# Lecture Notes in Civil Engineering

Volume 120

**Lecture Notes in Civil Engineering** (LNCE) publishes the latest developments in Civil Engineering - quickly, informally and in top quality. Though original research reported in proceedings and post-proceedings represents the core of LNCE, edited volumes of exceptionally high quality and interest may also be considered for publication. Volumes published in LNCE embrace all aspects and subfields of, as well as new challenges in, Civil Engineering. Topics in the series include:

- Construction and Structural Mechanics
- Building Materials
- Concrete, Steel and Timber Structures
- Geotechnical Engineering
- Earthquake Engineering
- Coastal Engineering
- Ocean and Offshore Engineering; Ships and Floating Structures
- Hydraulics, Hydrology and Water Resources Engineering
- Environmental Engineering and Sustainability
- Structural Health and Monitoring
- Surveying and Geographical Information Systems
- Indoor Environments
- Transportation and Traffic
- Risk Analysis
- Safety and Security

To submit a proposal or request further information, please contact the appropriate Springer Editor:

- Pierpaolo Riva at pierpaolo.riva@springer.com (Europe and Americas);
- Swati Meherishi at swati.meherishi@springer.com (Asia - except China, and Australia, New Zealand);
- Wayne Hu at wayne.hu@springer.com (China).

**All books in the series now indexed by Scopus and EI Compendex database!**

More information about this series at http://www.springer.com/series/15087

T. G. Sitharam · Raghuveer Rao Pallepati ·
Sreevalsa Kolathayar
Editors

# Seismic Design and Performance

## Select Proceedings of 7th ICRAGEE 2020

Springer

*Editors*
T. G. Sitharam
Indian Institute of Technology Guwahati
Guwahati, Assam, India

Raghuveer Rao Pallepati
Department of Civil Engineering
Indian Institute of Science
Bengaluru, Karnataka, India

Sreevalsa Kolathayar
Department of Civil Engineering
National Institute of Technology Karnataka
Mangalore, Karnataka, India

ISSN 2366-2557 ISSN 2366-2565 (electronic)
Lecture Notes in Civil Engineering
ISBN 978-981-33-4004-6 ISBN 978-981-33-4005-3 (eBook)
https://doi.org/10.1007/978-981-33-4005-3



This Springer imprint is published by the registered company Springer Nature Singapore Pte Ltd.
The registered company address is: 152 Beach Road, #21-01/04 Gateway East, Singapore 189721, Singapore

# Preface

The problems caused by devastating earthquakes in the past have forced the scientists and engineers to explore techniques on earthquake-resistant design of structures to safeguard lives and property during an earthquake. This book contains the latest research papers on review of seismic design codes, soil–structure interaction under dynamic loading, shallow and deep foundations, selected from the proceedings of 7th International Conference on Recent Advances in Geotechnical Earthquake Engineering and Soil Dynamics, 2021.

We thank all the staff of Springer for their full support and cooperation at all the stages of the publication of this book. We do hope that this book will be beneficial to students, researchers, and professionals working in the field of earthquake hazards. The comments and suggestions from the readers and users of this book are most welcome.

| | |
|---|---|
| Guwahati, India | T. G. Sitharam |
| Bengaluru, India | Raghuveer Rao Pallepati |
| Mangalore, India | Sreevalsa Kolathayar |

# Acknowledgements

We (editors) want to thank all the authors who have contributed to the book. We could bring this book out due to all the authors' timely contribution and cooperation. We thank and acknowledge the service of the following reviewers for their valuable time and efforts.

Ajay Chourasia, CSIR-CBRI
Amarnath Hegde, Indian Institute of Technology Patna
Amit Verma, IIT (BHU)
Anil Cherian, Strainstall
Anitha Kumari S. D., Ramaiah University of Applied Sciences
Arvind Kumar Jha, Indian Institute of Technology (IIT) Patna
Asha Nair, CMR Institute of Technology, Bengaluru
Babloo Chaudhary, NITK Surathkal
Bal Rastogi, Indian Society of Earthquake Science
Chittaranjan Birabar Nayak, Vidya Pratishthan's Kamalnayan Bajaj Institute of Engineering and Technology
Dauji Saha, Bhabha Atomic Research Center; Homi Bhabha National Institute
Deepankar Choudhury, Indian Institute of Technology Bombay
Dhanaji Chavan, IISc
Gopal Santana Phani Madabhushi, University of Cambridge
Govindaraju L., University Visvesvaraya College of Engineering, Bangaluru
Jagdish Sahoo, IIT Kanpur
Kalyan Kumar G., NIT Warangal
Karthik Reddy Konala S. K., IIT Hyderabad
Ketan Bajaj, Risk Management Solutions
Manas Kumar Bhoi, PDPU
Md Mizanur Rahman, University of South Australia
Padmanabhan G., Indira Gandhi Center for Atomic Research
Pradeep Kumar Singh Chauhan, CSIR-Central Building Research Institute, Roorkee
Premalatha Krishnamurthy, Anna University
Prishati Raychowdhury, IIT Kanpur

Purnanand Savoikar, Goa Engineering College
Rajib Saha, NIT Agartala
Rajib Sarkar, IIT (ISM) Dhanbad
Ramkrishnan R., Amrita Vishwa Vidyapeetham
Rangaswamy K., NIT Calicut
Ravi K., IIT Guwahati
Renjitha Varghese, National Institute of Technology, Calicut, Kerala, India
Sanjay Verma, Indian Geotechnical Society, Jabalpur Chapter
Sarat Kumar Das, Indian Institute of Technology (ISM) Dhanbad
Shreyasvi C., National Institute of Technology Karnataka
Snehal Kaushik, Girijananda Chowdhury Institute of Management and Technology, Guwahati.
Supriya Mohanty, Indian Institute of Technology (BHU),Varanasi
Surya Muthukumar, Amrita School of Engineering, Amrita Vishwa Vidhyapeetham
Vinay Srivastava, Retd. IIT ISM Dhanbad
Vipin K. S., Swiss Re
Visuvasam Joseph Antony, Vellore Institute of Technology

# Contents

# About the Editors

**Prof. T. G. Sitharam** is currently the Director of Indian Institute of Technology Guwahati, India. He is a KSIIDC Chair Professor in the area of Energy and Mechanical Sciences and Senior Professor at the Department of Civil Engineering, Indian Institute of Science, Bengaluru (IISc). He was the founder Chairman of the Center for Infrastructure, Sustainable Transport and Urban Planning (CiSTUP) at IISc, and is presently the Chairman of the AICTE South Western Zonal Committee, Regional office at Bengaluru and Vice President of the Indian Society for Earthquake Technology (ISET). Prof Sitharam is the founder President of the International Association for Coastal Reservoir Research (IACRR). He has been a Visiting Professor at Yamaguchi University, Japan; University of Waterloo, Canada, University of Dolhousie, Canada, and ISM Dhanbad, India, and was a Research Scientist at the Center for Earth Sciences and Engineering, University of Texas at Austin, Texas, USA.

**Dr. Raghuveer Rao Pallepati** is a Principal Research Scientist, Department of Civil Engineering, Indian Institute of Science, Bangalore and is involved in teaching, research and consultancy in the broad area of geotechnical engineering. He has taught courses on subsurface exploration and soil testing, earth retaining structures, behavior and testing of unsaturated soils, and fundamental of soil behavior for Master's and Doctoral students. His research interests are geotechnical instrumentation, slope stability analysis, numerical modelling, mechanics of unsaturated soils, contaminant transport through soil and reinforced earth structures. He has conducted several field and laboratory tests for design of foundations of different structures like buildings, turbo-generators and water tanks. He has 21 publications in journals and conference proceedings of national and international repute.

**Dr. Sreevalsa Kolathayar** pursued M.Tech. from IIT Kanpur, Ph.D. from Indian Institute of Science (IISc) and served as International Research Staff at UPC BarcelonaTech Spain. He is presently Assistant Professor in the Department of Civil Engineering, National Institute of Technology, Karnataka. Dr. Kolathayar has authored three books and over 65 research papers. His broad research areas are geotechnical earthquake engineering, geosynthetics and geonaturals, and water geotechnics. He is currently the Secretary of the Indian chapter of International Association for Coastal Reservoir Research (IACRR), and Executive Committee Member of Indian Society of Earthquake Technology. In 2017, The New Indian Express honored Dr. Kolathayar with 40 under 40—South India's Most Inspiring Young Teachers Award. He is the recipient of ISET DK Paul Research Award from Indian Society of Earthquake Technology, IIT Roorkee. He received "IEI Young Engineers Award" by The Institution of Engineers (India), in recognition of his contributions in the field of civil engineering.

# A Critical Review on Design Philosophies of Different Design Standards on Seismic Soil–Structure Interaction

**Vaibhav Mittal and Manojit Samanta**

**Abstract** Layout of philosophies of design standard plays an important role on the satisfactory long-term performance of the structure. Seismic soil–structure interaction (SSSI) is an important aspect, which determines the seismic response of the structure during any seismic disturbance. Various design codes around the world follow different approaches to incorporate the effects of SSSI on the design. This paper provides a critical review of the current design standard dealing with SSSI. The practicing provisions regarding the SSSI of different standards cover a wide range of structures supported over different soil types. Importance of SSSI, different criteria, design philosophies, nature of structure and soil conditions, etc., is systematically reported here. Areas of importance of different design standards regarding the SSSI for different types of soil and structure have been listed in detail. This paper may be useful for practicing professionals/engineers to identify various important provisions of different design standards regarding SSSI.

**Keywords** Seismic code · Monetary loss · Socioeconomic damages · Design provisions · Seismic soil–structure interaction

## 1 Introduction

In the recent decade, the huge losses in the socioeconomic are being caused due to occurrence of natural disasters. As per the report of Center for Research in Epidemiology of Disasters (CRED), earthquake is one of the fatal disasters, which adversely affects the society and economy of the country. As per this report, earthquake accounts for an economic loss of around $661 bn [1–3]. Non-availability of sufficient land

V. Mittal · M. Samanta (✉)
Geotechnical Engineering Group, Council of Scientific and Industrial Research - Central Building Research Institute (CSIR-CBRI), Roorkee 247667, India
e-mail: manojit@cbri.res.in

Academy of Scientific and Innovative Research (AcSIR), Ghaziabad 201002, India

V. Mittal
e-mail: vaibhavmittal2909@gmail.com

 
T. G. Sitharam et al. (eds.), *Seismic Design and Performance*, Lecture Notes in Civil Engineering 120, https://doi.org/10.1007/978-981-33-4005-3_1

and increased population results in construction of structures over soils having poor bearing characteristics. Generally, the structures are designed based on fixed-base conditions ignoring the soil characteristics. Seismic soil–structure interaction (SSSI) is a phenomenon, which accounts the response of structure on the soil and vice versa during a seismic event [4]. The waves originated from an earthquake source strike the bottom of the structure; the wave that enters the structure is transmitted wave ($E_1$), while the other, which reflects back to the soil, is reflected wave ($F_0$) as shown in Fig. 1. The transmitted wave travels up, strikes at the top of the structure, and returns after reflection resulting in the occurrence of SSSI [5, 6]. A part of the reflected wave travelling down transmits into the ground called the radiation wave ($R_1$), while the remaining part is reflected back to travel inside the structure which results in the vibration of the structure. The wave transmitted into the ground represents the radiation damping of the soil. The two types of interactions, namely kinematic and inertial, between soil and structure play an important role during SSSI. Due to these interactions between structure and ground during a seismic disturbance, the overall displacement of the structure increases.

The soil displacement due to earthquake ground motion is commonly known as free-field ground motion. Kinematic interaction occurs due to the inability of the embedded foundation to match with the free-field ground motion [4]. Considering, structure and foundation have no mass but stiffness; the equation of motion becomes:

$$[M_{\text{soil}}]\{\ddot{u}_{\text{kl}}\} + \left[K^*\right]\{u_{\text{kl}}\} = -[M_{\text{soil}}]\ddot{u}_{\text{b}}(t) \quad (1)$$

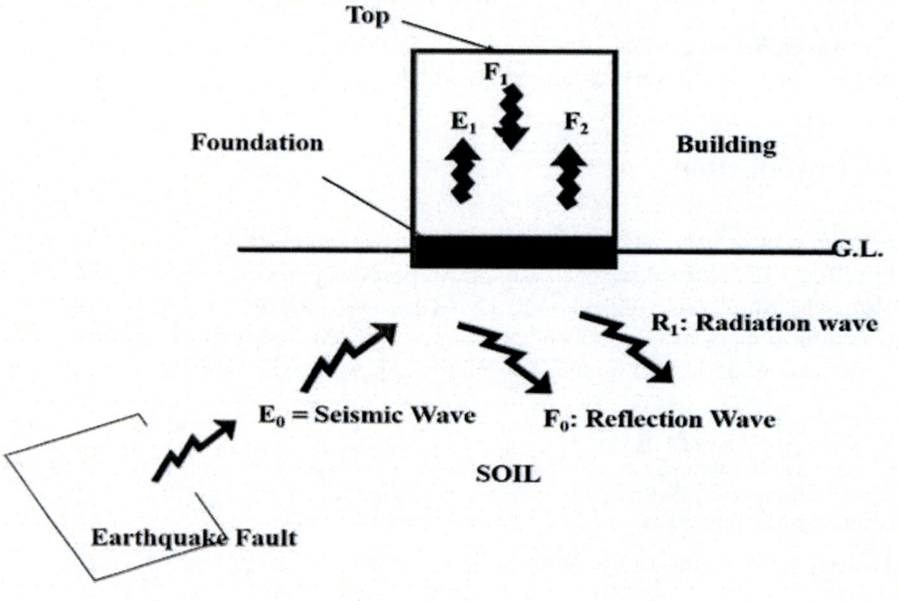


**Fig. 1** Interaction between ground and structure during an earthquake [5, 6]

where $[M_{soil}]$: mass matrix assuming the structure and foundation are massless

$\{u_{kl}\}$: foundation input motion

$[K^{*}]$: stiffness matrix

$\ddot{u}_b$: acceleration at the boundary.

While the inertial interaction is defined as the deformation in the soil due to transfer of inertial force to soil from the mass of the superstructure, the deformation due to inertial interaction can be computed using the equation:

$$[M_{soil}]\{\ddot{u}_{kI}\} + \left[K^{*}\right]\{u_{kI}\} = -[M_{Structure}]\{\ddot{u}_{kI}(t) + \ddot{u}_b(t)\} \quad (2)$$

where $[M_{Structure}]$: mass matrix assuming soil is massless.

While the right-hand side shows the inertial loading on the structure foundation system, which depends upon the base motion and foundation, input motion includes the kinematic interaction effect [4].

To incorporate the effects of SSSI, several analytical, experimental, numerical and field studies have been conducted in the past [7–19]. Analytical models such as Winkler p-y model, lumped-parameter models and cone models are vastly used to model the soil domain to study the effects of SSSI [7–9].

Several experimental studies have been conducted in the past to consider the SSSI phenomenon. Scaled-down physical model of a prototype is usually used to study the effects of SSSI. Shaking table test and centrifuge model testing are preferred for the scaled-time history simulation of SSSI on laboratory models [10–15]. To incorporate the boundary effects, laminar soil container sealed with absorbing layers of polymers is used. Several parameters such as natural frequency, drift, damping and base shear of the soil–structure system are measured using linear variable differential transformer (LVDT), accelerometers and strain sensors.

The discovery of modern computers, complexity of the SSSI problems and wider scope of numerical methods has significantly altered the computational techniques. Finite element method (FEM), finite difference method (FDM) and boundary element method (BEM) are widely used for the analysis of complex problems like SSSI [16–19]. The parameters like material nonlinearity, radiation damping, change in geometry, heterogeneous material conditions, etc., can be considered in the analysis easily using different numerical techniques [20–22]. Direct and substructure approaches are used to investigate the effects of SSSI numerically. In direct approach, both the structure and soil are modelled and analysed in single step. Solid finite elements and finite beam elements are utilized for the discretization of physical domain with special boundary elements. The special boundary elements are necessary to simulate the effects of unbounded soil medium which requires that the seismic energy should radiate away from the vibration source. The use of absorbing/transmitting boundaries prevents the seismic energy being reflected back into the problem domain. Several attempts are made in the recent past for the simulation of the effects of SSSI using this approach [23–25]. In the substructure approach, the SSI problem is divided into three distinct parts which also demonstrates the basic concept of substructure method of soil–structure interaction analysis. The three-step solution for SSI problems consists of determination of foundation input motion by solving the kinematic interaction problem. In the second step, determination of the frequency-dependent impedance functions describes the stiffness and damping characteristics of the soil–foundation interacting system. This step should account for the geometric and material properties of foundation and soil deposits and is generally computed using equivalent linear elastic properties for soil appropriate for the in situ dynamic shear strains. This step yields the so called soil springs. And lastly, computation of response of the real structure is supported on frequency-dependent soil springs and subjected to foundation input motion at the base of the spring [4]. Several researches have been carried out in the recent past to study the SSSI effects using this approach [26, 27]. It is found that the direct method is more superior to substructure approach due to its ability to accurately model and determine the complex nature of SSSI problem. All these studies found that during the seismic design of structures, ignorance of the effects of SSSI might shift the performance of the structures from life-safe to failure/collapse conditions [17, 18].

Several provisions are being recommended for the safety of structures/buildings, which influences the structural response during a seismic disturbance. The present study provides a critical review of the codal provisions followed across the world for incorporating the effects of SSSI during a design stage.

## 2 Case Studies for Seismic Soil–Structure Interaction of Structures

Several parameters like natural frequency, lateral displacement, interstorey drift, damping, etc., govern the structural performance when the effects of SSSI are considered. This section deals with advantageous or detrimental effects of SSSI on buildings, and bridges reported in the past studies.

Based on conventional theories, the effects of seismic soil–structure interaction are said to be beneficial. Nevertheless, consideration of the effects of SSSI during seismic design provides a better flexible structure. This consideration helps in obtaining the structure with larger natural period and damping ratio [6, 21, 28, 29]. The consideration of SSSI improved the performance in comparison with the corresponding rigid structure. Due to increased damping ratio, the capacity to resist the lateral load increases, resulting in improved performance. However, the increase in natural period is not always beneficial. It may increase the lateral displacement and drift of the structure [6]. The increase in natural period leads to resonance. In such conditions, the structure vibrates with a longer time period. This also results in increase in ductility demand of the structure [6]. The permanent deformation of structure increases, and an increased chance of soil failure is observed. The collapse of pile supported bridge structures during Loma Prieta Earthquake, 1989 and Kobe earthquake, 1995 manifests the detrimental effectsof non-consideration of SSSI (as shown in Figs. 2 and 3 respectively) [30, 31]. Mylonakis and Gazetas [30] observed that the amplitude of vibration is increased due to modification of seismic wave in the alluvial deposits across bridge piers. This increment leads to the collapse of the 18

**Fig. 2** Fallen deck at Pier E9 of the East Bay crossing during Loma Prieta Earthquake, 1989 [31]

**Fig. 3** Hanshin Expressway collapse during Kobe Earthquake, 1995 [30]

piers of the expressway, which shows the detrimental effects of non-inclusion of SSSI at the initial design stage.

## 3 Summary of Different Codal Provisions

Different guidelines and procedures are followed by various countries to incorporate the effects of SSSI. Under the recent trends and considerations provided by the design codes, the reduction in design base shear is considered when SSSI effects are included. The reduction in base shear of the structures for its elastic response was first proposed in ATC 3-06 [32] guidelines as described in the past studies [33, 34]. Later, it was also observed that SSSI affects the inelastic response of the structure.

American Standard, "Minimum design load and associated criteria for buildings and other structures (ASCE/SEI 7-16) [35, 36]" provides the guidelines regarding the inclusion of effects of SSSI on structures over flexible soil on the basis of shear wave velocity and SPT-N values. The provisions provided in this code are based on the National Earthquake Hazards Reduction Program (NEHRP) [37, 38]. The soil has been classified into different groups. The properties of different groups of soils are described in Table 1. The effects of SSSI are considered while designing the structures which are supported on sites C, D, E and F. It is recommended to use the following methods for the consideration of effects of SSSI [35, 36].

I. **Equivalent Lateral Force Procedure/Linear Static Procedure**

In the equivalent lateral force procedure or linear static procedure, the base shear of the fixed-base structure is computed. The base shear depends on effective seismic weight of the structure and seismic response coefficient. The determined seismic response coefficient is dependent on response modification factor, design spectral response acceleration parameter and importance factor for the structure. The design spectral response acceleration parameter can have a value equal to 1.0. When the

**Table 1** Site classification (as per ASCE/SEI 7-16) [35, 36]

| Site class | $V_s$ | $N_{ch}$ | $S_u$ |
|---|---|---|---|
| A. Hard rock | >5000 ft/s | NA | NA |
| B. Rock | 2500–5000 ft/s | NA | NA |
| C. Very dense soil and soft rock | 1200–2500 ft/s | >50 blows/ft | >2000 lb/ft$^2$ |
| D. Stiff soil | 600–1200 ft/s | 15–50 blows/ft | 1000–2000 lb/ft$^2$ |
| E. Soft clay soil | <600 ft/s | <15 blows/ft | <1000 lb/ft$^2$ |
| F. Soils requiring site response analysis | Very high plastic soils, soils vulnerable to potential failure, peat and organic clays | | |

effects of SSSI are considered with this procedure, base shear is reduced. The reduction in base shear depends on effective viscous damping ratio of the soil–structure system and foundation damping coefficient.

II. **Linear Dynamic Analysis**

In a linear dynamic analysis, the design spectral response acceleration ($S_a$) is calculated for a fixed-base structure. This parameter depends on fundamental period of the structure for short- and long-range transition periods. When the structures are supported on sites C, D, E and F, this analysis is performed using modified general design response spectrum or site-specific response spectrum. When the effects of SSSI are accounted through this method, the inclusion of kinematic interaction is not permitted. Under modified general design response spectrum method, the spectral response acceleration parameter is modified using a factor ($B_{SSSI}$) quantified for the damping ratios other than 0.05. While considering the effects of SSSI using the site-specific response spectrum analysis, probabilistic ground motions approach and deterministic ground motions approach are recommended.

III. **Nonlinear Response History Analysis**

When the effects of kinematic interaction are dominant, nonlinear response history analysis with site-specific response spectrum is recommended. The kinematic interaction effects are recommended to be estimated by the product of response spectral modification factors for base slab averaging and embedment. It is recommended that the value of this product shall not be less than 0.7. It is also recommended to consider the foundation damping effects through direct incorporation of soil hysteretic damping and radiation damping in the mathematical model of the structure.

Indian Design standard, "Criteria for earthquake resistant design of structures (IS 1893: 2016)" provides the guidelines for the consideration of the effects of SSSI based on the shear wave velocity and SPT-N value of the soil [39–42]. It has classified the existing soil into three major categories. Type A corresponds to rock or hard soils, while Type B and C represent medium or stiff soils and soft soils, respectively. When the structures are supported on a rock or a rock-like material, the effects of SSSI are not considered. It considers the four main attributes, i.e., adequate ductility, robust structural configuration, a minimum elastic lateral stiffness and lateral strength of the buildings for the seismic design of structures. The methods like equivalent static method and dynamic analysis method which include response spectrum, modal time history and time history method are suggested for the seismic design of structures. Equivalent static method is used to quantify the design lateral force for the buildings which are lesser than 15 m in height, while dynamic analysis method is recommended for all the buildings having more than 15 m height. It considers the quantification of seismic design forces for structures which are only founded on rocks or soils which do not settle, while the structures supported on soil, which have shear wave velocity less than 750 m/s or SPT-N value less than 10, require consideration of SSSI. But, the above-mentioned methods do not consider the influence of foundation type, foundation geometry, soil damping, kinematic and inertial interaction effects which affect the response of the structure during the seismic disturbance thereby ignoring the SSSI effects.

New Zealand design standard [43] provides the guidelines about the seismic design of structures. It considers the soil having shear wave velocity less than 150 m/s as very soft. The buildings and infrastructures in such soil are susceptible to damage during an earthquake event. Use of structural performance and ductility factor in the design stage has been suggested for the prevention/reduction of earthquake damage. The ductility factor represents the change in seismic response of structure undergoing plastic deformation, while structural performance factor represents energy dissipation, total structural capacity and structural demand corresponding to peak ground acceleration.

European design standard [44, 45] presents the guidelines regarding the consideration of seismic soil–structure interaction based on $P - \delta$ effects, unconfined compressive strength and shear wave velocity of the soil. It suggested to consider the SSSI for structures resting on soils having average shear wave velocity less than 100 m/s and unconfined compressive strength in the range 10–20 kPa. It considers the effects of SSSI to be detrimental for the structures subjected to significant $P - \delta$ effects (second order). Although, they identify structures for which SSSI should be considered, it does not stipulate the guidelines for enumerating the effects of SSSI.

National Standard of the People's Republic of China [46] provides the guidelines regarding the seismic design of structures. It suggests the consideration of SSSI for tall structures supported over soil having shear wave velocity less than 150 m/s. It also recommends for the reduction in the horizontal seismic shear forces assumed for the rigid base. The reduction of horizontal shear forces is being done based on the aspect ratio (height/width) of the structures. Although, the standard provides the information about the structures where SSSI is important, but did not provided specific guidelines to consider the same in the design.

Australian standard [47] provides the guidelines for the seismic design of structures. They suggest that the soil having SPT-N value less than 6 is considered to be soft. It provides the similar guidelines as provided by the New Zealand design standard. To incorporate the effects of SSSI and reduce the earthquake damage, the use of structural performance and ductility factor during the design stage is recommended.

Japan Society of Civil Engineers [48] provides guidelines regarding the consideration of SSSI. It provides the suggestion to consider the dynamic interaction between the structure and the ground for the design of bridge abutments, retaining walls, underground structures and foundations structures. Depending on the type and characteristics of the structures and the ground, the effects of SSSI may be included or excluded for remaining other structures. During an earthquake, the influence of surface layers depends on the location and position of engineering bed ground surface. The layer with a shear wave velocity of 400 m/s or more (generally, the SPT N value of 50 or more for sandy soil and 30 or more for cohesive soil) is referred to be the engineering bedrock surface. Considering the nonlinear effects of soil, use of coupled analysis is suggested if the effects of SSSI are predominant, while seismic coefficient method and non-linear spectra methods can be used in initial design phases, if the effects of SSSI are not predominant.

Mexican design standard [49, 50] emphasizes to consider the effects of SSSI based on probabilistic and deterministic approaches during the design stage of the structures. It suggested damping and interaction factors for the consideration of SSSI. The damping factor depends on the target damping of the structure, structural natural period and damping ratios. The damping factor allows modification from 5% initial damping to accommodate the SSSI. Changes in lateral displacement due to foundation's rotation and modification in base shear is suggested to consider the SSSI effects. Modified base shear is computed using either static analysis or modal analysis procedure. In static analysis procedure, the modified base shear depends on total weight of the structure. The values of interaction factor can neither be greater than 1.25 nor be less than 0.75. In modal analysis procedure, the base shear is modified based on the fundamental mode of vibration. The changes in lateral displacement of the structure are also accounted and are dependent on lateral stiffness and effective height of the structure, depth and rocking stiffness of the foundation and the overturning moment of the fixed-base structure.

## 4 Comparison of the Available Guidelines

As per the review of different design standards provided in the above section and depending upon the prevailing soil conditions across the world, different design standards recommend different values for SPT-N and shear wave velocities of the soil for the consideration of SSSI effects. For a range of SPT-N value less or equal to 15 blows per feet and shear wave velocity less than 750 m/s, effects of SSSI are recommended to be considered during the seismic design by ASCE 7-16 [35, 36] and Indian standards [39–42]. However, the other design standards recommend the consideration of SSSI for the soil showing SPT-N value less than 15 and shear wave velocity less than 150 m/s generally. During the seismic design, the P-delta effects of the structure is an important parameter for the consideration of SSSI as provided by the Euro code and Chinese code. The real structures are flexible in nature and are subjected to vertical loads in day-to-day life. When they are subjected to lateral loads such as seismic loads, the lateral displacement of the structure may increase, resulting abrupt changes in shear, overturning moment, ground motions, etc. However, other design standards mostly consider the shear wave velocity and SPT-N value of soil for the consideration of SSSI effects. Hence, for the implementation of SSSI during the seismic design, consideration of P-delta effects with shear wave velocity of soil should be paid attention.

The lateral displacement of the structure quantified during a seismic disturbance increases due to consideration of SSSI. The Mexican design standard [49, 50] considers the modification in lateral displacement due to the rotation and depth of the foundation during a seismic disturbance. However, the other design standards do not consider the effects of foundation rotation and its depth. ASCE standard has provided the guidelines regarding the consideration of foundation damping effects through direct incorporation of soil hysteretic damping and radiation damping in the mathematical model of the structure. Foundation type and its geometric properties influence the seismic response of the structures significantly. Foundation–soil stiffness also plays an important role in the seismic response of the structure. Hence, response of structure due to foundation type should also be included in the initial design stages.

Base shear is the estimation of the maximum lateral force at the base of the structure during a seismic disturbance. Different design standards suggest modification/reduction in base shear for considering the effects of SSSI. Among various standards outlined in the above section, ASCE 7-16 appears to be the most developed in the stipulation of SSSI in design practice. Mexican design standard also suggests the modification in base shear based on probabilistic and deterministic approach. However, for the modification of the base shear, it considers the effects of SSSI on the basis of structural ductility, a parameter which is ignored in all building codes. It is clear from various standards that they have outlined the conditions to include SSSI during design but still lack in evaluation and assessment of SSSI in design practice. Such a gap between the structures and soil conditions can change the performance of the structure from life-safe to collapse conditions.

## 5 Conclusions

The critical review on design philosophies of different design standards regarding SSSI leads to following broad conclusions:

i. The present paper describes different aspects of SSSI and its influencing parameters, different design guidelines for inclusion in the initial design stage.
ii. Consideration of the effects of SSSI on the structures designed on the basis of conventional design philosophies shows that the natural period of system increases, resulting in enhancement of lateral displacement and interstorey drift that may shift the performance the structure from life-safe to collapse/nearly collapse conditions.
iii. Considering the effects of SSSI in the initial design phases, the damping of the system is enhanced, which decreases the structural demand. This decrement leads to an economical design of the soil–structure system.
iv. The present study provides the range of shear wave velocity of soil and SPT-N values recommended in various standards for the SSSI analysis of structures resting on soil deposits.
v. ASCE 7-16 provides the detailed guidelines regarding the effects of SSSI and procedure to consider it in the initial design stage. They recommended the modification in base shear with equivalent lateral force procedures excluding the kinematic interaction effects. However, inclusion of effects of kinematic interaction is recommended with nonlinear response history procedure.

vi. Mexican design standard provides the detailed guidelines for the consideration of SSSI effects. They recommended the modification in base shear to incorporate the SSSI effects based on the structural ductility and modification in lateral displacement of the structure on the basis of foundation's depth and rotation.
vii. The other design standards recommend use of structural performance factor, ductility factor and reduction in base shear using modal analysis procedure in the initial design stage, during the seismic design of structures.

**Acknowledgements** The authors thank Director, CSIR-CBRI, for providing the environment and facilities. The authors also thank the anonymous reviewers for their valuable suggestions and advice.

## References

1. Center for Research on the Epidemiology of Disasters (CRED). https://www.statista.com/statistics/266325/death-toll-in-great-earthquakes/
2. Center for Research on the Epidemiology of Disasters (CRED). https://www.unisdr.org/2016/iddr/IDDR2018_Economic%20Losses.pdf
3. Daniell J, Khazai B, Wenzel F, Vervaeck A (2012) The worldwide economic impact of earthquakes. Paper No. 2038
4. Kramer SL. Geotechnical earthquake engineering. PGRPE. https://www.pearsoned.co.in
5. Miura K (2011) Dynamic soil structure interaction. https://iisee.kenken.go.jp/lna/?mod=view&cid=E0-190-2008#
6. Wolf JP (1985) Dynamic soil-structure interaction. In: Earthquake engineering & structural dynamics, vol 4. Prentice-Hall, Inc., Englewood Cliffs
7. Wolf JP (1994) Foundation vibration analysis using simple physical models. Prentice-Hall, Englewood Cliffs
8. Rha CS, Wallace JW, Tacigroglu E (2004) Analytical modelling of soil–structure interaction for bridge columns. In: 13th world conference on earthquake engineering, Vancouver, Paper No. 2388
9. Mohasseb S, Abdollahi B (2009) Soil–structure interaction analyses using cone models. JSEE 10(4)
10. Hokmabadi AS, Fatahi B, Samali B (2014) Physical modelling of seismic soil–pile–structure interaction for buildings on soft soils. American Society of Civil Engineers. https://doi.org/10.1016/(ASCE)GM.1943-5622.0000396, https://ascelibrary.org/doi/abs/10.1061/%28ASCE%29GM.1943-5622.0000396
11. Trombetta NW, Mason HB, Chen Z, Hutchinson TC, Bray JD, Kutter BL (2013) Nonlinear dynamic foundation and frame structure response observed in geotechnical centrifuge experiments. Soil Dyn Earthq Eng 50:117–133 (2013). ISSN 0267-7261
12. Ha JG, Lee S-H, Kim D-S, Choo YW (2014) Simulation of soil–foundation–structure interaction of Hualien large-scale seismic test using dynamic centrifuge test. Soil Dyn Earthq Eng 61–62:176–187 (2014). ISSN 0267-7261
13. Saha R, Haldar S, Dutta SC (2014) Influence of dynamic soil-pile raft-structure interaction: an experimental approach. Earthq Eng Eng Vib 14:625 (2015)
14. Ghandil M, Behnamfar F, Vafaeian M (2015) Dynamic responses of structure–soil–structure systems with an extension of the equivalent linear soil modelling. Soil Dyn Earthq Eng 80:149–162 (2016). ISSN 0267-7261
15. Hussien MN, Tobita T, Iai S, Karray M (2016) Soil–pile–structure kinematic and inertial interaction observed in geotechnical centrifuge experiments. Soil Dyn Earthq Eng 89:75–84 (2016). ISSN 0267-7261

16. Padron LA, Aznarez JJ, Maeso O (2009) Dynamic structure–soil–structure interaction between nearby piled buildings under seismic excitation by BEM–FEM model. Soil Dyn Earthq Eng 29:1084–1096. ISSN 0267-7261
17. Hokmabadi AS, Fatahi B, Samali B (2013) Assessment of soil–pile–structure interaction influencing seismic response of mid-rise buildings sitting on floating pile foundations. Comput Geotech 55:172–186. ISSN 0266-352X. https://doi.org/10.1016/j.compgeo.2013.08.011, (https://www.sciencedirect.com/science/article/pii/S0266352X13001353)
18. Tabatabaiefer SHR, Fatahi B, Samali B (2013) Seismic behavior of buildings considering dynamic soil–structure interaction. Int J Geomech ASCE 13:409–420. https://doi.org/10.1061/(ASCE)GM.1943-5622.0000231
19. Fatahi B, Van Nguyen Q, Xu R, Sun W-J (2018) Three-dimension response of neighboring buildings sitting of pile foundations to seismic pounding. Int J Geomech ASCE 4:04018007. ISSN: 1532–3641
20. Bowles JE (1996) Foundation analysis and design. In: Civil engineering series, 5th edn. McGraw-Hill International Editions, New York
21. Dutta SC, Roy R (2002) A critical review on idealization and modelling for interaction among soil-foundation-structure system. Comput Struct 80:1579–1594
22. Far H (2017) Advanced computation methods for soil-structure interaction analysis of structures resting on soft soils. Int J Geotech Eng 352–359. https://doi.org/10.1080/19386362.2017.1354510
23. Desai CS, Phan HV, Perumpral JV (1982) Mechanics of three dimensional soil–structure interaction. J EngMech 108(5):731–747
24. Fatahi B, Tabatabaiefar HR, Samali B (2011) Performance based assessment of dynamic soil–structure interaction effects on seismic response of building frame. In: Proceedings of georisk 2011—geotechnical risk assessment & management (geotechnical special publication no. 224). American Society of Civil Engineers (ASCE), Reston, pp 344–351
25. Fatahi B, Tabatabaiefar HR, Samali B (2014) Soil–structure interaction vs site effect for seismic design of tall buildings on soft soil. Geomech Eng 6(3):293–320
26. Carbonari S, Dezi F, Leoni G (2012) Nonlinear seismic behaviour of wall-frame dual systems accounting for soil–structure interaction. Earthq Eng Struct Dyn 41:1651–1672. https://doi.org/10.1002/eqe.1195
27. Yang XM, Chen Y, Yang BP (2008) Three-dimension dynamic soil–structure interaction analysis using the substructure method in the time domain. In: Proceedings of the 14th world conference on earthquake engineering, Beijing, 12–17 Oct 2008
28. Kausel E (2009) Early history of soil–structure interaction. Soil Dyn Earthq Eng 30:822–832. ISSN 0267-7261. https://doi.org/10.1016/j.soildyn.2009.11.001 (https://www.sciencedirect.com/science/article/pii/S0267726109001754)
29. Vishwajit A, Satish Kumar SR (2018) Seismic soil-structure interaction: a state-of-the-art-review. Structure 16:317–326. https://doi.org/10.1016/j.istruc.2018.10.009
30. Mylonakis G, Gazetas G (2000) Seismic soil structure interaction: beneficial or detrimental? J Earthq Eng 4(3):277–301. https://doi.org/10.1080/13632460009350372
31. Yashinsky M (1998) The Loma Prieta, California earthquake of October 17, 1989—highway systems. Professional Paper 1552-B. USGS, Washington. https://pubs.usgs.gov/pp/pp1552/
32. ATC (1978) Tentative provisions for the development of seismic regulations for buildings. ATC 3-06. Applied Technology Council, California
33. Velestos AS, Meek JW (1974) Dynamic behavior of building foundation system. Earthq Eng Struct Dyn 3(2):121–138
34. Velestos AS, Nair V (1975) Seismic interaction of structures on hysteretic foundations. J Struct Div 101(1):109–129
35. ASCE (2010) Minimum design loads for buildings and other structures. ASCE/SEI 7-10. American Society of Civil Engineers, Virginia
36. ASCE (2016) Minimum design loads for buildings and other structures. ASCE/SEI 7-16. American Society of Civil Engineers, Virginia

37. FEMA (2015) NEHRP recommended seismic provisions for new buildings and other structures. FEMA P-1050. Federal Emergency Management Agency, Washington
38. Khosravikia F, Mahsuli M, Ghannad MA (2017) Probabilistic evaluation of 2015 NEHRP soil-structure interaction provisions. J Eng Mech 143(9):04017065
39. IS 1893-1 (2016) Criteria for earthquake resistant design of structures—part 1: general provisions and buildings IS 1893-1. Bureau of Indian Standards, New Delhi
40. IS 1893-2 (2014) Criteria for earthquake resistant design of structures—part 2: liquid retaining tanks. IS 1893-2. Bureau of Indian Standards, New Delhi
41. IS 1893-3 (2014) Criteria for earthquake resistant design of structures—part 3: bridges and retaining walls. IS 1893-3. Bureau of Indian Standards, New Delhi
42. IS 1893-4 (2015) Criteria for earthquake resistant design of structures—part 4: industrial structures including stack-like structures. IS 1893-4. Bureau of Indian Standards, New Delhi
43. NZS 1170.5 (2004). tructural design actions—part 5: earthquake actions—New Zealand. NZS 1170.5. Standards New Zealand, Wellington
44. EN 1998-1 (2004) Eurocode 8: design of structures for earthquake resistance—part 1: general rules, seismic actions and rules for buildings. EN 1998–1. European Committee for Standardization, Brussels
45. EN 1998-5 (2004) Eurocode 8: design of structures for earthquake resistance—part 5: foundations, retaining structures and geotechnical aspects. EN 1998-5. European Committee for Standardization, Brussels
46. GB50011-2010 (2010) Code for seismic design of buildings. National Standard of the People's Republic of China
47. AS 1170.4-2007 (2007) Structural design actions—Australian Standard. Standards Australia, Sydney
48. JSCE 15 (2007) Standard specifications for concrete structures—design. JSCE 15. Japan Society of Civil Engineers, Tokyo
49. MOC-93, 1993 (1993) Manual of civil works design. Earthquake design. Institute of Electrical Research, Federal Electricity Commission
50. MOC-2008, 2008 (2008) Civil works design manual. Design by earthquake. Recommendations and comments. Institute of Electrical Research, Federal Electricity Commission

# A Critical Review of Indian Seismic Codes for Buildings in the Light of International Codes of Practice

P. Jayarajan

**Abstract** The field of earthquake-resistant design of structures is undergoing major transformations all around the world, and it is expected that the Indian seismic codes incorporate such developments from time to time to help engineers implement the latest methodologies in their design. The Indian seismic code IS 1893-Part 1 which covers general provisions and criteria for the earthquake-resistant design of building structures has been revised, and the 2016 edition is recently issued. The latest revision, though includes some improvements still lacks many important provisions required for earthquake-resistant design and detailing. Therefore, a critical review of the latest revision of seismic code is undertaken along with IS 800:2007 (section-12) and the adjunct code IS 13920:2016 based on the current state of the art and provisions in other major national codes. The paper highlights limitations and omissions in the latest Indian seismic codes and suggests important updates and revisions required to make current practice in the detailed engineering of earthquake-resistant structures at par with other international codes of practice.

**Keywords** Indian seismic codes · Base shear · Redundancy · Ductile detailing · Overstrength

## 1 Introduction

The protection of built environment against earthquake hazards is very crucial considering that almost 60% of Indian land is vulnerable to earthquakes. This requires that the Indian seismic codes shall be in place incorporating the latest advances in the field of earthquake-resistant design. However, it shall be noted that the progress made in the field of earthquake engineering and mitigation in India when compared to the developed world has been quite inadequate. The Indian seismic code IS 1893: Part 1 [1], though revised in 2016, based on the lessons learnt from Bhuj earthquake of

P. Jayarajan (✉)
National Institute of Technology, Calicut, Kerala, India
e-mail: Puttatt1@gmail.com


T. G. Sitharam et al. (eds.), *Seismic Design and Performance*, Lecture Notes in Civil Engineering 120, https://doi.org/10.1007/978-981-33-4005-3_2

2001 still needs to adopt to the modern state-of-the-art design and construction practices to ensure safety against seismic hazards. The objective of the present study is therefore to perform a critical review of the revised Indian seismic code and related documents and highlight potential areas requiring immediate remedial action. The review is done using the modern seismic codes of practice such as ASCE 7-16 [2] and Eurocode 8:2004 (EC8) [3]. The results of the review are presented under three heads, namely general provisions, RCC structures and steel structures. The important observations and conclusions are also presented.

## 2 Review Process

### *2.1 General Provisions*

The definition of design seismic action is based on seismic hazard maps which however are not prepared based on site-specific probabilistic seismic hazard analysis. In such a scenario, it would be impossible to define ground motions with different probabilities of exceedances with multiple performance target. Therefore, there is an urgent need to revise the seismic hazard map of the country based on probabilistic approach. Eurocode 8 (EC8) requires that the structures are deigned to meet the "No-collapse requirement" and "Damage limitation requirement." The first requirement represents the ultimate limit state (ULS) verified using seismic action with 10% probability of exceedance in 50 years, whereas the second forms the basis for serviceability limit state (SLS) using seismic action with 10% probability of exceedance in 10 years.

In Indian seismic code, the influence of local ground condition on the seismic action is represented through only three soil types (Type I, II and III) and therefore is to be necessarily reworked to adequately represent the variation of geotechnical profiles across the country. The soil classification is given on the basis of corrected standard penetration test (SPT) $N$-values. However, the soil classification shall be based on the average shear wave velocity in top 30 m ($v_{s,30}$) in line with modern international seismic codes. Further shear wave velocity ($v_s$) could capture the sensitivity of soil deposit in local amplification of seismic waves. Anbazhagan et al. [4] critically reviewed the geotechnical provisions in Indian seismic codes and highlighted the requirement of using the soil shear wave velocity for subsurface characterization in line with modern international codes. EC8 provides seven ground types A, B, C, D, E, S1 and S2 based on $v_{s,30}$, undrained cohesion ($c_u$) and plasticity index (PI). The code also highlights the requirement of special investigations for the seismic assessment in case of profiles S1 and S2. ASCE 7 provides six site classes, namely A, B, C, D, E and F, using same set of parameters as in EC8 and requires that site response analysis to be carried out for site class F. The shortcomings in the seismic site classification provisions of IS: 1893:2016 are dealt in detail [5].

It is well known that the soil nonlinearity alters the frequency content of earthquakes with larger surface wave magnitude ($M_s$) and results in spectra with lower spectral values for lower periods ($T < 1$ s) and higher values for larger periods ($T >$ 1 s). To take the above into account, EC8 provides two types spectra Type-1 and Type-2, the choice being dependent on the surface wave magnitude ($M_s$) of the dominant earthquake that contributes most to the seismic hazard in the site-specific probabilistic seismic hazard assessment. If $M_s$ is smaller than 5.5, Type-2 spectrum shall be used. Type-1 and Type-2 spectra as per EC8 are shown in Fig. 1. The effect of magnitude on the spectral accelerations toward the longer periods is evident. The amplification in response spectral values is also governed by earthquake magnitude in addition to site-specific soil properties [6]. However, such a magnitude-specific spectra definition is absent in IS 1893 as the same is not probabilistic. The design acceleration coefficient for different soil types as envisaged in IS 1893 is shown in Fig. 2. It is clear that the soil amplification in the short period range is totally neglected by the code. ASCE 7 defines the design response spectrum based on mapped maximum considered earthquake spectral acceleration values at periods of 0.2 and 1.0 s and adjusting them for site classes and acceleration values.

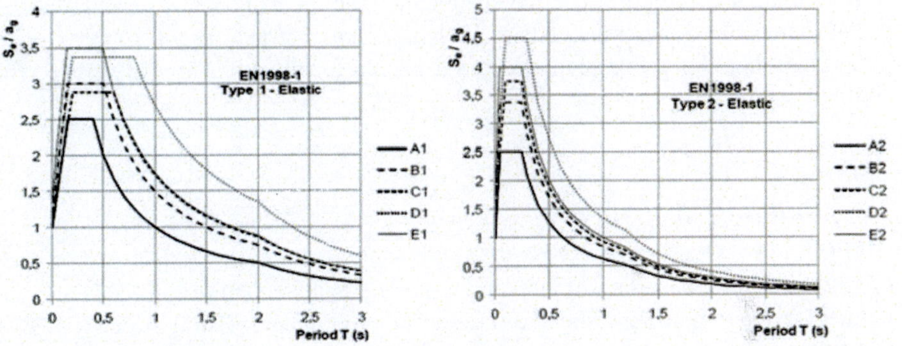


**Fig. 1** Spectral shapes for Type-1 and Type-2 seismic actions for various ground types (EC8)

**Fig. 2** Design acceleration coefficient (Sa/g) (IS 1893)

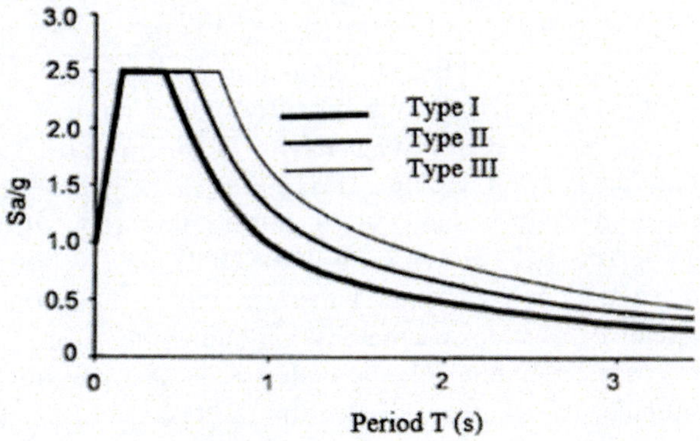

Seismic codes rely on the stable energy dissipation capacity and make use of the overstrength, ductility and redundancy of the structural system in the computation of seismic demands. This approach features the reduction of seismic forces obtained from an elastic analysis by the use of a factor identified as behavior factor ($q$) in EC8, response modification coefficient ($R$) in ASCE 7 and response reduction factor ($R$) in IS 1893. It shall be noted that the $q$ factor is not applied uniformly across all periods in EC8. Instead, for stiff structures with the period $0 < T < T_B$, a lower factor is employed computed using linear variation with $q = 1.5$ at $T = 0$ and full $q$ value at $T = T_B$. However, in Indian seismic codes, the same is applied uniformly across all time periods.

Partial safety factors of 1.20 and 1.50 in IS 1893 associated with earthquake actions in load combinations for the seismic design do not have any basis considering that the earthquake action (E) considered in the code itself represents an ultimate state and that dead and imposed loads represent the operating loads. Such combinations would obviously result in uneconomical designs and cannot be justified. It shall be noted that in EC8 and ASCE 7, a load factor of 1.0 is considered for earthquake actions in all combinations with other operating loads.

IS 1893 provides a detailed definition of a design acceleration spectrum for periods from 0 to 6.0 s specific to the equivalent static method. This is unwarranted as the use of such method as per the code is limited to regular structures with approximate period less than 0.4 s. For such structures, an ($Sa/g$) value of 2.50 can be used for computation of seismic action.

## 2.2 Building Structures

IS 1893 provides two values of response reduction factor ($R$) for reinforced concrete moment frame systems, a value of 3.0 for ordinary moment resisting frame (OMRF) and 5.0 for special moment resisting frames (SMRF) and similarly for buildings with reinforced concrete (RC) walls, an $R$ value of 3.0 for ordinary RC structural walls and 4.0 for ductile RC structural walls. Considering that only ductile systems are allowed in seismic zones III, IV and V, design engineers have only one option in each category, namely SMRF or ductile RC structural walls. The EC8 on the other hand classifies the RC buildings into two ductility classes, namely DCM (medium ductility) and DCH (high ductility) based on their hysteretic dissipation capacity and assign the respective behavior factor $q$ values. This provides design engineers to select the best ductility class suited to their structural system. For structural type including frames, dual system, and coupled walls, EC8 assigns basic $q$ values of 3.0 $\alpha_u/\alpha_1$ for DCM and 4.5 $\alpha_u/\alpha_1$ for DCH. The overstrength factor $\alpha_u/\alpha_1$ computed through a nonlinear static pushover analysis helps the design engineers to arrive at a realistic $q$ value and thereby to optimize the structural system. Such an approach is absent in Indian seismic codes. EC8 also requires that for buildings which are irregular in elevation, a reduction of 20% shall be made on the basic values of the behavior factor ($q_0$). Such a requirement is missing in Indian seismic codes. Further

IS 1893 does not provide $R$ values for structural systems such as coupled walls, torsionally flexible system and inverted pendulum systems.

It is normally observed that the approximate expressions for fundamental period of vibrations for different structural systems provided in various seismic codes result in conservative seismic demands. To arrive at a realistic demand, it is important that the code recommends expressions such as those based on Rayleigh method as typically pointed out in EC8. The computation of approximate periods for various structural systems such as steel eccentrically braced frames (EBF) and steel buckling-restrained braced frames (BRBF) is missing in the Indian code. The vertical distribution of base shears to different floor levels using a parabolic profile as given in IS 1893 with no specific reference to the structure period will unnecessarily increase the seismic demands such as the top story shears, roof displacements and base overturning moments. The linear distribution in EC8 and the period-specific distribution in ASCE 7 are worth noting in this case.

The effect of second-order effects ($P$-$\Delta$) in the estimation of seismic demand in every story is considered in EC8 through the use of inter-story drift sensitivity coefficient ($\theta$). The coefficient is computed for all story taking into account total gravity load at and above the story, story drift and story shear. As per EC8, the second-order effects need not be taken into account if $\theta \leq 0.10$. However, if $0.1 < \theta \leq 0.20$, the relevant seismic action shall be increased by a factor equal to $1/(1 - \theta)$. The value of $\theta$ in any case shall not exceed 0.30. ASCE 7 also uses almost a similar procedure using the stability coefficient ($\theta$). The seismic code IS 1893, however, does not have any such provisions for taking the second-order effects ($P$-$\Delta$ effects) into seismic demand calculations.

IS 1893 specifies that either linear dynamic time history analysis or response spectrum method is used to compute the design seismic force for all buildings except the regular buildings lower than 15 m in seismic zone II. It refrains from the use of nonlinear response history analysis. The code requires that when either time history or response spectrum method is used, the computed base shear shall not be less than the one computed using the approximate time periods stipulated in the code. This requirement does not have any basis and in addition eliminates the advantage derived from exact analysis. Though the code specifies that linear time history method shall be based on an appropriate ground motion, it does not provide any guidelines for the selection of ground motion records including the choice of range of periods required for checking compatibility with the design acceleration spectra. As the selection of proper ground motion records is an essential step in arriving at a robust estimation of structural response in both the linear and nonlinear response history analysis, the inclusion of the selection criteria shall be properly addressed in the code. Further, considering that the performance-based earthquake engineering (PBEE) is gaining acceptance, the code needs to urgently address this issue and provides relevant guidelines. The selection of recorded earthquake motions for seismic demand calculations is governed by earthquake magnitude, distance, site-specific soil profile and strong motion duration [7]. The PEER report [8] provides a detailed evaluation of ground motion selection and modification methods relevant to nonlinear response history analysis of structures. Detailed criteria for the selection of proper ground motion records for the nonlinear analysis are available in most international seismic codes. It shall be noted that as an alternate to linear method, EC8 also permits the use of nonlinear methods such as nonlinear static (pushover analysis) and nonlinear response history analysis.

ASCE 7 permits the use of nonlinear response history analysis to verify strength, stiffness and ductility of structures under maximum considered earthquake (MCE) shaking. The code highlights the importance of an independent peer review in the case of such analysis procedures. It states that the project-specific design criteria relevant to nonlinear analysis shall be approved by independent structural design reviewers. ASCE 7 provides detailed guidelines relevant to the use of target spectrum, selection criteria for ground motion records, detailed modeling criteria including modeling of element nonlinear hysteretic behavior and global acceptance criteria. As per the code, element nonlinear hysteretic behavior shall be consistent with ASCE 41-17 [9] or applicable laboratory test data. A typical generalized force–deformation relation for steel elements provided in ASCE 41 is shown in Fig. 3. The NIST report [10] provides recommendations for the modeling of structural components and their acceptance criteria in nonlinear analysis for various structural systems.

IS 1893 currently does not provide any guidelines for the nonlinear static (pushover analysis) analysis such as the load patterns to be adopted and the nonlinear modeling of members. Such an analysis provides the capacity curves and the hierarchy of hinge formation in various members of the structure. It helps the design engineers to obtain a realistic overstrength estimate of the designed structure and use higher values of force reduction factors eventually resulting in an efficient and economical structural design. BS EN 1998-1:2004 and ATC-40 [3, 11] provide detailed guidelines for pushover analysis including the load patterns for the lateral loads and target displacement required for obtaining the capacity curve.

As per international seismic codes, while the structure shall have sufficient strength and ductility to meet the "no-collapse requirement," it shall also satisfy the "damage limitation requirement" under a seismic action more frequent than the design one through a limitation of inter-story drifts. The story drift ratio limitation of 0.4% given in IS 1893 calculated corresponding to the design seismic action does

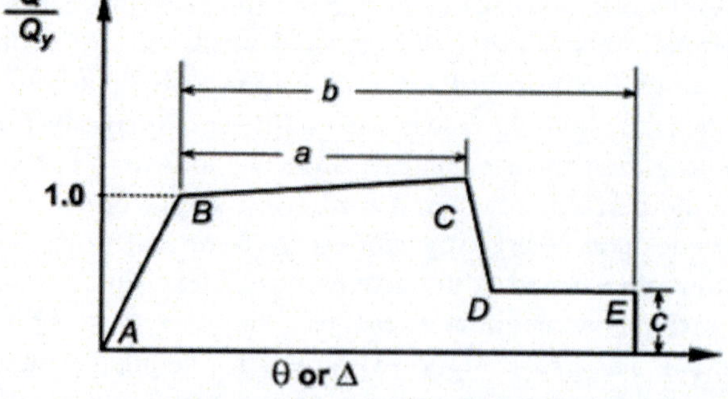


**Fig. 3** Generalized force–deformation relation for steel elements (ASCE 41)

not address the structure typology and the risk category. It is required that a separate serviceability state is defined considering earthquakes of smaller return period compared to the design seismic action and drift calculations performed accordingly. However, considering that the spectrum defined in the current code is not probabilistic, the definition of such a serviceability limit state would not be possible. EC8 defines the limiting inter-story drift ratio corresponding to a more frequent earthquake scenario based on the ductility of nonstructural elements attached to the structure. The limiting story drifts provided in ASCE 7 depend both on the risk category and the structure type; however, the story drifts computed by elastic analysis shall be increased by the deflection amplification factor to compare with the acceptable values. FEMA-445 [12] provides detailed guidelines for performance-based earthquake engineering to assess the performance capability of building structures.

Nonstructural elements such as curtain walls, mechanical appendages and equipment represent an important component of a building structure as their failure can cause injury to people and affect the critical facilities. As many of the nonstructural components are directly supported from various floors, it is important that they are verified for the corresponding floor spectra including the supports. Currently, IS 1893 does not provide any guidelines for the design of non-structural elements and their supports. Both EC8 and ASCE 7, on the other hand provide detailed guidelines for the seismic assessment and design of non-structural elements and their supports. While IS 1893 briefs on many issues relevant to RC frame buildings with unreinforced masonry including their modeling and irregularities, important requirements that qualify them as "interacting" elements are not addressed including the measure required against their out-of-plane failure.

The soil–foundation stiffness shall be properly incorporated into the structural model, where they are likely to affect the structural response and govern the design. The dynamic soil–structure interaction (DSSI) shall be considered especially in structures with significant second-order effects, structures with deep foundation or foundations supported on very soft soils. The analytical model in such cases shall incorporate vertical, horizontal and rotational foundation and soil flexibility. Detailed guidelines to include the SSI effects both in linear dynamic analysis and nonlinear response history analysis are provided in ASCE 7. Further, the calculations for the evaluation of soil hysteretic and radiation damping including the stiffness are provided in detail. IS 1893 does not provide any guidelines for including DSSI effects in the structural model.

The displacement-based approach which focuses on designing structures to achieve target displacement requires that a detailed definition of displacement spectra is available. Further, for structures with period larger than 2.0 s, such an approach would be more relevant as displacements would be more sensitive. However, IS 1893 does not provide guidelines for the definition of a displacement spectra. A detailed definition of displacement spectra for various periods is available in EC8. The elastic displacement response spectra as per EC8 for various ground types normalized by the ground displacement for ground type A are shown in Fig. 4. The attenuation equations for the construction of displacement design spectra for various magnitudes and distances are also available [13].

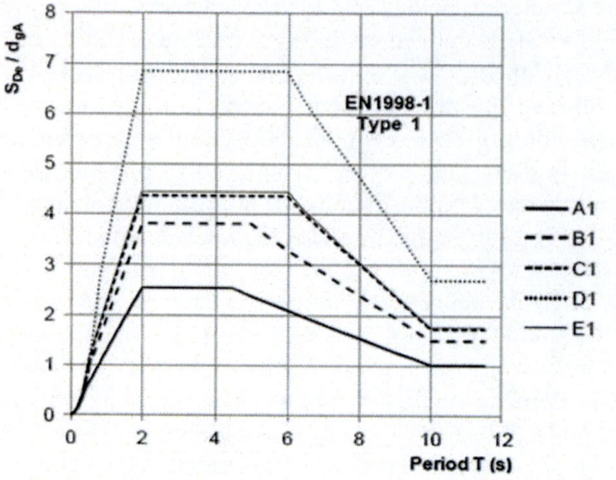


**Fig. 4** Normalized elastic displacement response spectra (EC8)

## 2.3 *RCC Structures*

The choice of a proper structural system for earthquake resistance in reinforced cement concrete buildings is mainly governed by the seismic demands, stiffness and redundancy requirements. A schematic representation of various earthquake-resistant structural systems widely used in reinforced concrete buildings is shown in Fig. 5. ACI 318 [14] is used as a basis for the formulation of most of the guidelines given in IS 13920: 2016 [15]; however, the omission of some of the critical requirements in ACI 318 is likely to result in over-conservative designs. The code IS 13920 provides criteria for the ductile design and detailing of RC structures, but applicable only for monolithic RC construction and not for precast RC structures. The absence

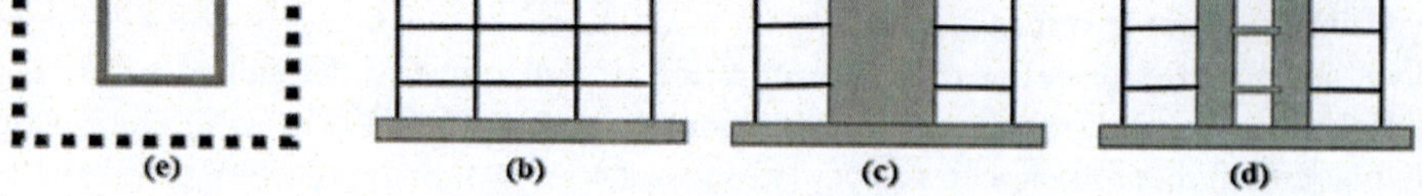


**Fig. 5** Earthquake-resistant systems in RCC buildings: **a** floor plan, **b** moment resisting frame, **c** structural shear wall, **d** coupled shear wall, **e** tube system

of detailing requirement to achieve a required level of ductility in case of precast RC structures results in uncertainty in their design and construction. The detailing provisions in EC8 relevant to precast RC structures are worth to note in such a case. The code also does not provide any design criteria for flat slab structures and instead refers to specialist literature leaving the scope for misuse. Further, it requires that the adequacy of design for such structures be verified by nonlinear dynamic analysis. This approach discourages the design engineers from using flat slab structures and makes the task more difficult in absence of any code specified criteria for nonlinear response history analysis such as the selection of ground motion records, hysteresis model and acceptance criteria of analytical results.

Currently, IS 13920 does not have any provision to limit the beam eccentricity measured with respect to column axis required to ensure efficient transfer of beam forces to supporting columns. EC8 requires that such an eccentricity shall be limited to $0.25b_c$, $b_c$ being the column width normal to beam axis. IS 13920 provides guidelines to achieve the required ductility levels for of primary seismic beams but does not suggest any parameters to ensure its achievement. EC8, on the other hand, uses the parameter curvature ductility factor ($\mu_\phi$) to assess the ductility levels in addition to providing prescriptive rules for its achievement. The limitation to spacing of links in the critical region to 100 mm in IS 13920 is likely to result in congestion. It shall be noted that the corresponding spacing is 150 mm in ACI 318. For the computation of shear strength in critical regions, IS 13920 requires that the concrete shear strength is ignored in the shear capacity calculations. However, ACI 318 very specifically mentions that such an over-conservatism in the computation of shear strength could be allowed only subject to conditions such as the level of shear forces induced by the seismic loading and the axial compression in beams. Further, while the Indian code uses a factor of 1.40 for beam moments required for computation of shears in critical region, ACI 318 requires that the computation of shears be based on a probable flexural strength computed on the basis of stress level of $1.25f_y$ in longitudinal tension reinforcement. The Indian code does not include the effect of vertical seismic actions in the computation of beam shear, whereas the same is considered in ACI 318.

In IS 13920, there exists lack of clarity as whether to choose the limiting aspect ratio for columns as 0.40 or 0.45. As per the code, the sum of nominal design strengths of column shall be at least 1.40 times the sum of the design strengths of beams at each beam-column joint. It shall be noted that the corresponding factor in ACI 318 is 1.20, whereas it is 1.30 in EC8, thereby introducing a higher estimate for overstrength without any basis. The Indian code specifies that the spacing of closed links in locations of lapped column bars shall not be more than 100 mm, whereas the same could be taken as 150 mm as per ACI 318. The required column shear demand in IS 13920 is computed on the basis of beam end moments at yielding, whereas ACI 318 computes it on the basis of maximum probable moments at the column ends. The requirement of special confining reinforcement is applicable to only columns, whereas IS 13920 mentions that it is applicable for both beams and columns and requires immediate correction. The minimum area of cross section of links or spiral forming the special confining reinforcement is given in ACI 318 as per the axial load levels and the concrete grade, whereas IS 13920 provides such a requirement without any consideration to the abovementioned factors. In addition, the philosophy for the computation of shear capacities of columns at locations of special confining reinforcement is missing in the Indian code.

The definition of effective width of joint shall be made in alignment with ACI 318, which clearly mentions that except where a beam spans into a wider column, effective joint width shall be taken as the overall width of column. The effective width and breadth are used interchangeably in IS 13920 and the figure for beam-column joint shall be reviewed in particular the shaded area. The criteria for the computation of shear demands across a joint as provided in IS 13920 corresponding to a stress level of $1.25f_y$ in the beam longitudinal bars are based on ACI 318 and the same philosophy shall also be applied uniformly in other cases of shear demand calculations as beams and columns.

ACI 318 specifies that two curtains of reinforcement shall be provided in slender walls with the in-plane aspect ratio greater than 2.0, to improve the lateral stability of compression zone. However, such a requirement is missing in IS 13920, wherein the provision for two curtains is based only on the shear demand and wall thickness. The computation of shear strength of structural walls in ACI 318 also reflects the importance of in-plane aspect ratio of walls. However, the shear strength computed in IS 13920 depends only on the concrete grade and reinforcement percentage and does not account for wall aspect ratio. Further, it does not provide an upper limit to the shear strength of coupling beams. ACI 318 requires the boundary elements to be provided only for walls with in-plane aspect ratio more than 2.0; however, such a requirement is missing in the Indian code. The horizontal extent of boundary elements measured from extreme compression fiber and the minimum width required for flexural compression zone are not furnished in IS 13920. Further, it does not provide the minimum dimension requirements applicable to rectangular transverse reinforcement for diagonally reinforced coupling beams.

## 2.4 *Steel Structures*

A number of seismic structural systems are used in steel structures, some of which are schematically presented in Fig. 6. The seismic code currently refers to section-12 of IS 800:2007 [16] for ductile detailing of steel members. However, section-12 of the code provides only limited guidance, in most cases with no basis. In the absence of proper guidelines, design engineers involved in the detailed engineering of steel structures in various critical industries such as petrochemicals, power, refineries and petrochemicals do not have any option other than depending on international codes of practice. It is felt that a dedicated document similar to IS 13920 for RC Structures is made available for seismic design and ductile detailing of steel structures. The codes AISC 341-16 [17] and EC8 are used as benchmarks for the review of steel structures.

The code does not provide an upper limit for yield stress of steel members to be used as a part of dissipative zone in seismic force resisting system. Partial safety

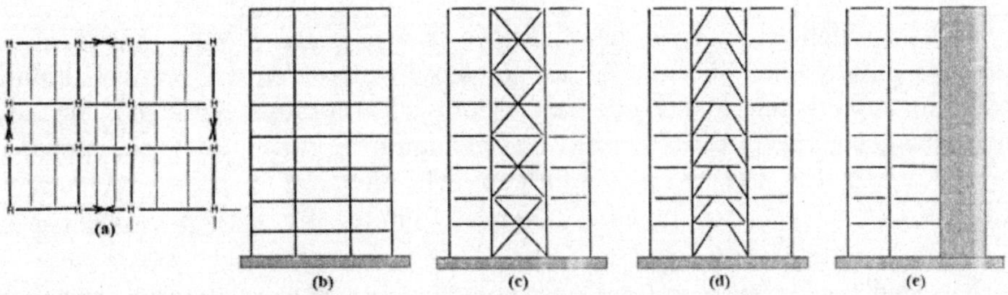


**Fig. 6** Earthquake-resistant systems in steel str.: **a** floor plan, **b** moment resisting frame, **c** concentrically braced frame, **d** eccentrically braced frame, **e** steel shear wall

factors of 1.20 and 1.50 associated with the earthquake loads for the limit state design of steel structures will result in highly conservative designs considering that most of the international codes use a load factor of 1.0 for earthquake loads. Further, a load factor of 2.50 assigned to earthquake loads for the verification of column members, bracing connections and connections of special moment resisting frames (SMRF) is highly unacceptable, first considering that such requirements do not have any basis, and secondly, structures designed as per these provisions would result in highly uneconomic designs. As clearly indicated in international codes (EC8, AISC 341), a higher load factor shall be considered only for columns located in non-dissipative zones which shall essentially remain elastic during the development of inelasticity in the connected members. However, the corresponding load factors shall be obtained on case-to-case basis based on the material overstrength and capacity ratios of connected members in dissipative zones. For bracing connections, the design forces shall be based only on the available material overstrength.

It is noted that a mismatch exists in *R* values recommended for ordinary moment frames (OMF) of steel structures, the value being 3 in IS 1893 and 4 in IS 800. To avoid such situations, it is recommended to provide the *R* values relevant to steel structures only in IS 800. Based on the ability of structure for energy dissipation through inelastic behavior, EC8 recommends two ductility classes DCM (Medium) and DCH (High) for steel structures, and the same is missing in IS 800. In the case of concentric braced frames (CBF), the values of *R* used in IS 800 are not specific to the configuration of bracings. It is well known that structures with active tension only diagonal members perform better than the one with tension and compression diagonal members. The above is reflected in EC8 wherein an *R* value of 4.0 is recommended for diagonal bracings and a lower value of 2.50 for V-bracings of DCH class. Recommendation for a specialist literature in the case of most commonly used V-type bracing could lead to its misuse. It shall be noted that the brace connecting beams in case of V-type bracings shall be suitably sized considering the unbalanced vertical seismic action effect after the buckling of compression diagonals. EC8 also allows the design engineers to use a higher value of behavior factor for moment resisting frames (MRF) and eccentrically braced frame (EBF) based on the structural overstrength and the sequence of plastic hinge formation in the structure obtained typically through a nonlinear static pushover analysis. The *R* values for moment resisting frames with concentric braces and those with masonry or reinforced infill walls are missing in IS 800. While the code provides very limited design guidelines for ordinary and special concentrically braced frames, the guidelines for eccentrically braced frames (EBF) are missing in the code. The reference to special literature for EBF, in fact, would discourage the engineers from the use of such a system with superior ductility.

The axial load levels to be used in the computation of column moment capacities to verify capacity ratios of columns to beams in SMRF are missing in the code. The code also does not provide any requirements for stability bracings required in highly ductile members. Detailed guidelines are required to be provided for both the panel zone doubler plates and continuity plates. In the case of column splices, guidelines for bolted column splices and seismic shear demand for column Web splices verification are not provided. Guidelines are to be included in the code for the ductile design of structural systems such as special truss moment frames (STMF), buckling-restrained braced frames (BRBF) and special plate shear walls (SPSW).

Design rules and detailing guidelines for dissipative structural behavior shall be provided in detail for all structural types including their connection in order to achieve the required ductility levels. The current level of detailing provided in IS 800 (Section-12) is highly incomplete and would result in undesired structural performance and misuse of several provisions. Therefore, it is very important that the current seismic provisions relevant to steel structures are revised at earliest in line with the international codes of practice such as AISC 341.

## 3 Conclusions

A critical review of Indian seismic codes for buildings is done taking various international codes of practice as the benchmark. The review shows that the recently revised Indian seismic codes for buildings are plagued by conservatism, empirical approach and omission of much-needed procedures and guidelines. It is strongly felt that the Indian seismic codes need to embrace the latest developments in earthquake engineering and detailing. While the ductile detailing of RCC structures is in place with enough scope for further improvement, earthquake detailing of steel structures is still an area of great concern. It is expected that a separate code dedicated to seismic design and detailing of structural steel buildings will be developed in near future in line with international codes. Further, seismic design requirements for structures with seismic isolation and damping systems are yet to be developed. The existing codes also do not cover many of the structural systems being formulated and widely used in the seismic regions of the developed world for the last three decades.

## References

1. IS 1893 (Part 1): 2016. Indian standard criteria for earthquake resistant design of structures, part 1: general provisions and buildings (sixth revision). Bureau of Indian Standards, New Delhi
2. ASCE 7-16. Minimum design loads and associated criteria for buildings and other structures. American Society of Civil Engineers (ASCE), Reston
3. BS EN 1998-1:2004. Eurocode 8, design of structures for earthquake resistance, part 1: general rules, seismic actions and rules for buildings. Brussels
4. Anbazhagan P, Prabhu G, Moustafa Sayed SR, Arifi Al Nassir S, Parihar A (2014) Provisions for geotechnical aspects and soil classification in Indian seismic design code IS-1893. Disaster Adv 7(3):72–89
5. Mittal RK, Rawat S, Tangirala A (2017) Shortcomings in the seismic site classification provisions of IS 1893-1:2016. In: Indian geotechnical conference
6. Zhao J, Zhang J, Irikura K (2005) Effects of earthquake magnitude and source distance on the response spectral amplification ratios of soft-soil sites. In: NZSEE conference
7. Katsanos EI, Sextos AG, Manolis GD (2010) Selection of earthquake ground motion records: a state-of-the-art review from a structural engineering perspective. Soil Dyn Earthq Eng 30:157–169
8. Haselton CB (2009) Evaluation of ground motion selection and modification methods: predicting median interstory drift response of buildings. PEER report 2009/01. Pacific Earthquake Engineering Research Center, Berkeley
9. ASCE 41-17. Seismic evaluation and retrofit of existing buildings. American Society of Civil Engineers (ASCE), Reston
10. NIST GCR 17-917-45 (2017) Recommended modelling parameters and acceptance criteria for nonlinear analysis in support of seismic evaluation, retrofit and design. Prepared by Applied Technology Council for National Institute of Standards and Technology, Gaithersburg
11. ATC-40 (1996) Seismic evaluation and retrofit of concrete buildings. Applied Technology Council, Redwood City
12. FEMA-445 (2006) Next-generation performance-based seismic design guidelines. Prepared by Applied Technology Council for Federal Emergency Management Agency, Washington
13. Bommer JJ, Elnashai AMRS (1999) Displacement spectra for seismic design. J Earthq Eng 1–32
14. ACI 318-19. Building code requirements for structural concrete. ACI Committee 318. American Concrete Institute
15. IS 13920: 2016. Ductile design and detailing of reinforced concrete structures subjected to seismic forces-code of practice (first revision). Bureau of Indian Standards, New Delhi
16. IS 800: 2007. Indian standard general construction in steel-code of practice (third revision). Bureau of Indian Standards, New Delhi
17. AISC 341-16. Seismic provisions for structural steel buildings. American Institute of Steel Construction (AISC), Chicago

# An Application to the Conditional Mean Spectrum Approach for Seismic Analysis of Coupled Dam–Foundation System

**Aniket R. Tidke and Shrabony Adhikary**

**Abstract** The nonlinear time-history analysis is the most suitable method to depict the seismic response of a concrete gravity dam. Also, ground motion selection is an important aspect of seismic analysis. This study focuses on the seismic response analysis of Koyna Dam–foundation system. The material nonlinearity is specified to the dam and foundation using the concrete damaged plasticity model and Mohr–Coulomb failure model, respectively. The 2D finite element model of the dam–foundation system is analyzed using ABAQUS. The infinite elements are used for the finite element analysis of the unbounded domain of foundation. The seismic analysis is conducted for ten ground motions that are selected using the conditional mean spectrum approach. Additionally, the ten ground motions are modified using the spectrum compatible method, and the effect of the horizontal component of earthquake ground motion is considered for the dynamic analysis. The damage indices are evaluated based on damage dissipation energy and crack length. Meanwhile, the damage index value shows the global damage states of the gravity dam, and it can be considered as a damage measure parameter for the seismic analysis. The importance of the conditional mean spectrum approach over the spectrum compatible method is also discussed. It is concluded that the earthquake frequency content significantly influences the seismic response of the gravity dam.

**Keywords** Koyna dam · Conditional mean spectrum · Ground motion characteristics

A. R. Tidke (✉) · S. Adhikary
Department of Civil Engineering, VNIT, Nagpur 440010, India
e-mail: aniket.tidke8@gmail.com

S. Adhikary
e-mail: shrabonyeq@gmail.com

 
T. G. Sitharam et al. (eds.), *Seismic Design and Performance*, Lecture Notes in Civil Engineering 120, https://doi.org/10.1007/978-981-33-4005-3_3

# 1 Introduction

Gravity dams serve a strong purpose like water storage, flood control, irrigation, power generation, etc. In India, many dams are constructed in recent years, as its demands have been increasing rapidly. However, the large number of gravity dams is situated in the seismic zone IV and V according to Indian seismic code IS 1893 (part 1)-2016 [1]. Earthquakes are very uncertain and cause a severe level of damages in the superstructures. Therefore, seismic analysis and design of dams are very important to examine for engineers and researchers. Many man-made structures are affected due to ground motions, viz. Bhuj, Koyna, Uttarkashi, etc. The 6.5 magnitude Koynanagar earthquake devastated the Koyna gravity dam, and cracks were observed on both faces of the dam [2]. Since then, the Koyna Dam is considered a benchmark problem to study the seismic analysis of gravity dams. In earthquake analysis of dams, interactions between two different materials play a significant role. Also, ground motion characteristics, viz. Amplitude, frequency content, duration, are strongly responsible for the damages in the appurtenant structure such as dams. Many of the existing dams were damaged due to various issues such as earthquakes, a foundation problem, poor construction, inadequate spillways, and unequal settlement. So, it is very important to study the static as well as dynamic response analysis of gravity dams. The primary objective of this study is to design the gravity dams for future earthquakes so that it can withstand during strong ground motions.

In this study, Koyna Dam–foundation system is adopted for the numerical investigation. The nonlinear time-history analysis is performed, and the concrete damaged plasticity (CDP) model is adopted to show cracks in the dam. The foundation is modeled using Mohr–Coulomb model. The upstream face of the Koyna Dam is considered as perfectly vertical. Earthquake ground motions are selected according to the conditional mean spectrum (CMS) approach, and the selected ground motions are again matched with IS 1893 (part 1)—2016 [1] zone IV spectrum using spectrum compatible method. Then, the two sets of ground motions are used as input motion for the time-history analysis. Various damage measure parameters are also adopted here to characterize the earthquake ground motions. Additionally, damage indices, viz. local damage index and global damage index are also evaluated based on damage dissipation energy and crack length.

# 2 Materials and Methodology

## *2.1 Finite Element Analysis of the Koyna Dam–Foundation System*

The finite element analysis of the Koyna Dam–foundation system is performed. Figure 1 shows the FEM model of the Koyna Dam–foundation system, and it is modeled in the FEM software, Abaqus. The dam and foundation domains are modeled

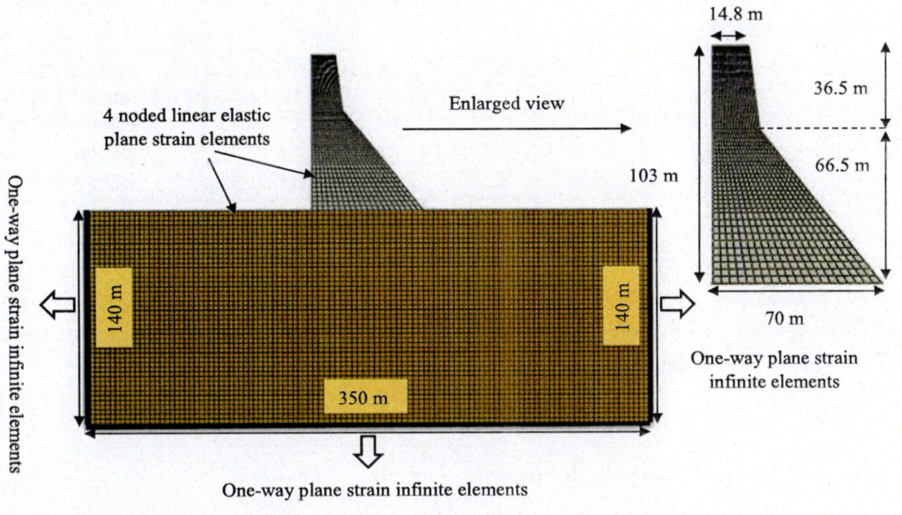


**Fig. 1** FEM model of coupled Koyna Dam–foundation system

with four-noded linear elastic plane strain elements. The concrete damaged plasticity (CDP) model is used for the dam, and Mohr–Coulomb failure model is used for the foundation. The dam–foundation system is analyzed for empty reservoir condition, so the effect of hydrodynamic pressure and uplift pressure is not considered for the dynamic analysis. The Koyna Dam is assumed to be resting on a 350 × 140 m foundation block. The horizontal component of ground motion is applied to the foundation base as an input motion for the seismic analysis, and the nodes of the foundation base are fixed along the vertical direction. For dam–foundation interaction, the nodes along dam base are tied with the nodes along the foundation. The one-way plane strain infinite elements are used for the far end of the foundation domain to allow way propagation effect. Table 1 shows the linear material properties of the dam and foundation, whereas Table 2 shows the material properties of the CDP model for the dam. For Mohr–Coulomb failure model, the plasticity and cohesive properties are assigned appropriately. For numerical simulation, an implicit integration scheme is used.

The frequency analysis of the coupled dam–foundation system is performed before the dynamic analysis. The natural frequencies (Hz) for the first five modes are 2.7950 Hz, 4.1113 Hz, 6.7435 Hz, 7.1781 Hz, and 7.8181 Hz, respectively. Using

**Table 1** Linear material properties of dam and foundation

| Sr. No. | Parameters | Dam | Foundation |
|---|---|---|---|
| 1 | Elastic modulus ($E$) (MPa) | 31,027 | 62,054 |
| 2 | Poisson's ratio ($\mu$) | 0.2 | 0.33 |
| 3 | Mass density ($\rho$) (kg/m$^3$) | 2643 | 3300 |

**Table 2** Nonlinear material properties of the dam superstructure (CDP model)

| Dilation angle ($\psi_c$) | Compressive initial yield stress ($\sigma_{co}$) (MPa) | Compressive ultimate stress ($\sigma_{cu}$) (MPa) | Tensile failure stress ($\sigma_{to}$) (MPa) | Flow potential eccentricity (e) | Ratio of the initial equibiaxial to the uniaxial compressive yield stress |
|---|---|---|---|---|---|
| 36.31° | 13 | 24.1 | 2.9 | 0.1 | 1.16 |

the first and second mode natural frequency, the Rayleigh damping coefficients are evaluated and assigned to the dam as well as foundation.

## 2.2 *Ground Motion Selection*

In recent years, a lot of studies are presented on the ground motion selection procedure. Every earthquake prone-region has different geological conditions. So, it is very important to select the input ground motions that contain the suitable ground motion characteristics which matches to the seismic site conditions. Baker [3] proposed the efficient tool for ground motion selection called the conditional mean spectrum (CMS) approach. The probabilistic seismic hazard analysis (PSHA) was used to develop the CMS approach. This CMS approach gives the mean spectrum which can be used as the target spectrum for ground motion selection. The procedure for spectrum calculation is presented earlier [4]. Jayaram et al. [5] presented the algorithm for the ground motion selection in which the ground motions are matched with target spectrum mean and variance. Later, the algorithm proposed by Jayaram et al. [5] was improved [6]. Bernier et al. [7] used the CMS method for the fragility assessment of gravity dam, which was originally proposed by Jayaram et al. [5].

In this study, the input ground motions are selected using the CMS approach, and then, the selected ground motions are again matched with the target response spectrum (IS 1893 (part 1)—2016) [1] using spectrum compatible method. The mean spectrum is generated using the MATLAB code developed by Baker [3]. The CMS gives the target spectrum, conditioned on the appearance of the target spectral acceleration at certain period of interest Sa($T_i$). For CMS, the first mode period is considered as a period of interest. The modal analysis of dam–foundation system is done in Abaqus using Lanczos iteration method. Thus, the first mode natural frequency is 2.7950 Hz, so the fundamental period of the system is calculated as 0.3578 s.

The other important parameters such as magnitude, site to fault distance, fault type, and average shear-wave velocity to 30 m depth ($Vs_{30}$) for Koyna (1967) earthquake are considered appropriate for this study. Figure 2a shows the conditional mean spectrum that is used for target spectrum conditioned on the appearance of target spectral acceleration (0.6 g) at period of interest (0.3578 s). The ten ground motions

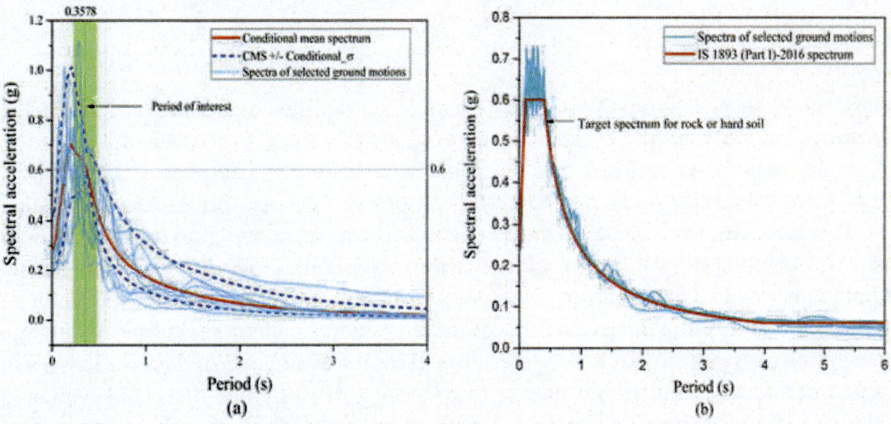


**Fig. 2** Selected ground motions for the seismic analysis using: **a** conditional mean spectrum approach, **b** spectrum compatible method

are selected from PEER ground motion database [8], and these ground motions are shown in Table 3. Figure 2b shows the ten selected ground motions that are matched with the target spectrum (IS 1893 (part 1)—2016) [1] for rock or hard soil. It is obvious that the earlier ground motions have different ground motion characteristics, whereas the later ground motions contain nearly the same peak ground acceleration (PGA) and frequency content but different durations.

**Table 3** Selected ground motions using conditional mean spectrum approach

| No. | Seismic event | Date | Station | ($M$) | Distance to fault (km) ($R_{jb}$) | $Vs_{30}$ (m/s) | PGA ($H$) | Total duration ($T$) |
|---|---|---|---|---|---|---|---|---|
| TH1 | Koyna | 11/12/1967 | Koyna dam | 6.3 | 13 | 800 | 0.41 | 10 |
| TH2 | Parkfield | 28/06/1966 | Temblor-Pre | 6.19 | 15.96 | 527.92 | 0.28 | 30.39 |
| TH3 | Landers | 1992 | Joshua Tree | 7.28 | 11.03 | 379.32 | 0.27 | 21.93 |
| TH4 | Coyote Lake | 6/8/1979 | Gilroy Array # | 5.74 | 3.11 | 663.31 | 0.26 | 26.83 |
| TH5 | New Zealand—02 | 2/3/1987 | Matahina Dam | 6.6 | 16.09 | 551.3 | 0.25 | 24.6 |
| TH6 | Kocaeli, Turkey | 17/08/1999 | Izmit | 7.51 | 7.21 | 811 | 0.23 | 29.99 |
| TH7 | Irpinia, Italy-01 | 23/11/1980 | Brienza | 6.9 | 22.56 | 561.04 | 0.22 | 34.98 |
| TH8 | Northridge—01 | 17/01/1994 | LA—Chalon Rd | 6.69 | 20.45 | 740.05 | 0.22 | 31 |
| TH9 | Santa Barbara | 13/08/1978 | Cachuma Dam Toe | 5.92 | 27.42 | 465.51 | 0.21 | 13.18 |
| TH10 | Northern Calif—07 | 7/6/1975 | Cape Mendocino | 5.2 | 34.73 | 567.78 | 0.21 | 29.14 |

## 2.3 Damage Indices

The damage indices show the structural damage and often considered as the damage measure parameter in the seismic analysis of gravity dam. Moreover, the selection of proper response quantities for the evaluation of damage indices is also important. A lot of studies were presented by various researchers on the damage index and the associated response quantities. The seismic response quantities of gravity dam can be crest displacement, crest acceleration, damage dissipation energy, etc. Alembagheri and Ghaemian [9] adopted the maximum deformation, and damage dissipation energy for the evaluation of damage indices. The crack length, damage dissipation energy and drift ratio were adopted for the evaluation of damage indices in local, intermediate, and global state of gravity dam [10]. Alembagheri [11] evaluated damage indices of the pine flat dam, Koyna Dam, and Folsom dam based on cumulative inelastic area—demand capacity ratio (CIA-DCR) response curve. Zhang and Wang [12] proposed a new type of local and global damage index for gravity dam based on damage dissipation energy and crack length. Various researchers were used the damage index proposed by Zhang and Wang [12] for the seismic analysis of the gravity dam [13–16].

In this study, the damage indices (local damage index and global damage index) are evaluated based on damage dissipation energy and crack length in the dam. Initially, the local damage index at base of the dam ($\mathrm{DI_L}$) and at neck of the dam is evaluated based on crack length, and finally, the global damage indices are evaluated based on local damage index ($\mathrm{DI_L}$) and damage dissipation energy. The local damage index ($\mathrm{DI_L}$) at different paths and global damage index ($\mathrm{DI_G}$) is evaluated using Eq. (1) and (2), respectively. Figure 3 shows the tensile cracks in neck and base of the dam. The compression crushing and tensile cracking can be the two main failures occurred in the concrete, and it can be depicted using CDP model in the numerical investigation.

$$\mathrm{DI_{Li}} = \frac{l_i}{L_i} \tag{1}$$

$$\mathrm{DI_G} = \frac{\sum_{i=1}^{n} \mathrm{DI}_{\mathrm{L}i} \times E_i}{\sum_{i=1}^{n} E_i} \tag{2}$$

Where $l_i$ is the crack occurred in the $i$th path and $L_i$ is the total length of crack path $i$ which is expected to occur. $\mathrm{DI}_{\mathrm{L}i}$ is the local damage index for the $i$th crack path, and $E_i$ is the damage dissipation energy in the $i$th crack path and $n$ is the number of crack path in the dam.

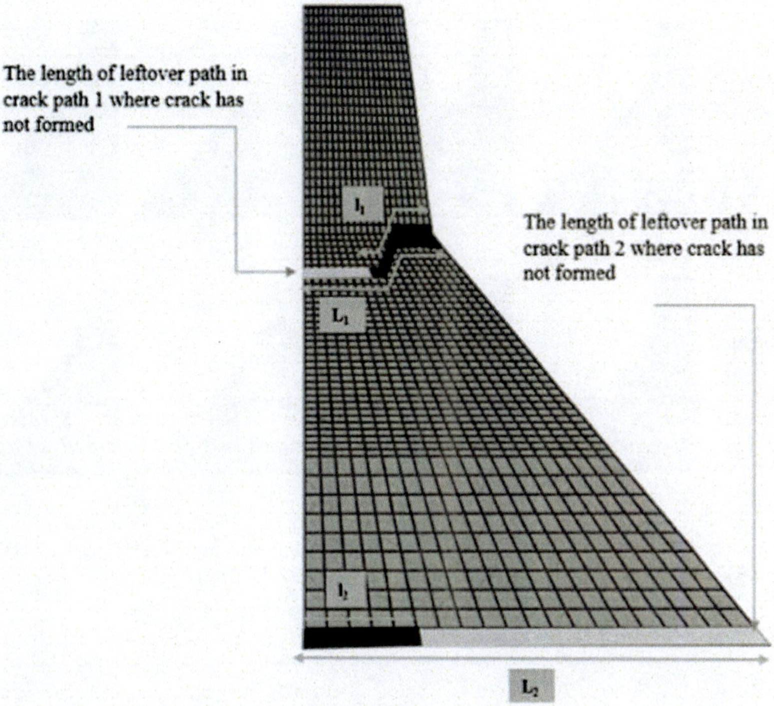


**Fig. 3** Crack formation in the gravity dam

## 3 Numerical Analysis Outcomes

The coupled Koyna Dam–foundation system was analyzed for the effect of different ground motions. The conditional mean spectrum approach and spectrum compatible method were adopted for the ground motion selection. The total ten ground motions were selected, and the nonlinear time-history analysis was performed by considering the material nonlinearity. In this section, the numerical analysis results are presented and discussed.

Figure 4 shows the failure profile of the Koyna Dam for the considered ground motions. Figure 4a, b shows the failure profile of the dam for TH1 and TH2. Similarly, Fig. 4c, d shows the failure profile of the dam for TH7 and TH8. The failure profile is shown for recorded and artificial accelerograms. It is observed that the maximum level of damages in the dam is seen for TH2 (Fig. 4b). The PGA of TH1 (0.41 g) is more than TH2 (0.28 g), even though more level of damages are seen for TH2 than TH1. Moreover, as a result of TH2, high amount of damage dissipation energy is imparted in the dam body. The effect of frequency content is significant; in addition, the ground motion characteristics, viz. PGA, frequency content, and duration of TH2, can be a primary cause of tensile damages. Moreover, TH7 and TH8 contain the same PGA and nearly same duration, but the extent of failure in the dam is fairly different

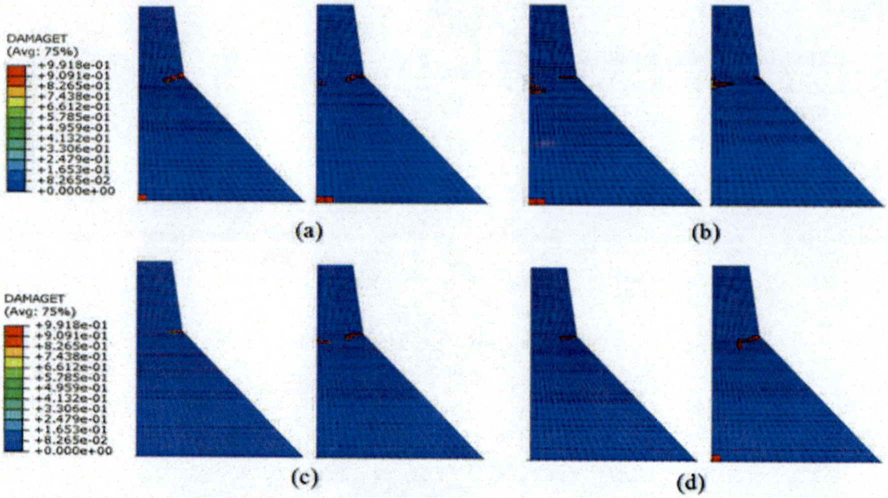


**Fig. 4** Failure profile of the dam for the considered ground motions: **a** TH1 (RA), TH1 (AA), **b** TH2 (RA), TH2 (AA), **c** TH7 (RA), TH7 (AA), **d** TH8 (RA), TH8 (AA)

(Fig. 4c, d). For the same ground motion, more damages are imparted as a result of the artificial accelerogram than recorded accelerogram except TH2. So, it can be said that the TH2 contains high-frequency content than the other ground motions presented.

Figure 5 shows the horizontal displacement response at dam crest for the considered ground motions, viz. TH1, TH2, TH7, and TH8. For the recorded accelerogram, peak crest displacement for TH2 is observed as 4.21 cm followed by TH1 (3.25 cm), TH8 (3.01 cm), and TH7 (2.64 cm). The artificial time histories contain the same PGA and frequency content, but different ground motion durations. Moreover, the difference in the duration of all ground motion is very less. The peak crest displacements for artificial time histories, viz. TH1, TH2, TH7, and TH8, are found as 3.85 cm, 3.79 cm, 3.62 cm, and 3.46 cm, respectively. The peak crest displacement response for all the considered ground motions is almost same. For the artificial ground motions, the effect of duration on the seismic response is insignificant.

The local and global damage index are also evaluated based on damage dissipation energy and crack length. Figure 6a shows the local damage index for neck and heel of the dam, while Fig. 6b shows the global damage index for the entire dam. The mean of local damage index ($DI_L$) for neck (RA), neck (AA), base (RA), and base (AA) is 0.025, 0.035, 0.326, and 0.495, respectively. The mean of global damage index ($DI_G$) for RA and AA is 0.294 and 0.429, respectively. The difference in between the mean $DI_L$ between RA and AA is very less; however, the difference between the mean $DI_G$ of RA and AA is significant. Moreover, all the artificial ground motions produce nearly the same response; nevertheless, the difference in between the response results of the recorded and artificial accelerograms is significant. For TH5 and TH9, the global damage index shows a zero value for recorded

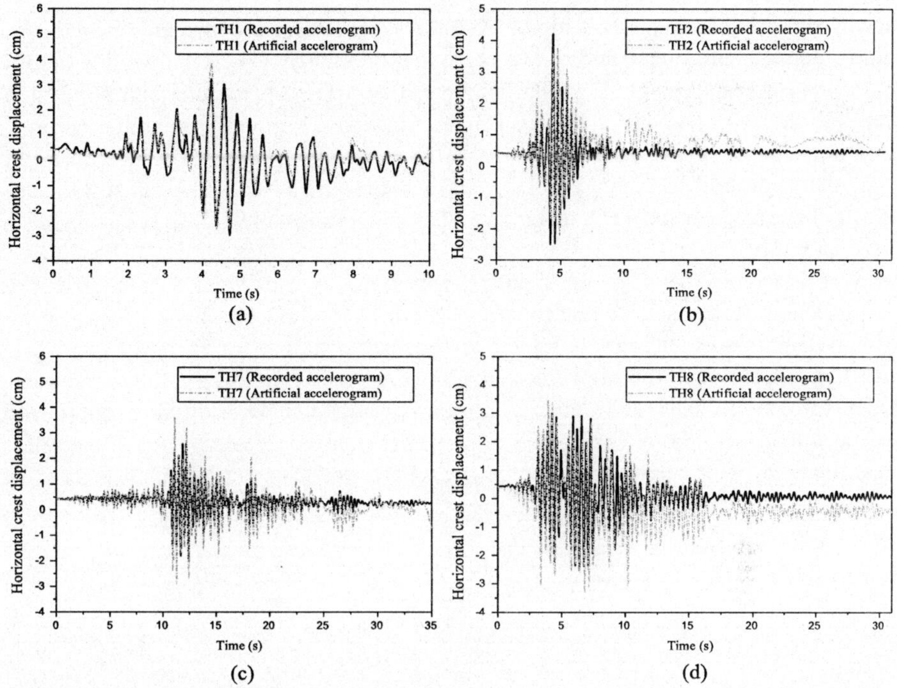


**Fig. 5** Comparison of horizontal crest displacement response for recorded and artificial accelerogram: **a** TH1, **b** TH2, **c** TH7, **d** TH8

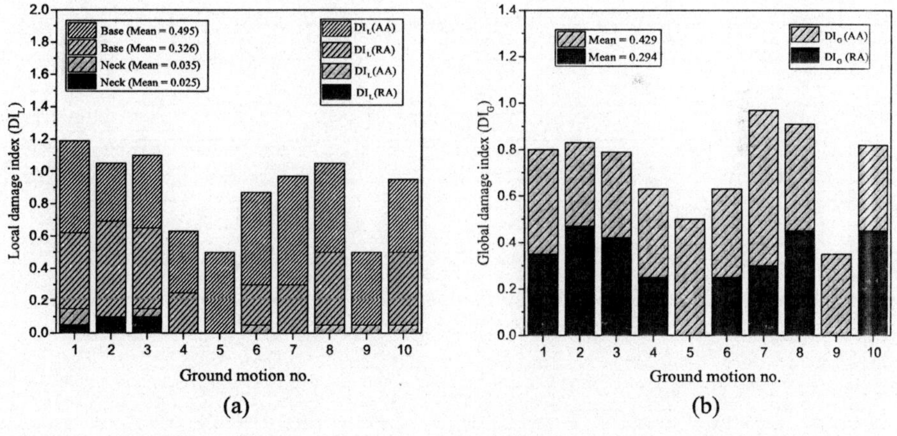


**Fig. 6** Damage indices: **a** local damage index ($DI_L$), and **b** global damage index ($DI_G$)

accelerogram. It is interesting to note that the superstructure withstands even during high-intensity earthquake and shows slight to moderate level of damages due to low-intensity earthquakes. Thus, global damage index ($DI_G$) parameter can be used as damage measure parameter for the seismic analysis. Figure 7 shows the comparison between the peak responses of the various damage measure parameters for the considered ground motions. For the recorded time histories, the peak response is mostly observed for the TH1, TH2, and TH3, and the least response is observed for TH5 and TH9.

The difference between the peak responses for the artificial accelerograms is insignificant. It is observed that the peak response for TH1 (RA) is as a result of its high peak ground acceleration. However, the peak response for TH2 (RA) and TH3 (RA) is as a result of the high-frequency content. It is also observed that the mean value of damage dissipation energy, horizontal crest acceleration, horizontal crest displacement, and the normal stress at heel for recorded accelerogram is quite less than the artificial accelerogram. Moreover, the recorded accelerogram gives an exact

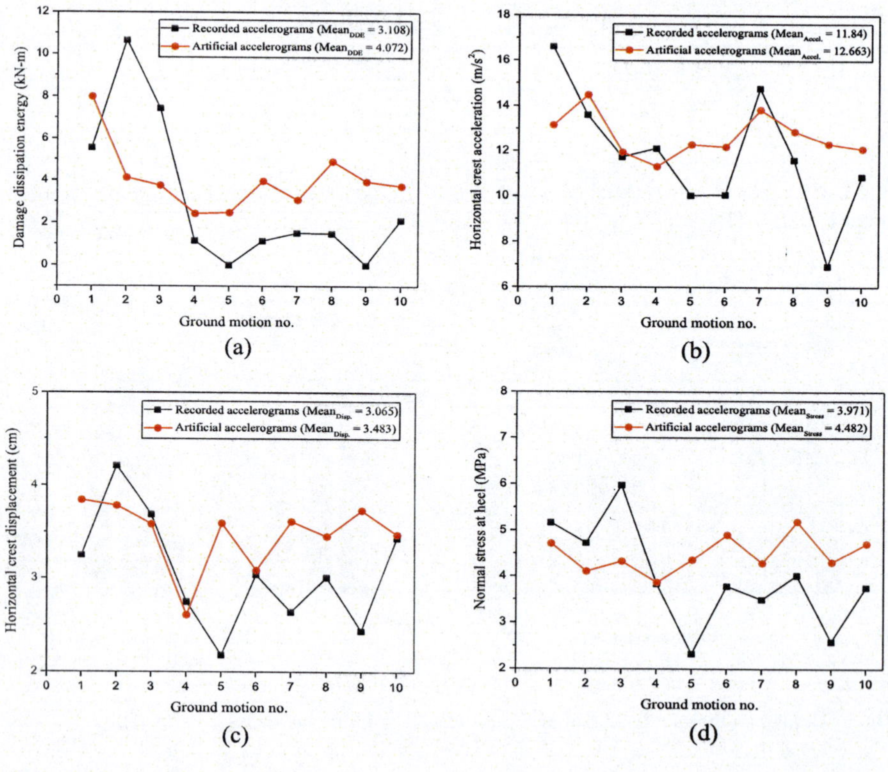


**Fig. 7** Comparison of peak seismic response in between recorded and artificial accelerogram: **a** damage dissipation energy, **b** horizontal crest acceleration, **c** horizontal crest displacement, **d** normal stress at heel

seismic response than the artificial accelerogram, as recorded accelerogram contains different ground motion characteristics, viz. PGA, frequency content, duration.

## 4 Conclusions

The work was done on the nonlinear time-history analysis of the Koyna Dam–foundation system. The nonlinear material behavior of the dam and foundation is simulated using the damaged plasticity model and Mohr–Coulomb model, respectively. The conditional mean spectrum approach was employed for the ground motion selection, and the numerical investigation was done. Based on the failure investigation results, the following conclusions are presented.

1. The maximum damage dissipation energy is imparted in the dam for TH2 (Fig. 7a), even if the PGA of TH2 (0.28 g) is quite less than TH1 (0.41 g). Moreover, the PGA and duration of TH5 are more than TH9, still the seismic responses (crest displacement and heel stress) as a result of TH5 are very less. It is concluded that the PGA and duration of ground motion influence the seismic response of the gravity dam. However, the effect of frequency content is predominant for the greater damages in the dam.
2. It is advisable to consider the actual ground motion for the time-history analysis, since the ground motion shows different characteristics. Consequently, the conditional mean spectrum is a suitable method for ground motion selection.
3. The global damage index is known to be a good damage measure parameter, as it shows the overall damage in the superstructure.
4. The seismic failure in the dam can be better represented by the damage measure parameters such as damage dissipation energy and horizontal crest displacement (Fig. 7a, c).

## References

1. IS 1893 (Part 1)-2016 Criteria for earthquake resistant design of structures: general provisions and buildings. Bureau of Indian Standards, Delhi
2. Nuss LK, Matsumoto N, Hansen KD (2012) Shaken, but not stirred—earthquake performance of concrete dams. United States Society on Dams
3. Baker JW (2011) Conditional mean spectrum: tool for ground-motion selection. J Struct Eng 137(3):322–331
4. Baker JW, Cornell CA (2006) Spectral shape, epsilon and record selection. Earthq Eng Struct Dyn 35(9):1077–1095
5. Jayaram N, Lin T, Eeri M, Baker JW (2010) A computationally efficient ground—motion selection algorithm for matching a target response spectrum mean and variance. Earthq Spectra 27(3):1–25
6. Baker JW, Lee C (2018) An improved algorithm for selecting ground motions to match a conditional spectrum. J Earthq Eng 22(4):708–723

7. Bernier C, Monteiro R, Paultre P (2016) Using the conditional spectrum method for improved fragility assessment of concrete gravity dams in Eastern Canada. Earthq Spectra 32(3):1449–1468
8. Peer Strong Motion Database (2000) https://peer.berkeley.edu/smcat
9. Alembagheri M, Ghaemian M (2013) Seismic assessment of concrete gravity dams using capacity estimation and damage indexes. Earthq Eng Struct Dyn 42(1):123–144
10. Hariri-Ardebili MA, Saouma V (2015) Quantitative failure metric for gravity dams. Earthq Eng Struct Dyn 44(3):461–480
11. Alembagheri M (2016) Earthquake damage estimation of concrete gravity dams using linear analysis and empirical failure criteria. Soil Dyn Earthq Eng 90:327–339
12. Zhang S, Wang G (2013) Effects of near-fault and far-fault ground motions on nonlinear dynamic response and seismic damage of concrete gravity dams. Soil Dyn Earthq Eng 53:217–229
13. Wang G, Zhang S, Zhou C, Lu W (2015) Correlation between strong motion durations and damage measures of concrete gravity dams. Soil Dyn Earthq Eng 69:148–162
14. Wang G, Wang Y, Lu W, Yan P, Zhou W, Chen M (2016) A general definition of integrated strong motion duration and its effect on seismic demands of concrete gravity dams. Eng Struct 125:481–493
15. Wang G, Wang Y, Lu W, Yan P, Zhou W, Chen M (2017) Damage demand assessment of mainshock-damaged concrete gravity dams subjected to aftershocks. Soil Dyn Earthq Eng 98:141–154
16. Wang C, Hao H, Zhang S, Wang G (2018) Influence of ground motion duration on responses of concrete gravity dams. J Earthq Eng 1–25

# Tsunami Resilient Foundation for Breakwater: Centrifuge Model Tests

**Babloo Chaudhary, Hemanta Hazarika, Akira Murakami, and Kazunori Fujisawa**

**Abstract** Many coastal protection structures collapsed due to the past earthquakes and tsunamis. For example, several breakwaters damaged during the 2011 Great East Japan Earthquake and Tsunami in Japan. Due to the failure of the breakwaters, the tsunami waves could not be blocked by the breakwaters. Thus, the tsunami entered in the coastal areas; and imposed deep devastation there. It was found that the breakwaters damaged mainly due to their foundation failures. In order to mitigate such damage of breakwater caused by earthquake and tsunami, new techniques were developed by the authors for breakwater foundation. In the technique, gabions and sheet piles are used in breakwater foundation. Effectiveness of the developed foundations model of breakwater were evaluated by conducting centrifuge model tests. It was observed that the developed models could mitigate damage, and make the breakwater resilient against earthquake and tsunami-induced damage.

**Keywords** Breakwater · Foundation · Earthquake · Tsunami · Damage · Resilient

## 1 Introduction

Breakwater is an offshore structure which is constructed to protect the ports and harbors from destructive effects of sea waves, typhoons and tsunamis. It provides calm water for safe navigation, anchorage of ships and other harbor activities at port and harbor. But in the past few decades, it was found that breakwaters are vulnerable to tsunami. For example, many breakwaters were collapsed by the 2004 Indian Ocean Earthquake and the 2011 off the Pacific Coast of Tohoku Earthquake and subsequent tsunamis. The world deepest breakwater at Kamaishi port (Iwate Prefecture), Japan

B. Chaudhary (✉)
National Institute of Technology Karnataka (NITK), Surathkal, India
e-mail: babloomit@gmail.com

H. Hazarika
Kyushu University, Fukuoka, Japan

A. Murakami · K. Fujisawa
Kyoto University, Kyoto, Japan



T. G. Sitharam et al. (eds.), *Seismic Design and Performance*, Lecture Notes in Civil Engineering 120, https://doi.org/10.1007/978-981-33-4005-3_4

was one of them which collapsed and failed to block the tsunami. The breakwater collapsed mainly due to failure of its foundation [1–7]. The mound was scoured by the overflowing tsunami. The caissons slid over the mound and sank into the sea. Due to failure of the breakwater, the tsunami entered into the coastal areas, and led to catastrophic losses for structures and population living near the sea coast [1–3]. There were some other breakwaters in Japan which were also collapsed mainly due to the failure of their foundations during the earthquake and tsunami. Therefore, safety of breakwater is very important for smooth operation of costal activities and human lives during such compound geo-disaster caused by earthquake and tsunami.

There are many earthquake prone coastal areas in the world which are susceptible to earthquake and subsequent tsunami in the future. In order to mitigate the devastating damage triggered by a tsunami in the future, breakwater foundation should be made enough strong to bear these destructive forces. For this purpose, countermeasure need to be established for foundation of breakwater so that it can make the breakwater tsunami resistant. Hence, it is utmost important to develop a reinforcing model for breakwater foundation which can mitigate the damage of the breakwater triggered by tsunami.

Some attempts have been made to develop a resilient breakwater foundation against tsunami. For example, Oikawa et al. [8] installed a row of steel piles in the harbor side of the mound to increase stability of the foundation during tsunami. Seepage through rubble mound is one of the main reasons of failure of a breakwater during tsunami. Ueda et al. [9] proposed to use membrane for covering the seaside mound to reduce seepage beneath the mound during tsunami. Kikuchi et al. [10] installed wall of piles to reduce destructive effects of tsunami. However, all those models were developed to increase the stability of the breakwater foundation against only tsunami. They were not aimed to mitigate the earthquake induced damage of breakwater foundation. Generally, an earthquake precedes a tsunami. Reduction in height of caisson due to settlement (caused by earthquake that precedes tsunami) is very significant, because maintaining the crown is extremely important for reducing the run-up height of tsunami waves.

Actually, strong earthquake often precedes tsunami. Pore water pressure is generated in the foundation by the earthquake which may cause of deformation of mound. It may result in excessive settlement and lateral displacement of the caisson. The effective height of the breakwater may reduce due to the settlement caused by the earthquake. If tsunami strikes the breakwater in this condition, breakwater may not block the tsunami waves, and tsunami can easily overflow the breakwater. Therefore, earthquake induced settlement of the breakwater should be prevented or reduce up to a great extent. This phenomenon makes it a compound disaster caused by both the earthquake and tsunami (caused not by tsunami only). Hence, for a practical solution against tsunami, the countermeasure should be effective against both the earthquake and tsunami. Chaudhary et al. [11] performed a study to determine stability of a breakwater under actions of an earthquake and tsunami. But, no countermeasure has been suggested. To the end, new reinforcing model were developed by Hazarika et al. [12] for foundation of breakwater to make the breakwater resilient against both earthquake and tsunami-induced forces. Sheet piles and gabions are used as reinforcing countermeasures for the foundation in order to reduce damage of breakwater caused by earthquake and tsunami. Chaudhary et al. [13–20] and Hazarika et al. [21] performed shaking table tests, hydraulic model tests to evaluate the effectiveness of the developed reinforced foundation models subjected to earthquakes and tsunamis. But, all the tests were conducted under 1 g gravitational field. To see the effect of prototype stress for model tests, centrifuge model tests were performed for the developed reinforced foundation. Performance of the reinforced model is described against earthquake and tsunami based on the results obtained from the centrifuge model tests.

## 2 Newly Developed Resilient Foundation Breakwater

New reinforcing techniques have been developed by Hazarika et al. [12] for breakwater foundations which can make the breakwaters resilient against both the earthquake and tsunami-induced damage. As countermeasures, gabions and two rows of steel sheet piles are used. In the developed model, a number of gabions are used to cover the entire mound beside breakwater as shown in Fig. 1. The purpose of using gabion is to protect the rubble mound from scouring and toe erosion during tsunami. It can reduce lateral displacement of breakwater due to friction between rubble mound and gabions during earthquake and tsunami. Owing to these properties, lateral displacement of breakwater can be prevented or reduced to a certain extent during earthquake and tsunami. In addition to the gabions, steel sheet piles are installed at both the edges of the breakwater. The sheet piles can restrict lateral deformation of foundation soils due to their bending characteristics during high excess pore water pressure generation (or liquefaction) caused by earthquake in foundation ground. These sheet piles act as cut off walls, and prevent seepage of seawater beneath the breakwater during tsunami. The sheet piles can prevent scouring of the foundation soils caused by overflowing tsunami (both leading wave and backrush). Such multifunctional measures can prevent any bearing capacity failure of the foundation during the earthquake and tsunami. In the following sections, effectiveness

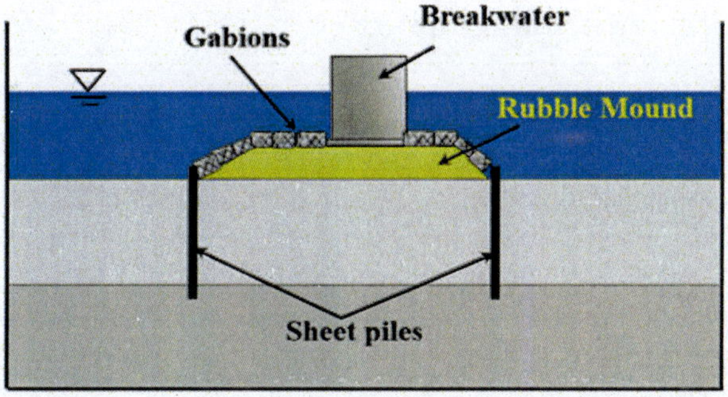


**Fig. 1** Reinforced foundation for a resilient breakwater

of the reinforcing model is described against earthquake and tsunami based on the results of centrifuge model tests.

## 3 Centrifuge Model Tests

A series of centrifuge model tests were conducted for model breakwaters under 25 g gravitational field. The centrifuge model tests were performed at Kyoto University, Japan. Earthquake loadings and tsunami overflow tests were carried out separately for the model breakwaters. The acceleration, pore water pressure, water pressure, horizontal displacement and settlement were monitored during the tests. In order to examine the effectiveness of the reinforced foundation, comparisons are made between the unreinforced and reinforced foundation.

### *3.1 Model Description and Test Procedures*

In the study, the breakwater at Miyazaki port (Miyazaki Prefecture), Japan which is likely to be affected by the predicted Nankai Earthquake, was chosen as a prototype. About 2 m seismic subsidence is predicted in the Kochi area of Shikoku Island, Japan by the Nankai earthquake [22]. Prototype to model ratio adopted in this case was 225. Generalized scaling law developed by Iai et al. [23] was used for the tests. Soil box was made of steel, except for the front face, which was made of acrylic plate. The box type model caisson (breakwater) was made of aluminum' and filled with sand and lead balls to adjust its weight (specific gravity 2.3) and center of gravity. Foundation soils, which was prepared using Toyoura sands, consist of two layers: a loose upper layer of relative density, $D_r = 60\%$, and a dense lower layer of relative density, $D_r = 90\%$. To observe the deformation of foundation soil, red colored membranes were used in the front wall of soil box, and fixed in such a way that these could move or deform along with the soil during the earthquake.

### *3.2 Instrumentations*

Two different models of breakwater foundation were used in the tests (i) Conventional breakwater foundation (unreinforced foundation) and (ii) Breakwater foundation reinforced with gabions and two rows of sheet piles (reinforced foundation). Instrumentation for the teste is shown in Fig. 2. Displacement gauges were used to monitor settlements at two locations (V1 and V2) and horizontal displacements at two points (H1 and H2) of the caisson. In addition, accelerations (A1-A6), water pressures (W1-W4) and pore water pressures (P6-P10) were monitored at different locations of the models.

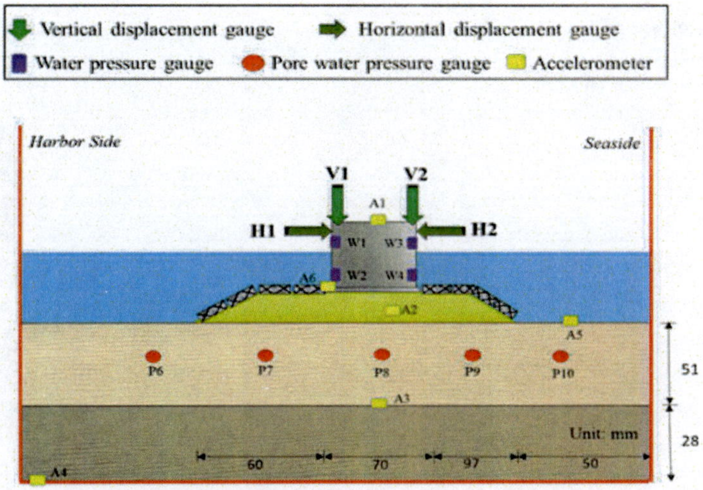


**Fig. 2** Instrumentations for the centrifuge model tests

### Earthquake Loadings and Tsunami Overflow

Sinusoidal accelerations waves with frequency of 83 Hz and time of 1.4 s were used as earthquake loadings. Two foreshocks (0.1 g and 0.2 g) and one main shock (0.4 g) were considered for the earthquake loadings. Accelerations amplitudes were 2.5 g (0.1 g in the prototype scale) and 5.0 g (0.2 g in the prototype scale) for the first and second foreshock respectively. The acceleration amplitude was 10 g (0.4 g in the prototype scale) for the main shock.

For the tsunami overflow test, fresh models were prepared, and tests were conducted under 25 g gravitational field. Tsunami waves were generated by dam break method.

# 4 Results and Discussions

## 4.1 *Earthquake Loadings*

To observe the effects of different acceleration amplitudes of the earthquake loadings, average settlement and average horizontal displacement of the caisson are compared. The comparisons are made for the unreinforced foundation, and shown in Figs. 3 and 4 respectively. It can be seen that the settlement and horizontal displacement of the caisson increase with the amplitude of acceleration. Moreover, the settlement and horizontal displacement increased to a high value for the main shock.

In order to evaluate effectiveness of the reinforcing countermeasures, average settlement and average horizontal displacement of the caisson is compared between

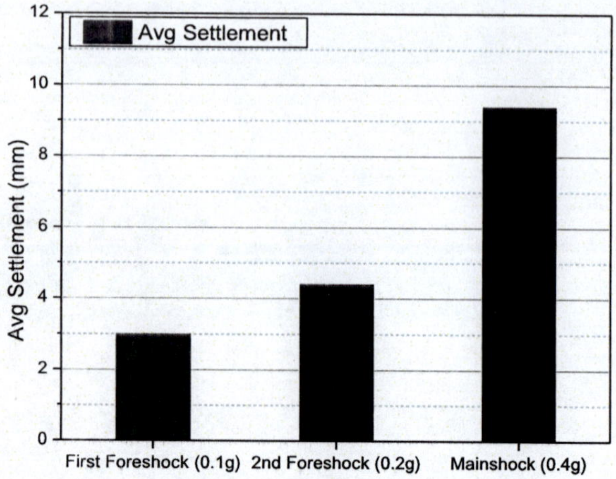


**Fig. 3** Settlement of the caisson during the foreshocks and main shock

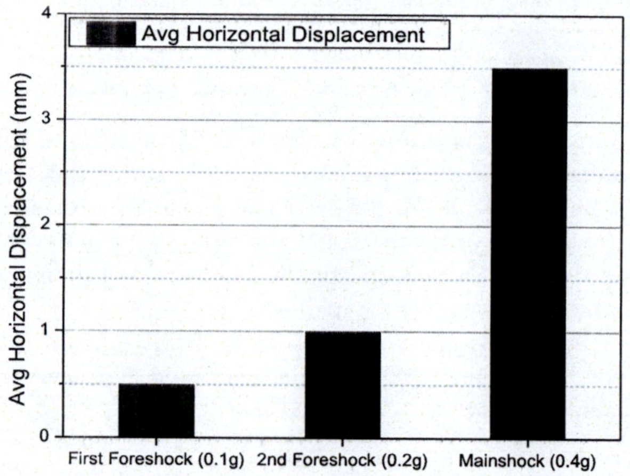


**Fig. 4** Horizontal displacement of the caisson during the foreshocks and main shock

the reinforced and unreinforced foundations during the main shock. The comparison for average settlement of the caisson is shown in Fig. 5. It can be observed from this figure that the average settlement for the unreinforced foundation is 9.4 mm, which could be reduced to 4.6 mm (51% reduction) by the reinforced foundation. The excessive ground deformation or lateral flow of soils were the main cause of settlement of the unreinforced case, the deformation was restricted by the sheet piles of the reinforced foundation.

Figure 6 shows comparison of the average horizontal displacement of the caisson for the unreinforced and reinforced models during the main shock. It can be noted that the average horizontal displacement for the unreinforced foundation is 3.5 mm, which was reduced to 1.8 mm (49% reduction) by the reinforced model. Therefore, it can be said that the reinforced foundation can reduce average settlement and average horizontal displacement of a caisson significantly during earthquake loading.

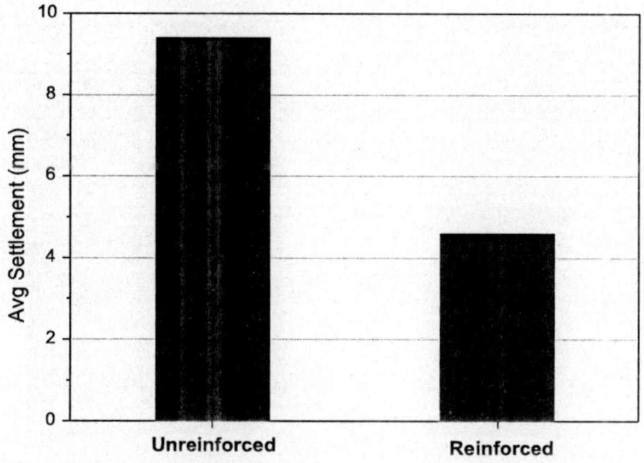


**Fig. 5** Average settlement of the caisson during the main shock

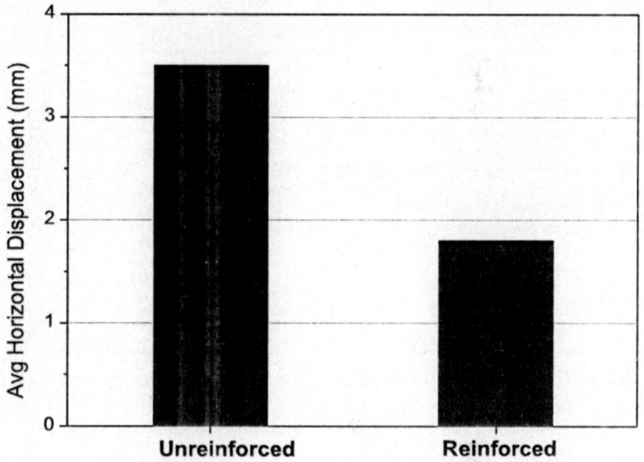


**Fig. 6** Average horizontal displacement of the caisson during the main shock

To see the effects of the reinforced foundation on excess pore water pressure (EPWP) generated by the earthquake, EPWP in the foundation soils, at P8, during the main shock is compared between the reinforced and unreinforced foundations. It can be seen in Fig. 7 that the EPWP is less for the reinforced foundation compared with that for the unreinforced foundation.

## *4.2 Tsunami Overflow Tests*

To examine effectiveness of the reinforced foundation during tsunami, settlements and the average horizontal displacements of the caisson are compared during the tsunami overflow tests.

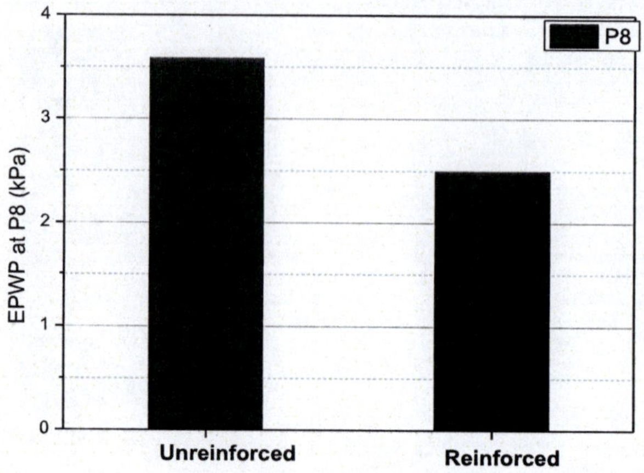


**Fig. 7** Excess pore water pressure at P8 during the main shock

Comparison of average settlements of the caisson during the tsunami overflow tests is shown in Fig. 8. It can be seen that unreinforced foundation led to complete collapse of the breakwater. However, the reinforced breakwater exhibits very little settlement during the tsunami overflow. The reinforced foundation provided a resilient breakwater during the tsunami. In the case of unreinforced foundation, tsunami impact forces on the caisson, scouring of rubble mound and seepage beneath the caisson were the main reasons of failure of the breakwater during the tsunami. However, the gabions were found very effective to prevent scouring of the rubble mound. Due to heavier weight of the gabion (compared to a rubble), it could not be scoured by the overflowing tsunami. The sheet piles prevent seepage beneath the mound. Thus, the multifunctional reinforcing elements mitigated the tsunami-induced damage of the breakwater.

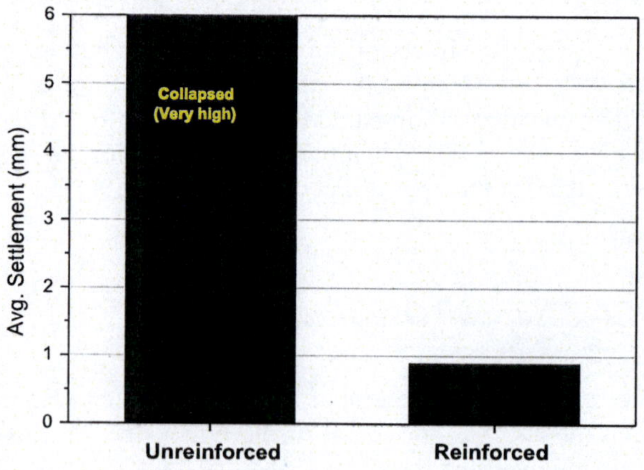


**Fig. 8** Average settlement of the caisson during the tsunami overflow tests

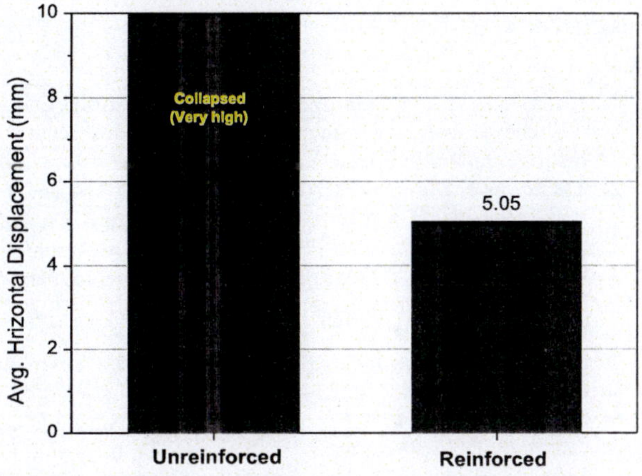


**Fig. 9** Average horizontal displacement of the caisson the tsunami overflow tests

Figure 9 shows average horizontal displacement of the caisson during the tsunami. The unreinforced breakwater toppled towards the harborside and collapsed. However, the tsunami overflow tests.

## 5 Conclusions

Centrifuge model tests were conducted to evaluate the performance of newly developed reinforcing foundation for breakwater subjected to an earthquake and tsunami. The following conclusions are made based on this study.

1. During the earthquake loadings, lateral deformation of the foundation soils was one of the main causes of damage of the conventional/unreinforced breakwater. The sheet pile of the reinforced foundation could restrict the deformation of foundation ground.
2. During the tsunami overflow tests, tsunami impact forces on the caisson was the main reason of collapse of the unreinforced breakwater. The harborside gabions could protect the caisson from the tsunami impact forces.
3. During tsunami, seepage flow took place through the rubble mound and foundation soils. The ceepage flow reduced bearing capacity of the foundation soils of the breakwater which causes for instability of the breakwater during the tsunami. The sheet piles of the reinforced foundation were found effective in reducing the seepage to a greater extent.
4. Scouring of harborside rubble mound by the overflowing tsunami waves was one of the reasons of failure of the unreinforced foundation. The gabions protected the rubble mound from the scouring caused by the overflowing tsunami, and thus can prevent scouring induced damage of breakwater during tsunami.

## References

1. Arikawa T, Sato M, Shimosako K, Tomita T, Yeom GS, Niwa T (2013) Failure mechanism and resiliency of breakwater under tsunami. Technical Note No. 1269, Port and Airport Research Institute, Japan (in Japanese)
2. Takahashi S, Kuriyama Y, Tomita T, Kawai Y, Arikawa T, Tatsumi D, Negi T (2011) Urgent survey for 2011 Great East Japan earthquake and tsunami disaster in ports and coasts part I (tsunami). Technical note 1231, Port and Airport Research Institute, Japan, pp 1–9
3. Kazama M, Noda T (2012) Damage statistics (Summary of the 2011 off the Pacific Coast of Tohoku Earthquake damage. Soil Found 52(5):780–792
4. Hazarika H, Kasama K, Suetsugu D, Kataoka S, Yasufuku N (2013) Damage to geotechnical structures in waterfront areas of northern Tohoku due to the March 11, 2011 tsunami disaster. Indian Geotech J 43(2):137–152
5. Hazarika H, Kataoka S, Kasama K, Kaneko K, Suetsugu D (2012) Composite ground disasters caused by the earthquake and tsunami in Aomori, Iwate Prefecture, northern. Geotech Eng J, Special Issue on 2011 Great East Japan Earthquake 7(1):13–23, (in Japanese)
6. Hara T, Okawara M, Osumi T, Yamanaka M, Ishihara Y, Tsunekawa Y, Okamura M, Uzuoka R (2012) Damages in south-central coastal area of Iwate prefecture in 2011 off the Pacific Coast of Tohoku Earthquake. Japanese Geotech J 7(1):25–36
7. Sugano T, Nozu A, Kohama E, Shimosako KI, Kikuchi Y (2014) Damage to coastal structures. Soils Found 54(4):883–901
8. Oikawa S, Kikuchi Y, Kawabe S, Mizuno R, Moriyasu S, Tanaka R, Takenaka S (2014) An experimental study for reinforcing construction of breakwater with structural steel. In: Proceedings of the special symposium of japanese geotechnical society on the Great East Japan Disaster, Tokyo, Japan, pp 703–709. (in Japanese)
9. Ueda K, Iai S, Tobita T (2016) Centrifuge model tests and large deformation analyses of a breakwater subject to combined effects of tsunami. Soil Dyn Earthquake Eng 91:294–303
10. Kikuchi Y, Kawabe S, Takaenaka S, Moriyasu S (2015) Horizontal loading experiments on reinforced gravity type breakwater with steel walls. Japan Geotech Soc Spec Publ 2(35):1267–1272
11. Chaudhary B, Hazarika H, Ishibashi I, Abdullah A (2017) Sliding and overturning stability of breakwater under combined effect of earthquake and tsunami. Ocean Eng 136:106–116
12. Hazarika H, Chaudhary B, Monji N, Ishikura R, Kasama K, Hara T, Yamazaki N, Noda T, Yamada S (2015) Resilient breakwater foundation against level II earthquake and tsunami. IN: Proceedings of the 6th international geotechnical symposium on disaster mitigation in special geoenvironmental conditions, Chennai, India, pp 35–46
13. Chaudhary B, Hazarika H, Murakami A, Fujisawa K (2019) Development of resilient breakwater against earthquake and tsunami. Int J Geomech ASCE 19(1):1–17
14. Chaudhary B, Hazarika H, Murakami A, Fujisawa K (2018) Geosynthetic-sheet pile reinforced foundation for mitigation of earthquake and tsunami induced damage of breakwater. Geotext Geomembr 46:597–610
15. Chaudhary B, Hazarika H, Murakami A, Fujisawa K (2018) Countermeasures for enhancing the stability of composite breakwater under earthquake and subsequent tsunami. Acta Geotech 13(4):997–1017
16. Chaudhary B, Hazarika H (2018) Centrifuge modelling for stability evaluation of a breakwater foundation subjected to an earthquake and a tsunami. Ocean Eng 148:169–181
17. Chaudhary B, Hazarika H, Nishimura N (2017) Effects of duration and acceleration level of earthquake ground motion on the behavior of unreinforced and reinforced breakwater foundation. Soil Dyn Earthquake Eng 98:24–37
18. Chaudhary B, Hazarika H, Murakami A, Fujisawa K (2018) Mitigation of earthquake induced damage of breakwater by geogrid reinforced foundation. Mar Georesour Geotechnol 36(7):827–840

19. Chaudhary B, Hazarika H, Pasha SMK (2017) Countermeasures for breakwater foundation subjected to foreshocks and main shock of earthquake loading. Mar Georesour Geotechnol 36(3):308–322
20. Chaudhary B, Hazarika H, Nishimura N (2017) Effects of reinforcement on the performance of breakwater foundation subjected to earthquake loadings. Int J Geotech Eng 11(2):186–197
21. Hazarika H, Hara T, Nishimura K, Yamasaki N, Monji N, Chaudhary B, Ishikura R, Kasama K (2016) Fundamental study on seismic resistant behavior of caisson type breakwater foundation reinforced by steel sheet pile and gabion. J Japan Assoc Earthquake Eng 16(1):184–204 (in Japanese)
22. Council Central Disaster Management (2003) Report of the 16th committee meeting. Investigation Committee on Tonankai Earthquake and Nankai Earthquake, Cabinet Office, Government of Japan (in Japanese)
23. Iai S, Tobita T, Nakahara T (2005) Generalised scaling relations for dynamic centrifuge tests. Geotechnique 55(5):355–362

# Comparative Study on Behavior of Vertical and Batter Piles in Lateral Spreading Soil

**J. S. Rajeswari and Rajib Sarkar**

**Abstract** Liquefaction induced lateral spreading of soil is reported to be one of the primary causes of failures of pile supported bridges and structures during major earthquakes (1964 Niigata, 1989 LomaPrieta and 1995 Kobe). The present study investigates the performance of single piles subjected to lateral spreading displacement through probabilistic approach. Beam on non-linear Winkler foundation method has been adopted to evaluate the pile response. The prediction models for pile response (maximum bending moment) are developed using response surface methodology for piles of different batter angles (0°, 10° and 20°) and slenderness ratios (15, 20 and 25). The probability of failure of various soil-pile systems is then evaluated using Monte Carlo technique. The results indicate that positive batter piles are effective in restricting the bending moment compared to vertical piles, and the behavior of piles subjected to lateral spreading is majorly controlled by the thickness of surface crust overlying the liquefiable soil.

**Keywords** Vertical pile · Batter pile · Lateral spreading

## 1 Introduction

Post liquefaction site observations indicated failure of many pile supported structures resulting from lateral spreading of surface crust due to liquefaction of underlying soil. This is due to the large passive pressure exerted by moving soil blocks on piles passing through layered ground [1]. The intensity of lateral spreading depends on the complex interaction of various factors such as soil layering, slope of ground, location of water table, and so on [2, 3]. Neglecting the consideration of these displacements in the design of piles in seismic areas may lead to huge economic losses.

J. S. Rajeswari · R. Sarkar (✉)
IIT (ISM) Dhanbad, Dhanbad 826004, India
e-mail: rajib@iitism.ac.in

J. S. Rajeswari
e-mail: rajeswarijs1993@gmail.com



T. G. Sitharam et al. (eds.), *Seismic Design and Performance*, Lecture Notes in Civil Engineering 120, https://doi.org/10.1007/978-981-33-4005-3_5

Due to higher lateral stiffness of batter piles compared to same diameter vertical piles, research has been carried to investigate the potentiality of batter piles under seismic conditions. Though studies have indicated undesirable pile performance under purely kinematic loading, favorable response was observed when subjected to combined (inertial + kinematic) loads [4]. Literature review indicated that the studies predominantly reported the behavior of batter piles embedded in non-liquefiable soil [5–7]. The efficiency of batter piles in liquefiable and laterally spreading ground is still a matter of concern. Wang and Orense [8] observed that negative batter piles develop lesser bending moment at the pile head whereas positive batter piles develop lesser bending moment at the interface between liquefied and non-liquefied soil compared to vertical piles. Tazoh et al. [9] reported the beneficial response of batter piles in controlling the horizontal displacement of foundation when subjected to liquefaction-induced lateral flow.

Traditional factor of safety approach in pile design has been replaced by standard probabilistic-based techniques in recent years owing to its capability to adequately treat various uncertainties [10–12]. The present study utilizes conventional pseudo-static approach coupled with response surface methodology and Monte Carlo technique for assessing the performance of single piles subjected to lateral spreading displacements. The behavior of batter piles (angles 10° and 20°) is compared to vertical piles (i.e., 0° batter angle) based on the probability of failure of soil-pile systems considering the performance criteria.

## 2 Details of Modeling

### 2.1 Problem Statement

Single, fixed head vertical, and batter piles having diameter $D$ and length $L$ subjected to lateral spreading displacements are considered for the present study. Two different batter angles $\alpha$—10° and 20° are analyzed to study the effect of batter on the performance of pile. The batter pile is considered to be positive if the pile is inclined in the direction of loading and negative if the pile is inclined opposite to the direction of loading. The soil medium comprises of liquefiable layer of thickness $H_{\mathrm{L}}$ embedded between crust layer of thickness $H_{\mathrm{NL}}$ and base non-liquefiable layer as shown in Fig. 1. The water table is located at the interface between top crust and middle liquefiable soil. The inertial effects are neglected considering its insignificance in the post liquefaction stage [13].

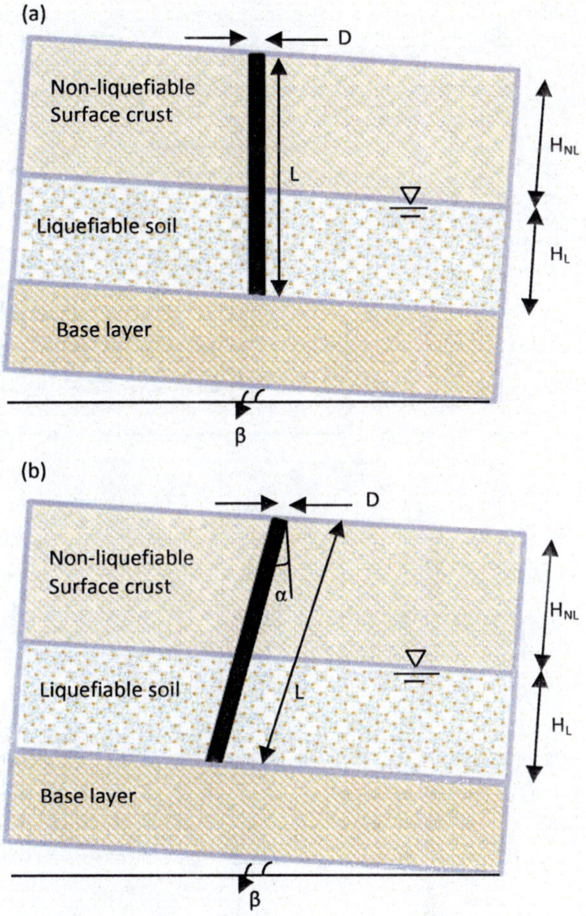


**Fig. 1** Schematic diagram of the soil-pile system for **a** vertical pile and **b** batter pile

## 2.2 *Numerical Model*

A simplistic approach known as *Beam on Nonlinear Winkler Foundation* (*BNWF*) is adopted for modeling the soil-pile system. In this approach, the non-linear behavior of soil is simulated using spring elements attached along the discrete lengths of the pile. *p-y* curves are used to represent the spring elements, where *p* is the lateral resistance per unit length of the pile and *y* is the relative soil-pile displacement. Similarly, *t-z* and *q-z* curves represent spring elements modeling the shaft and tip resistance, respectively. All the models are developed in finite element programmable software *OpenSees*. The pile sections are modeled using elastic beam-column elements, and zero length elements are used to model the non-linear springs. Figure 2a shows the typical soil-pile system modeled in *OpenSees*.

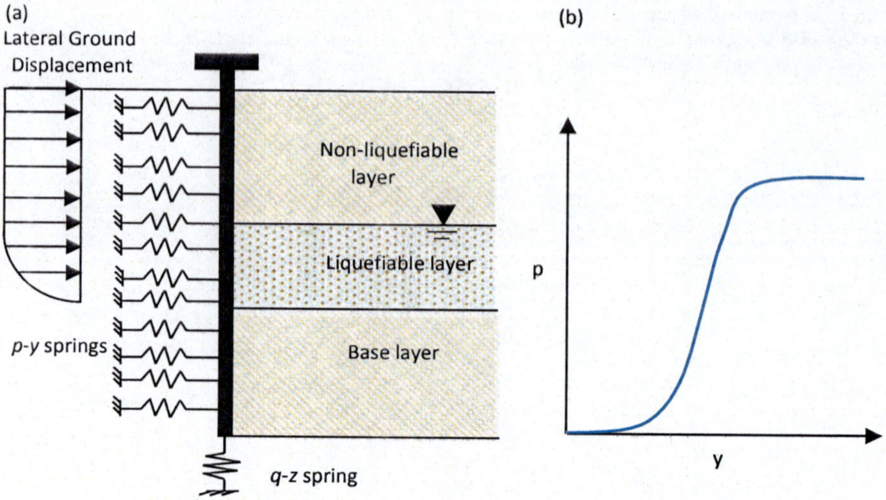


**Fig. 2** **a** Typical soil-pile system model, **b** typical convex *p-y* curve used for liquefiable soil

***p-y* springs**. The ultimate lateral resistance ($p_u$) along the depth of the pile and displacement at which 50% of the ultimate resistance is mobilized is computed as per Hansen and Christensen [14] and Boulanger et al. [15], respectively.

**Modification for batter piles**. The modification factor $\psi$ as suggested by Zhang et al. [16] is adopted to account for the effects of batter, and $\psi$ is proportional to the ratio of passive earth pressure coefficient considering batter pile to that of vertical pile. The ultimate lateral resistance per unit depth of batter pile ($p_{ub}$) is given as

$$p_{ub} = \psi p_u \tag{1}$$

***p-y* curves for liquefiable layer**. Generally, *p-y* curves for liquefiable soil are obtained by reducing the corresponding *p-y* curves for non-liquefiable soil. These *p-y* curves are concave downward in shape. But *p-y* curves back-calculated from experimental studies conducted on liquefiable soil exhibits convex downward shape indicating negligible stiffness at small displacements and higher stiffness with shearing contrary to concave *p-y* curves [17, 18]. Since this conflicting shape of *p-y* curves can have significant effect on pile behavior, the present study adopts convex *p-y* curves for liquefiable soil. The curves are developed using the practical procedure proposed by Lombardi et al. [19] and Fig. 2b shows the typical convex *p-y* curve used for liquefiable soil.

***t-z* springs**. The ultimate vertical resistance per unit depth of the pile can be calculated as per Kulhawy [20] and displacement at which 50% of ultimate capacity is mobilized is obtained as the ratio of ultimate vertical resistance ($t_u$) to initial tangent ($k_f$) [21].

***q-z* springs**. The ultimate tip resistance ($q_u$) for spring element at the pile tip is calculated from Meyerhof [22]. The displacement at which 50% of the ultimate resistance is mobilized ($z_{50}$) is taken as one fourth of the critical deflection (5% of the pile diameter) [23].

## 2.3 Ground Displacement

The free-field ground displacement $U_{gr}$ is computed using the equation proposed by Hamada et al. [24]. The profile is assumed to have cosine variation within the liquefiable layer with zero value at the interface of liquefiable soil and base layer and $U_{gr}$ at the top which remains constant throughout the upper crust as shown in Fig. 2a.

## 2.4 Selection of Input Parameters and Consideration of Variability

Piles are made of M30 concrete with 0.5 m diameter and three different slenderness ratios—15, 20, and 25. The base sand is assumed to have a relative density of 85%. Based on extensive literature review, the ground geometry and properties of surface crust and liquefiable soil are taken as random parameters and are summarized in Table 1 [2, 25–32].

**Table 1** Statistics of random parameters

| Sl. No | Random variables | Unit | Symbol | Probability distribution function | Mean | Standard deviation or variation |
|---|---|---|---|---|---|---|
| 1 | Thickness of surface crust | m | $H_{NL}$ | Uniform | 3.0 | 0–6 |
| 2 | Relative density of surface crust | % | $Dr_{NL}$ | Normal | 77.5 | 0.05 |
| 3 | Liquefied layer thickness | m | $H_L$ | Log normal | 2.0 | $\sigma_{ln} = 0.7$ |
| 4 | Relative density of liquefied sand | % | $Dr_L$ | Normal | 45.0 | 0.12 |
| 5 | Ground slope | % | $\beta$ | Normal | 2.75 | 0.75 |

## 3 Application of Response Surface Methodology and Monte Carlo Technique

The pile performance is evaluated based on maximum bending moment developed along the length of the pile. $2^k$ factorial design is adopted along with analysis of variance (ANOVA) $F$-test to study the effect of various random parameters and to develop approximate prediction models for pile response. Each random variable is fixed at two levels—lower limit and upper limit based on their mean and standard deviation, and consequently, $2^k$ analyses are carried out for each case considered (vertical pile, 10° and 20° positive and negative batter piles) [33].

The probability of failure of systems ($P_f$) is then evaluated using Monte Carlo technique considering maximum bending moment as

$$\text{Probability of failure, } P_f = \frac{\text{Number of runs exceeding plastic moment capacity of pile}}{\text{Total number of runs}} \tag{2}$$

The Monte Carlo technique allows the generation of $N$ (decided based on acceptable error and confidence level) number of random parameters [34]. The response is then computed for each combination of parameters using the prediction model built for each case. Subsequently, the probability of failure is calculated using Eq. (2).

## 4 General Behavior of Batter Piles in Laterally Spreading Soil

The effect of batter on the behavior of piles in laterally spreading soil is presented in this section. Figure 3 shows bending moment profile and lateral displacement profile for vertical and 10° positive and negative batter piles. The bending moment profiles indicate promising response of positive batter pile in lateral spreading soil. This may be because at the interface, positive batter piles experience lesser stiffness contrast owing to the lesser change in the soil resistance across the boundary. On the other hand, negative batter pile tend to aggravate bending moment response significantly compared to vertical pile. It is also worth noting from Fig. 3b that irrespective of pile batter, the piles are displaced to same extent when subjected to lateral spreading displacement. These results are confirmed by Wang and Orense [8].

## 5 Effect of Random Parameters on Pile Response

The influence of various parameters on pile response in lateral spreading soil is presented in this section based on the results of factorial experiments. Figure 4 shows the effect of thickness of surface crust, thickness of liquefiable layer, slope of

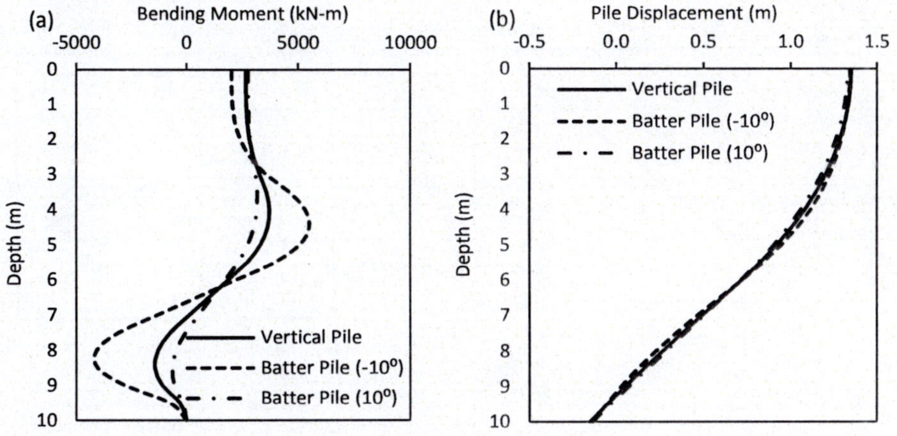


**Fig. 3** **a** Bending moment profile and **b** lateral pile displacement profile for vertical and 10° positive and negative batter piles

ground, relative density of liquefiable layer, and relative density of surface crust on maximum bending moment of vertical pile. Since the parameters indicated similar influence on batter piles, the results are not presented to avoid repetition.

It can be observed from the plots that thickness of surface crust has greatest influence on pile response in laterally spreading soil and maximum bending moment increases with the crust thickness. This is followed by the thickness of liquefiable layer. With the increase in liquefiable layer thickness, maximum bending moment increases. The increase in the ground slope and relative density of liquefiable sand aggravates the pile response, whereas increase in the relative density of surface crust is observed to have insignificant effect on pile bending moment.

# 6 Effect of Batter on Probability of Failure of Piles in Laterally Spreading Soil

The prediction models for pile response considering different batter angles and slenderness ratios are developed using the results of factorial experiments. The probability of failure of system is then computed considering the failure criteria as explained in Sect. 3. Due to the lack of space, results are presented only for slenderness ratio of 20. Figure 5 shows the effect of various random parameters on probability of failure $P_f$ for piles of different batter angles. The probability of failure of negative batter piles is observed to be highest and it increases with increase in batter angle as opposed to positive batter piles which displayed commendable performance even at lower batter angles.

The probability of failure of piles $P_f$ increases abruptly to maximum as crust layer thickness reaches a value of 1 m and thereafter remains constant for all the piles.

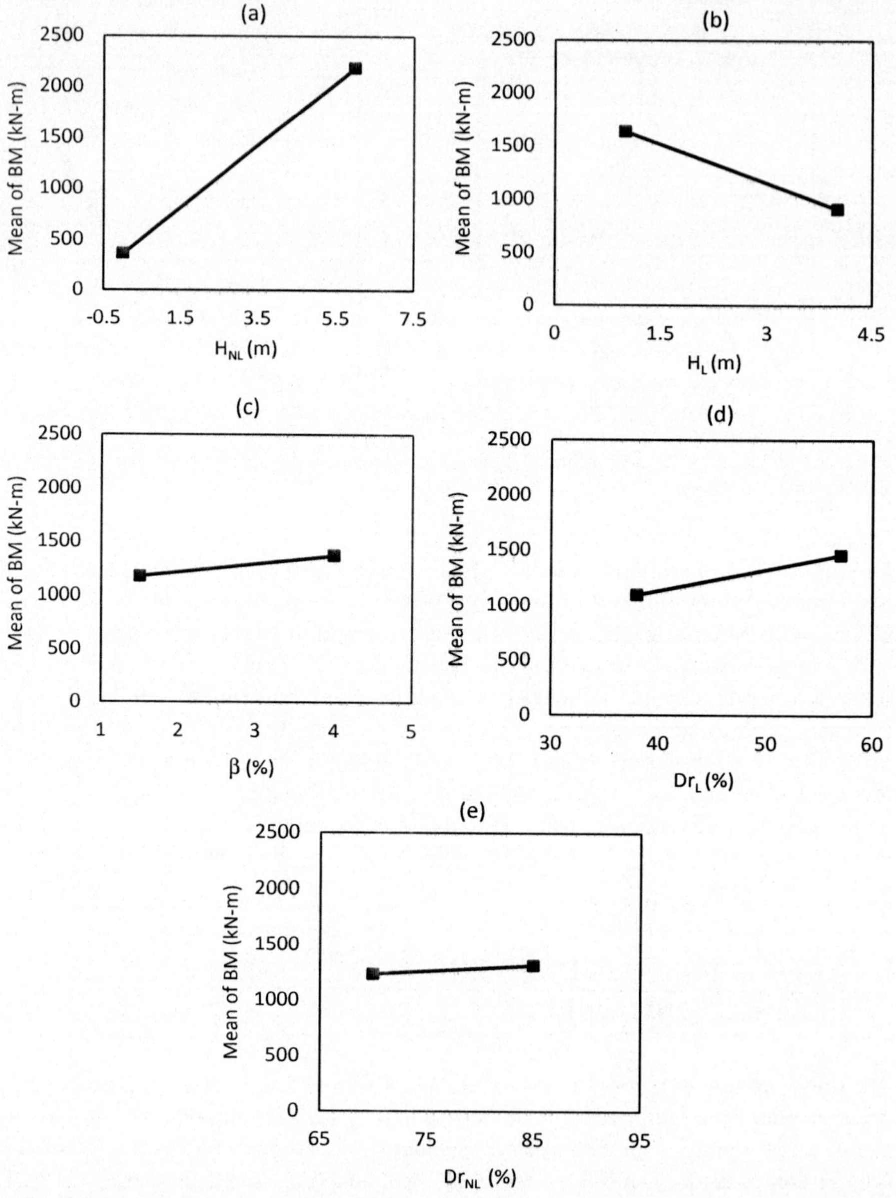


**Fig. 4** Effect of **a** thickness of surface crust, **b** thickness of liquefiable layer, **c** slope of ground, **d** relative density of liquefiable layer, and **e** relative density of surface crust on the maximum bending moment of vertical pile

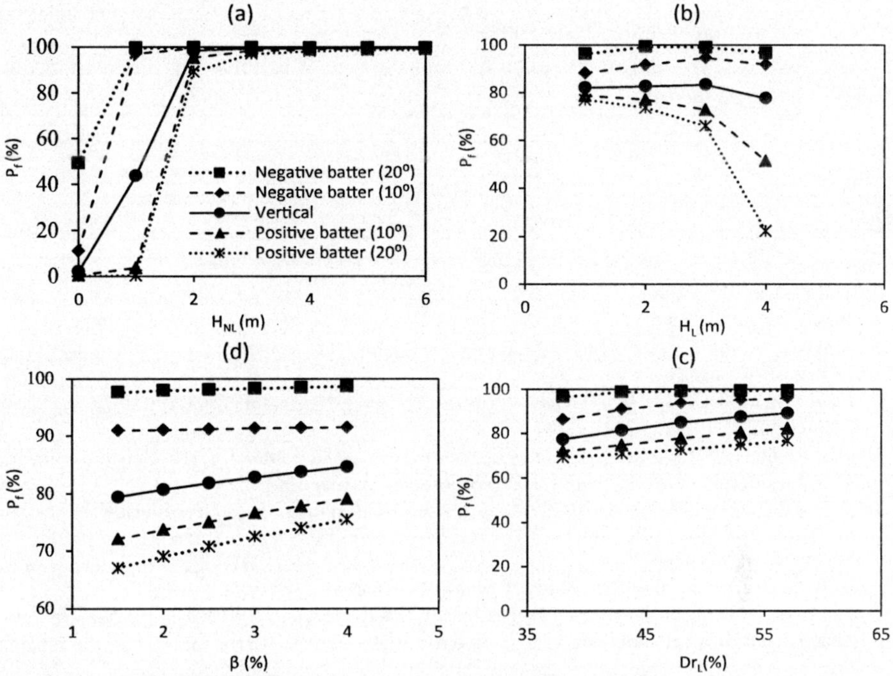


**Fig. 5** Effect of various random parameters **a** $H_{NL}$, **b** $H_L$, **c** $Dr_L$ and **d** $\beta$ on probability of failure $P_f$ of piles for different batter angles

This indicates the significance of crust layer thickness on the behavior of piles in laterally spreading soil. $P_{f1}$ reduces with increase in thickness of liquefiable layer whereas increases with slope of the ground as expected. The relative density of sand is observed to have minor effect on $P_{f1}$.

## 7 Conclusions

The present study focusses on the behavior of single fixed head vertical and batter piles in laterally spreading ground conditions. The effect of various soil and geometrical parameters on pile response is investigated. The probability of failure of piles of different batter angles is also evaluated using Monte Carlo technique utilizing the prediction models developed adopting response surface methodology. The major conclusions from the study are following:

- Positive batter piles are advantageous in restricting bending moment throughout the depth of pile and its efficiency increases with increase in batter angle.

- The pile displacement is independent of batter angle.
- The thickness of surface crust is the major parameter influencing pile response in lateral spreading soil.

## References


1. Madabhushi G, Knappett J, Haigh S (2010) Design of pile foundations in liquefiable soils. Imperial College Press
2. Youd TL, Perkins DM (1987) Mapping of liquefaction severity index. J Geotech Eng 113(11):1374–1392
3. Fiegel GL, Kutter BL (1994) Liquefaction-induced lateral spreading of mildly sloping ground. J Geotech Eng 120(12):2236–2243
4. Giannakou A, Gerolymos N, Gazetas G, Tazoh T, Anastasopoulos I (2010) Seismic behavior of batter piles: elastic response. J Geotech Geoenviron Eng 136(9):1187–1199
5. Sarkar R, Roy N, Serawat A (2018) A three dimensional comparative study of seismic behaviour of vertical and batter pile groups. In: Geotechnical and geological engineering, pp 1–19
6. Sadek M, Isam S (2004) Three-dimensional finite element analysis of the seismic behavior of inclined micropiles. Soil Dyn Earthq Eng 24(6):473–485
7. Ghorbani A, Hasanzadehshooiili H, Ghamari E, Medzvieckas J (2014) Comprehensive three dimensional finite element analysis, parametric study and sensitivity analysis on the seismic performance of soil–micropile-superstructure interaction. Soil Dyn Earthq Eng 58:21–36
8. Wang S, Orense RP (2017) Numerical simulation of inclined piles in liquefiable soils. In: Alexander GJ, Chin CY (eds) 20th NZGS geotechnical symposium, Napier
9. Tazoh T, Sato M, Jang J, Taji Y, Gazetas G (2010) Seismic behavior of batter pile foundation: kinematic response. In: International conferences on recent advances in geotechnical earthquake engineering and soil dynamics, p 41
10. Eloseily KH, Ayyub BM, Patev R (2010) Reliability assessment of pile groups in sands. J Struct Eng 128(10):1346–1353
11. Puła W, Różański A (2012) Reliability of rigid piles subjected to lateral loads. Arch Civ Mech Eng 12(2):205–218
12. Farag R (2014) Probabilistic pseudostatic analysis of pile in laterally spreading ground: two layer soil profile. Ain Shams Eng J 5(2):343–354
13. Tokimatsu K (1999) Performance of pile foundations in laterally spreading soils. In: Proceedings of 2nd international conference earthquake geotechnical engineering, Lisbon, vol 3, pp 957–964
14. Hansen JB, Christensen NH (1961) The ultimate resistance of rigid piles against transverse forces. Bulletin No. 12. Geoteknisk Institute, Copenhagen
15. Boulanger RW, Kutter BL, Brandenberg SJ, Singh P, Chang D (2003) Pile foundations in liquefied and laterally spreading ground during earthquakes: centrifuge experiments and analyses. Rep. UCD/CGM-03/01. Center for Geotechnical Modeling, University of California at Davis, Davis
16. Zhang L, McVay MC, Lai PW (1999) Centrifuge modelling of laterally loaded single battered piles in sands. Can Geotech J 36(6):1074–1084
17. Wilson DW, Boulanger RW, Kutter BL (2000) Observed seismic lateral resistance of liquefying sand. J Geotech Geoenviron Eng 126(10):898–906
18. Ashford SA, Rollins KM (2002) TILT: the treasure island liquefaction test (No. UCSD/SSRP-2001/17). Department of Structural Engineering, University of California, San Diego
19. Lombardi D, Dash SR, Bhattacharya S, Ibraim E, Wood DM, Taylor CA (2017) Construction of simplified design p–y curves for liquefied soils. Géotechnique 67(3):216–227

20. Kulhawy FH (1991) Drilled shaft foundations. In: Fang H-Y (ed) Foundation engineering handbook, 2nd edn. Van Nostrand Reinhold, New York
21. Mosher RL, Dawkins WP (2000) Theoretical manual for pile foundations (No. ERDC/ITL-TR-00-5). Engineer Research and Development Center
22. Meyerhof GG (1976) Bearing capacity and settlement of pile foundations. J Geotech Geoenviron Eng 102:195–228
23. Vijayvergiya VN (2019) Load-movement characteristics of piles. In: Proceedings of the ports 77 conference, New York
24. Hamada M, Towhata I, Yasuda S, Isoyama R (1987) Study on permanent ground displacement induced by seismic liquefaction. Comput Geotech 4(4):197–220
25. Cubrinovski M, Ishihara K (2004) Simplified method for analysis of piles undergoing lateral spreading in liquefied soils. Soils Found 44(5):119–133
26. Dobry R, Abdoun T (2001) Recent studies on seismic centrifuge modeling of liquefaction and its effect on deep foundations. In: Fourth international conference on recent advances in geotechnical earthquake engineering and soil dynamics
27. Seed HB, Idriss IM (1967) Analysis of soil liquefaction: Niigata earthquake. J Soil Mech Found Div 93(3):83–108
28. Ghasemi-Fare O, Pak A (2016) Numerical investigation of the effects of geometric and seismic parameters on liquefaction-induced lateral spreading. Soil Dyn Earthq Eng 89:233–247
29. Rahmani I, Zahmati A, Hamed AA (2013) Seismic behavior analysis of piles by considering lateral spreading. Electron J Geotech Eng 18:2989–2996
30. Ziotopoulou K, Montgomery J (2017) Numerical Modeling of earthquake-induced liquefaction effects on shallow foundations. In: Proceedings of 16th world conference on earthquake engineering
31. Brandenberg SJ, Zhang J, Kashighandi P, Huo Y, Zhao M (2011) Demand fragility surfaces for bridges in liquefied and laterally spreading ground. Pacific Earthquake Engineering Research Center
32. Dai SH, Wang MO (1992) Reliability analysis in engineering applications. Van Nostrand Reinhold, New York
33. Myers RH, Montgomery DC, Anderson-Cook CM (2016) Response surface methodology: process and product optimization using designed experiments. Wiley
34. Baecher GB, Christian JT (2005) Reliability and statistics in geotechnical engineering. Wiley

# Influence of Different Seismic Loadings on the Closely Spaced Interfering Footings Embedded in Cohesionless Foundation Medium

**Anupkumar G. Ekbote and Lohitkumar Nainegali**

**Abstract** In some circumstances, the foundations are laid down adjacent to each other in order to fulfil structural requirements or scarcity of land/construction space. Under such situations, the footings do not behave like that of an isolated footing which the foundation theory suggests; instead, they behave in a group. This alteration in characteristic behaviour is attributed due to the phenomenon of interference of footings. In the present study, the response/behaviour of two closely spaced strip footing embedded in cohesionless soil medium subjected to earthquake excitations of different magnitude and acceleration is explored by estimating settlement behaviour and the stresses (vertical and shear stresses) below the footing base. The analyses have been performed using finite element software, ABAQUS, by discretizing the domain with 2D continuum linear plane strain elements. The soil is modelled to follow the Mohr–Coulomb failure criterion. A constant critical damping ratio of 5% is considered to determine Rayleigh's coefficient ($\alpha$ and $\beta$). It has been observed that the settlement and the shear stresses at the base of the footings are enhanced due to the interference effect; in contrast, normal stresses are observed to vary little or no variation with the change in spacing between the footings.

**Keywords** Seismic analysis · Footing interference · Embedded footings · Strip footings · Finite element analysis · Infinite elements

## 1 Introduction

The response of foundations and serviceability is an essential pre-requisite for the design of any structures since foundations are the integral part that has to transfer the loads to the underlying soil stratum. In some circumstances, the foundations are

A. G. Ekbote · L. Nainegali (✉)
Department of Civil Engineering, Indian Institute of Technology (ISM), Dhanbad, Dhanbad 826004, India
e-mail: lohitkumar@iitism.ac.in

A. G. Ekbote
e-mail: anupge@gmail.com

 
T. G. Sitharam et al. (eds.), *Seismic Design and Performance*, Lecture Notes in Civil Engineering 120, https://doi.org/10.1007/978-981-33-4005-3_6

laid down adjacent to each other in order to fulfil structural requirements or scarcity of land/construction space or uncontrolled urbanization. Under such situations, the footings do not behave like that of an isolated footing which the classical foundation theory suggests. Instead, the footings behave in a group with alteration in the ultimate bearing capacity, settlement, failure mechanism, rotational characteristic, etc. This alteration in characteristic behaviour is attributed due to the phenomenon of interference of footings. This makes the problem of footing interference of significant importance given foundation design and particularly under seismic loading conditions. The literature on static analysis on interfering footings is vast. The spark was given by Stuart [15], who first observed the phenomenon of interference by conducting a series of experiments along with the limit equilibrium analysis. In a line, several researchers carried the study adopting different methodologies such as a method of stress characteristics [7], upper bound and lower bound limit analysis [6]; finite difference method [10]; finite element method [11, 12] and small-scale laboratory tests [3, 14]. It can be observed from the said literature that the interference has a significant effect on the bearing capacity, settlement, and rotational characteristics as well as on the failure mechanism. The bearing capacity is observed to be increasing with a decrease in spacing between the footings, and in contrast, the settlement increases with a decrease in spacing between the footings. The studies on the dynamic response (either due to machine excitation or due to seismic loading) of interfering shallow foundations are scarce. Ghosh [2] performed the seismic interaction between two nearby square foundations using finite difference analysis through FLAC, from which they concluded that settlement of closely spaced footings under seismic condition is found to be higher than that of single isolated footing. Swain and Ghosh [16] conducted the large-scale model tests on closely spaced square foundations subjected to machine vibrations inducing harmonic loading on one of the footings (active) and the other as a passive footing. They presented results in terms of transmission ratio with frequency. Javdanian [4] examined the dynamic bearing capacity of adjacent shallow strip foundations by applying cyclic loading, which is located on sandy soil using FLAC.

The numerous studies performed on the closely spaced footings are for the case of static loading conditions. The reports on the dynamic response of adjacently placed footings are seldom, and this makes the motivation in the present study to fill the gap in the literature. The objective of the study is to assess the response/behaviour of two closely spaced strip footing embedded in cohesionless soil medium subjected to earthquake excitations of different magnitude and acceleration. The footings are placed close to each other at a clear spacing, *S*, which are assumed to be rigid, rough and embedded in homogenous, isotropic, semi-infinite soil medium. The attempt is made here to estimate settlement behaviour and the stresses (normal and shear stresses) below the footing.

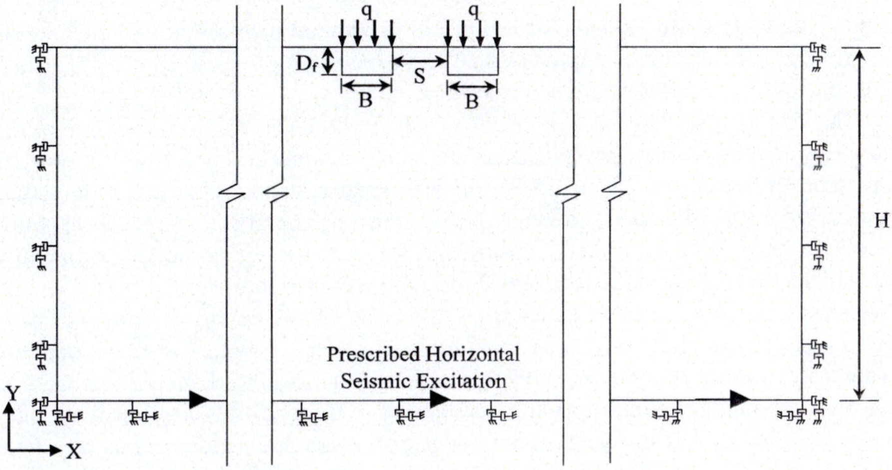


**Fig. 1** Typical geometry of the problem

## 2 The Objective of the Problem

A two-dimensional plane strain problem is considered to study the response of seismic loading on two closely spaced strip footings of width, *B* well-thought-out to be embedded at depth $D_f$ in the dry homogenous, isotropic, semi-infinite cohesionless foundation medium. The footings are placed at a clear spacing, *S*, which are assumed to be rigid, rough and carry the uniform static load, *q*; Fig. 1 represents the typical geometry of the model considered in the analysis. The study aims to estimate the response due to different horizontal seismic excitations prescribed at a depth, *H*, from the base of the footings, in order to analyse the settlement behaviour and stresses developed below the footings.

## 3 Methodology

The present study is carried out using the commercially available finite element software, ABAQUS, by discretizing the problem domain using two-dimensional linear continuum plane strain elements (CPE4R). The relatively finer mesh is adopted at the vicinity of footings, and the size of the elements increased gradually towards the boundary of the problem domain. The maximum size of the element is determined from the criterion that the element size in the direction of wave propagation should be restricted to 1/8th to 1/10th of the shortest wavelength [5, 9]. According to Zhang et al. [17], the shortest wavelength is corresponding to the maximum frequency of 15 Hz, since most of the recorded earthquake ground motion does not contribute significant energy, and this corresponds to the maximum length of element as 0.73 m.

Hence, the maximum element size of 0.5 m is adopted near the boundary, and the finer element of 0.2 m is adopted at the vicinity of the footings to capture the stress concentration around the corner of footings.

The seismic excitation is prescribed at a bedrock level, $H$, from the base of the footings. In the present analysis, the assumption is made that the bedrock level exists at a depth of $H = 50$ m. The boundary condition such as standard fixities (both vertical and horizontal displacements restricted for bottom horizontal boundary whereas horizontal displacements restricted for vertical side boundaries of a domain) is generally considered during the static loading. During the seismic loading condition, it should be taken care that the boundaries allow the most of the energy to transmit away from soil domain and restrict any unnatural reflection of waves inside the domain. Therefore, infinite elements (CINPE4) coupled with radiation dampers (dashpots as absorbing boundary condition) at each node in both horizontal and vertical directions are provided at the vertical side edges of a domain which acts as free-field soil columns [8, 13]. The coefficients of dashpots were obtained from $\rho.v_s.A$ and $\rho.v_p.A$, where $\rho$, $v_s$, $v_p$ and $A$ are mass density, shear wave velocity, compression wave velocity and tributary surface area of the corresponding node, respectively.

The footings of width, $B = 1$ m, are considered to be placed at an embedment depth, $D_f/B = 1$, in the analysis by varying the clear spacing, $S$, between the footings. The footings are considered to behave as a linear elastic material having Young's modulus, $E = 25$ GPa, and Poisson's ratio, $\nu = 0.2$. The soil is modelled to behave as linear elastic, perfect plastic obeying Mohr–Coulomb constitutive law having mass density, $\rho = 1600$ kg/cm$^3$; static Young's modulus, $E_s = 32$ MPa; Poisson's ratio, $\nu = 0.3$; soil friction angle, $\phi = 30°$; cohesion, $c = 2$ kPa. The change in soil stiffness during dynamic loading which occurs due to densification of soil due to shaking is determined from the empirical relation as provided by Alpan [1]; thus, dynamic Young's modulus, $E_d = 149.5$ MPa is used. Furthermore, material damping is provided in terms of Rayleigh damping coefficients ($\alpha$ and $\beta$) calculated from the equation below (Eq. 1) and by considering the critical damping ratio, $\xi = 5\%$, corresponding to the natural frequency, $\omega_i$.

$$\xi i = \frac{\alpha}{2.\omega i} + \frac{\beta.\omega i}{2} \tag{1}$$

In the present analysis, as mentioned earlier, horizontal seismic excitation is applied at the level of bedrock. Initially (Prior to the seismic loading), earth pressure at rest, which is geostatic stresses are generated then a uniformly distributed static load, $q$ in the range of working load is applied simultaneously on both footings. The working load of 300 kPa is applied, which is obtained by dividing the ultimate bearing capacity obtained from the static analysis by a factor of safety 2.0. Thereafter, the soil domain is applied with different seismic excitations as presented in Table 1.

**Table 1** Prescribed horizontal seismic excitation

| Name of the input excitation | Magnitude of input excitation | Acceleration versus time history of input excitation |
|---|---|---|
| El-Centro | 7.1 | 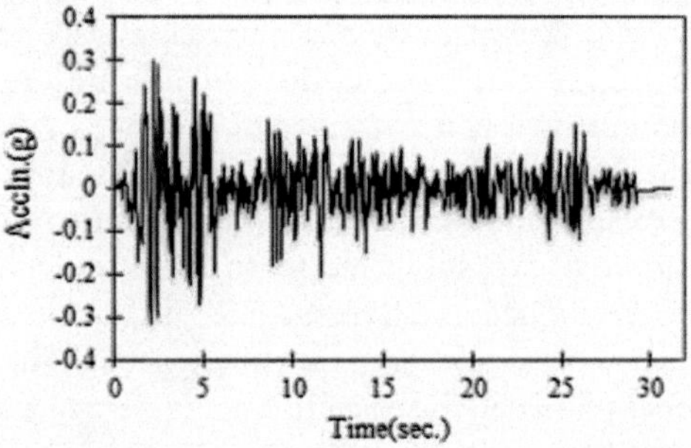 |
| Loma Gilroy | 6.9 | 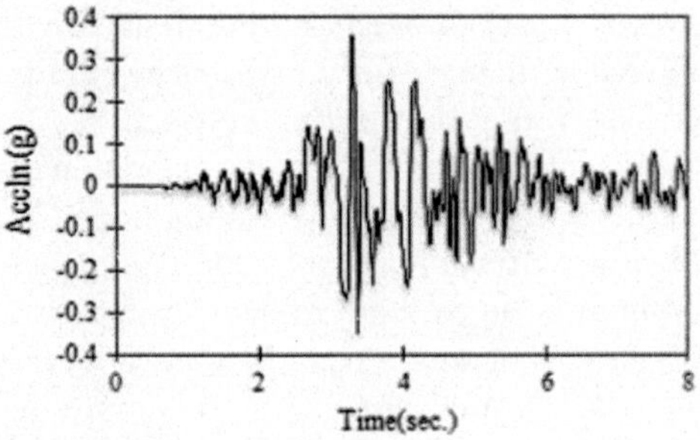 |
| Northridge | 6.7 | 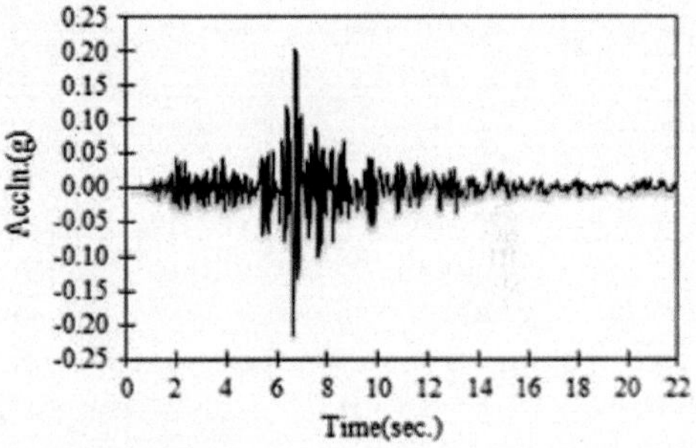 |
| Parkfield | 6.0 | 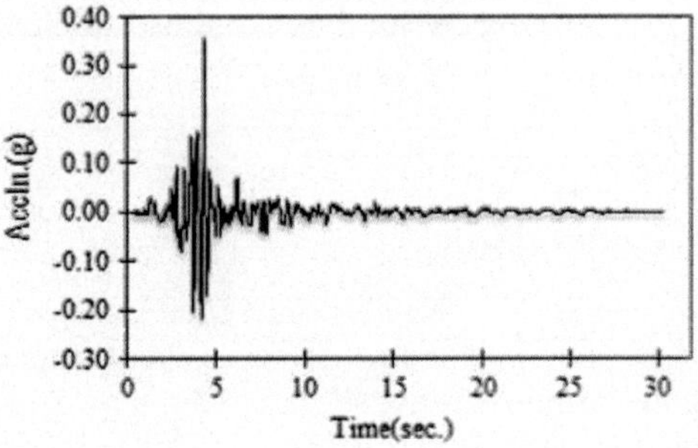 |

## 4 Results and Discussions

The seismic excitation is prescribed at the level of bedrock, which is followed by geostatic stress and the static working load, $q$, on closely spaced embedded footings. The analysis at hand is divided into different steps, in which geostatic stress is generated in 1 s, the static load is applied for 10 s, and seismic excitation is applied for a prescribed time interval of acceleration versus time as mentioned in Table 1. Figure 2 represents the time versus settlement for single and two footings placed at different $S/B$ subjected to different horizontal seismic excitations. It can be seen that the induced settlement due to seismic excitation occurs from 11 s onwards, hence at duration where maximum accelerations sudden increase in the settlement is observed. For all the cases, the settlement due to static loading is observed to the same magnitude, but the settlement due to seismic excitation is of different magnitudes. However, it can be seen that the trend of variation of settlement for given loading is the same; that is, for close spacing the induced settlement is maximum which decreases within the spacing between the footings. The percentage increase in the settlement due to applied seismic excitation in comparison with static loading is given in Table 2. The maximum variation in the settlement occurs for a case when $S/B =$ 0.25, and the seismic induced excitation is observed to be maximum for the greater magnitude of an earthquake. However, the seismically induced settlement for lesser earthquake magnitude (Parkfield) is not lesser (Table 2); this is because the peak ground acceleration for Parkfield is more than the Northridge. The difference in a seismically induced settlement is 10.5, 16.3 and 11.6% for Loma Gilroy, Northridge and Parkfield in comparison with El-Centro seismic loading.

The maximum normal stress (S22) and the maximum shear stress (S12) developed at the base of the footing are evaluated by averaging the stresses (S22/S12) induced in different soil elements below the footing base. In Table 3, developed normal stress is given at the end of dynamic loading. From the observation of normal stress, it can be seen that although the magnitude of normal stress for two footings case is small than that of single footing case for all the seismic loading, no significant variation occurs; that is, a change in the stress is approximately about 2–5%. Similarly, the shear stress (S12) is estimated, and the same is given in Table 4. It can be observed that the significant variation in the shear stress is observed when the footings are in close spacing. Also, the magnitude of shear stress decreases with an increase in spacing.

## 5 Conclusion

The seismic interference effect of closely spaced embedded strip footings placed in a semi-infinite, homogenous medium is carried out by the finite element method. From the analysis, it can be concluded that a significant increase in settlement occurs for the footings in close spacing. The settlement induced due to seismic excitation is

**Fig. 2** Time versus settlement for different horizontal accelerations

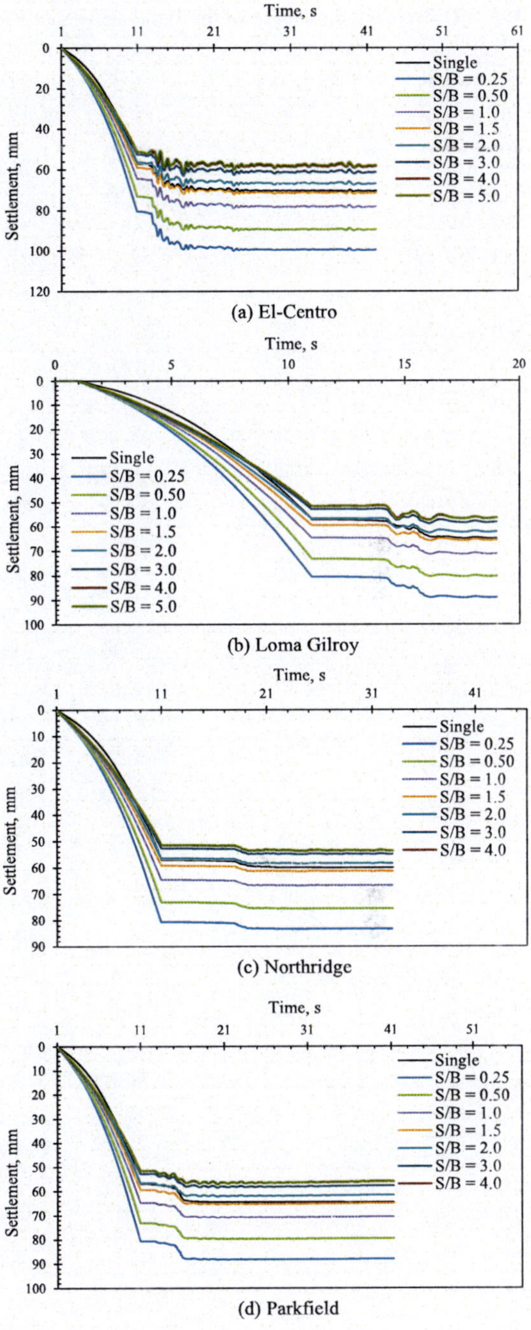

**Table 2** Percentage increase in the settlement due to seismic excitation from static loading for different horizontal accelerations

| Spacing between the footings (*S*/*B*) | El-Centro | Loma Gilroy | Northridge | Parkfield |
|---|---|---|---|---|
| Single footing | 23.32 | 13.57 | 5.21 | 13.15 |
| 0.25 | 23.49 | 10.49 | 3.37 | 9.22 |
| 0.50 | 22.14 | 9.71 | 2.22 | 8.48 |
| 1.0 | 21.15 | 10.23 | 3.46 | 9.01 |
| 1.5 | 20.65 | 11.03 | 3.73 | 9.38 |
| 2.0 | 18.56 | 10.61 | 3.57 | 8.86 |
| 3.0 | 16.23 | 10.93 | 4.31 | 9.36 |
| 4.0 | 18.28 | 11.06 | 4.24 | 8.70 |
| 5.0 | 18.41 | 11.12 | 4.30 | 8.53 |

**Table 3** Normal stress (S22, kPa) developed below the footing

| Spacing between the footings (*S*/*B*) | El-Centro | Loma Gilroy | Northridge | Parkfield |
|---|---|---|---|---|
| Single footing | 320.52 | 318.07 | 319.13 | 319.53 |
| 0.25 | 317.63 | 308.52 | 305.67 | 308.86 |
| 0.50 | 314.96 | 305.84 | 302.17 | 307.82 |
| 1.0 | 321.12 | 312.14 | 306.44 | 312.64 |
| 1.5 | 320.01 | 311.21 | 305.88 | 311.81 |
| 2.0 | 317.18 | 308.26 | 303.42 | 309.14 |
| 3.0 | 314.70 | 306.95 | 304.17 | 308.42 |
| 4.0 | 313.74 | 306.84 | 304.64 | 308.11 |
| 5.0 | 313.24 | 306.68 | 305.2 | 308.17 |

**Table 4** Shear stress (S12, kPa) developed below the footing

| Spacing between the footings (*S*/*B*) | El-Centro | Loma Gilroy | Northridge | Parkfield |
|---|---|---|---|---|
| Single footing | 6.26 | 4.36 | 2.3 | 4.45 |
| 0.25 | 13.93 | 9.74 | 6.32 | 7.73 |
| 0.50 | 17.1 | 12.46 | 10.3 | 12.6 |
| 1.0 | 17.94 | 17.11 | 13.55 | 15.68 |
| 1.5 | 15.21 | 15.10 | 11.42 | 14.03 |
| 2.0 | 11.53 | 11.26 | 7.34 | 10.07 |
| 3.0 | 7.47 | 8.01 | 4.01 | 6.07 |
| 4.0 | 7.08 | 8.26 | 4.75 | 7.17 |
| 5.0 | 6.73 | 8.07 | 5.07 | 7.28 |

observed to be more in the close spacing, that is, at $S/B = 0.25$, approximately 42% increase in the settlement noted in comparison with single footing. The evaluation of normal and shear stress from the analysis reveals that for footings in close spacing, the normal stress below the footings is observed to be more or less of similar magnitude as that of single footing. In contrast, the shear stress at close spacing is found to be more than that of single footing, and maximum shear stress occurs at the spacing of $S/B = 1.0$.

## References

1. Alpan I (1970) The geotechnical properties of soils. Earth Sci Rev 6(1):5–49
2. Ghosh P (2011) Seismic interference effect of two nearby square footings. In: Geo-frontiers 2011: advances in geotechnical engineering, pp 352–361
3. Gupta A, Sitharam TG (2018) Experimental and numerical investigations on interference of closely spaced square footings on sand. Int J Geotech Eng 1–9
4. Javdanian H (2018) Behavioral interference of vibrating machines foundations constructed on sandy soils. Int J Eng 31(4):548–553
5. Kuhlemeyer RL, Lysmer J (1973) Finite element method accuracy for wave propagation problems. J Soil Mech Found Div
6. Kumar J, Bhattacharya P (2013) Bearing capacity of two interfering strip footings from lower bound finite elements limit analysis. Int J Numer Anal Methods Geomech 37(5):441–452
7. Kumar J, Ghosh P (2007) Ultimate bearing capacity of two interfering rough strip footings. Int J Geomech 7(1):53–62
8. Lysmer J, Kuhlemeyer RL (1969) Finite dynamic model for infinite media. J Eng Mech Div 95(4):859–878
9. Lysmer J, Udaka T, Tsai C, Seed HB (1975) FLUSH—a computer program for approximate 3-D analysis of soil-structure interaction problems (No. PB-259332). California University
10. Mabrouki A, Benmeddour D, Frank R, Mellas M (2010) Numerical study of the bearing capacity for two interfering strip footings on sands. Comput Geotech 37(4):431–439
11. Nainegali LS, Basudhar PK, Ghosh P (2013) Interference of two asymmetric closely spaced strip footings resting on non-homogeneous and linearly elastic soil bed. ASCE Int J Geomech 13(6):840–851
12. Nainegali L, Basudhar PK, Ghosh P (2018) Interference of strip footings resting on nonlinearly elastic foundation bed: a finite element analysis. Iran J Sci Technol Trans Civ Eng 42(2):199–206
13. Nielsen AH (2006) Absorbing boundary conditions for seismic analysis in ABAQUS. In: ABAQUS users' conference, May 2006, pp 359–376
14. Salamatpoor S, Jafarian Y, Hajiannia A (2019) Bearing capacity and uneven settlement of consecutively constructed adjacent footings rested on saturated sand using model tests. Int J Civ Eng 17(6):737–749
15. Stuart JG (1962) Interference between foundations, with special reference to surface footings in sand. Geotechnique 12(1):15–22
16. Swain A, Ghosh P (2015) Experimental study on dynamic interference effect of two closely spaced machine foundations. Can Geotech J 53(2):196–209
17. Zhang Y, Conte JP, Yang Z, Elgamal A, Bielak J, Acero G (2008) Two-dimensional nonlinear earthquake response analysis of a bridge-foundation-ground system. Earthq Spectra 24(2):343–386

# Experimental Analysis of Seismic Cloak Foundation by Using Elastoplastic Material

**Anup Shelar, Varsha Yadav, Mirajj Thaker, and Dhruval Suriya**

**Abstract** The consequences of earthquakes cannot be underemphasized as more than 60% of the land area falls under seismic zones having intensity VII and above (MMI scale). This paper aims at having a comparative analysis of various parameters of two models that tested on a shake table and field. Model 1 has circular concentric ring of elastoplastic material at the foundation and Model 2 has a conventional circular foundation.

The proposed structure has circular raft foundation which is surrounded by concentric layers of elastoplastic material. Thus, the structure acts as a monolithic structure and moves in accordance with the foundation during seismic activity. By controlling the stiffness and elasticity of concentric rings, waves that travel through the "cloak" will dissipate into small fluctuations due to the effect of pressure and density. The cloak has the ability to absorb and alleviate the intensity of waves approaching it from any direction due to its circular shape.

The effect of seismic wave on footing results in the generation of impact force of high intensity which comes laterally or longitudinally from the surrounding soil and rocks. To mitigate the impact of seismic forces, the foundation can be surrounded by a circular concentric ring of elastoplastic material laid in different geometrical formations. This may absorb and divert the impact of forces generated by seismic waves.

**Keywords** Elastoplastic material · Seismic wave · Concentric ring · Shake table · Seismic cloak

---

A. Shelar · V. Yadav
Parul University, Vadodara, India
e-mail: anupok3@gmail.com

V. Yadav
e-mail: varshayadav179@gmail.com

M. Thaker (✉) · D. Suriya
Universal College of Engineering, Mumbai, India
e-mail: mirajjketanthaker1@gmail.com

D. Suriya
e-mail: dhruvalsuriya76@gmail.com



T. G. Sitharam et al. (eds.), *Seismic Design and Performance*, Lecture Notes in Civil Engineering 120, https://doi.org/10.1007/978-981-33-4005-3_7

# 1 Introduction

Earthquake is amongst the catastrophic natural calamities where an abrupt movement or tremor is initiated under earth's crust causing major damage to mankind. Sebastien Guenneau at the University of Liverpool says "The seismic waves of an earthquake fall into two main groups—body waves that propagate through the earth and surface waves that travel only across the surface. Although controlling body waves would be too complex, the controlling surface waves are within the ability of conventional engineering" [1]. The elastodynamic cloak or invisibility cloak that was created by the English mathematician, William Parnell, is said to divert shock waves, leaving a building almost untouched by earthquake. The rubberized material should be wrapped around the base or the key components of any structure. By doing such, the building is expected to be spared from the severe damages brought about by an earthquake [2]. Seismic cloaks are currently being tested as a means of creating protective barriers capable of rerouting seismic energy away from aboveground structures. Seismic cloaking involves the modification of soil and other ground materials surrounding a building to deflect or redirect the force created by an earthquake. This innovation revolves around the theory that seismic waves pass energy between the potential energy stored in the planet's crust and the kinetic energy within the seismic wave itself.

The new theoretical cloak comprises a number of large, concentric rings made of plastic fixed to the Earth's surface. The stiffness and elasticity of the rings must be precisely controlled to ensure that any surface waves pass smoothly into the material, rather than reflecting or scattering at the material's surface. When waves travel through the cloak they are compressed into tiny fluctuations in pressure and density that travel along the fastest path available. By tuning the cloak's properties that path can be made to be an arc that directs surface waves away from an area inside the cloak. To design an earthquake-resistant structure, engineers need to reinforce the structure and counteract an earthquake's forces. Since earthquakes release energy that pushes a building from one direction, the strategy is to have the building pushed from the opposite way. Instead of just counteracting forces, researchers are experimenting with ways buildings can deflect and reroute the energy from earthquakes altogether. Dubbed the "seismic invisibility cloak", this innovation involves creating a cloak of 100 concentric plastic and concrete rings in and burying it at least three feet beneath the foundation of the building [3]. Nowadays, advanced techniques are being developed to resist earthquake which moreover focuses on enhancing superstructures. The root cause is a failure of foundation, protecting which shall safeguard the entire structure. The developing countries or economically backward countries cannot afford to install such techniques as they are uneconomical and require high maintenance.

The proposed technique as shown in Fig. 1 shows the conceptual diagram that consists of circular raft foundation cloaked with elastoplastic material in concentric manner which can absorb and resist the intensity of waves approaching from any direction due to its circular shape. The seismic waves that enter the rings can be

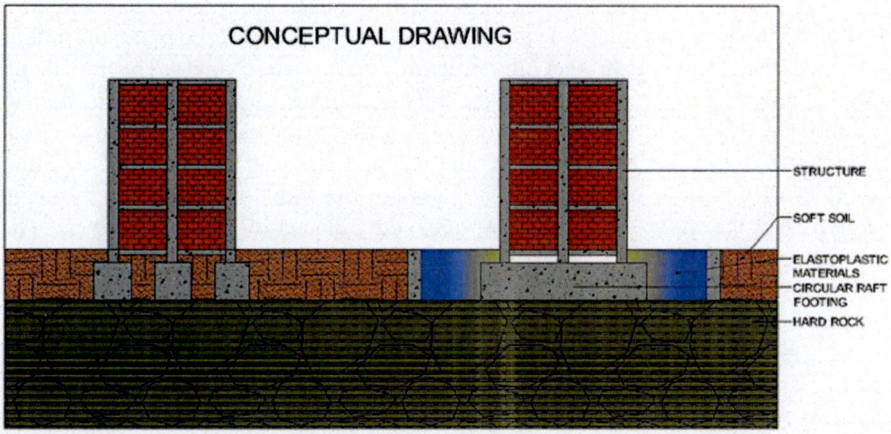


**Fig. 1** Conceptual drawing of seismic cloak foundation

forced to move to the outer ring and then channel away from the building. It is easy to install and has low cost as compared to other techniques. This technique can be used for any type of substructure. This technique will prevent the building from collapsing suddenly during earthquakes thus preventing fatal injuries in and around the building. The testing has been carried out by performing field test and making a prototype.

## 2 Literature Review

Noon [4] conducted a model testing at Arizona State University civil Engineering laboratory that simulated Tohoku Earthquake by using base isolation system. The system comprised of two bearing outer laminated rubber bearing (LRB) and inner high damping rubber bearing (HDRB) placed alternately in 18 layers. The bearings were then connected on each inner hexagonal side with 4 shock transferring rubber springs to absorb the axial force in the central high damping rubber bearing which would increase with the increased magnitude and decrease with decreased magnitude of earthquake. By this, the system was able to self-stabilize the shock. The shock transferring rubber spring would direct the required amount of force to HDRB or LRB to decrease 60% energy transfer which was better as conventional systems decreased by merely 40%. It could also achieve self-stabilization in 0.5 s by about 28%.

Mohammad et al. [5] have simulated broadband cloaking of electromagnetic waves by using elastic cloaking in thin plates in concentric manner. The design consisted of a cloak with 10 concentric layers of 6 materials namely flurocarbon elastomer, granular moulding powder, SABIC innovative plastic, epoxy and polyester. These materials greatly reduced the flexural plane waves having range from 30 to

150 hertz. However, this method cannot be implemented on a large scale, the author suggested design shield to protect infrastructures from seismic waves. The broadband waves of only larger wavelength than the thickness of the plate could be obstructed.

The paper [3] demonstrated earthquake-resistant structure using seismic cloak foundation. The author had developed the technique to test the ability of resisting the effect of vibration on structure by cloaking the individual footings by using materials like thermocol, wire mesh, sponge, etc., placed in concentric layers. The test was carried out by cyclic impact hammering that created shock waves throughout the structure. The graphical results of proposed and conventional structures were compared in terms of acceleration and gyro effect. This technique could de-intensified the wave to half of its magnitude. However, a major shortcoming of the research was that the failure of single footing could demolish entire structure. Furthermore, the impact shocks applied were non-uniform as hammering was done manually.

Tsang et al. [6] had proposed rubber-soil cushion for earthquake-resistant structure. The paper applied seismic isolation method by using rubber-soil mixture as cushion. It reduced the level of shaking of structure not just in horizontal direction by 60–80% but also in vertical direction by 80–90%. The research is significant as it utilized scrap tyres from stockpiles, which are environmentally friendly and inexpensive, thus, benefiting the developing countries. The method was presented by simulating a series of numerical models. Also, a parametric study that included nonlinear site response, soil resonance effect, liquefaction, ground settlement and environmental effect was carried out to evaluate the effectiveness of cushions.

Kashiwa et al. [7] had studied the method of earthquake response reduction of direct foundation using composite geomaterial which consisted of soil slurry, cement, rubber chips and fibres. The author discussed the mechanism and properties of the compound by performing unconfined compression test and cyclic simple shear test. The shaking table test comprised of three conditions namely nonslip-foundation having soil adhered by epoxy resin to the, slip-foundation having soil not adhered to base and isolation-foundation which floated by 10 mm by geodynamic magnet. Acceleration and Fourier spectrum from top of above three model conditions were compared. Results showed maximum amplification ratio become small in order of isolation < slip < nonslip. The maximum amplification also followed the same sequence. The earthquake response of the building decreased when the input energy from the bottom of foundation was decreased. Geomaterial that was used to increase the absorption of seismic wave energy of surrounding soil. However, the earthquake response capacity decreases in case of large earthquakes. Horizontal response of the building decreased and rocking component of the foundation increased.

## 3 Shaking Table Test

There are several different experimental techniques that can be used to test the response of structures and soil or rock slopes to verify their seismic performance. One of which is the use of an earthquake shaking table. Shake table provides an experimental platform that simulates earthquake motion to verify seismic performance of building structure. It is a device for shaking scaled slopes, structural models or building components with a wide range of simulated ground motions.

As to test our model, we developed a prototype in which we stimulate the ground motions as shown in Figs. 2 and 3. The prototype consists of two structures (Model 1 and Model 2) that represented proposed structure and conventional structure, respectively. Model 1 represents circular foundation with concentric rings of elastoplastic material, whereas Model 2 represents conventional footing as shown in Fig. 4.

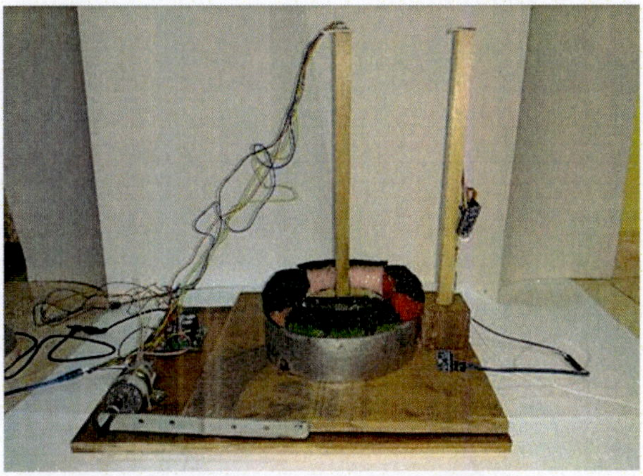

**Fig. 2** Front view of prototype

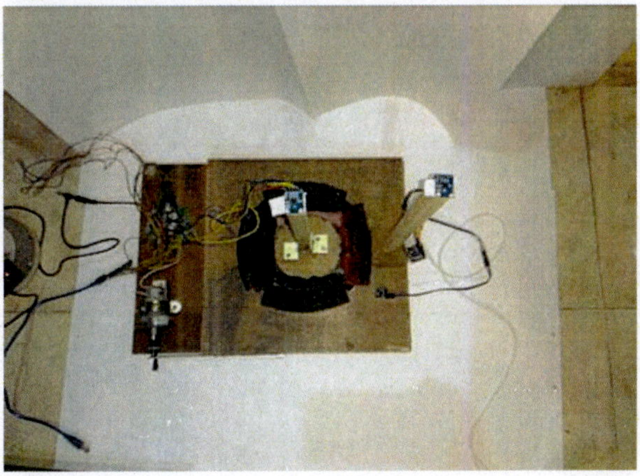

**Fig. 3** Top view of prototype

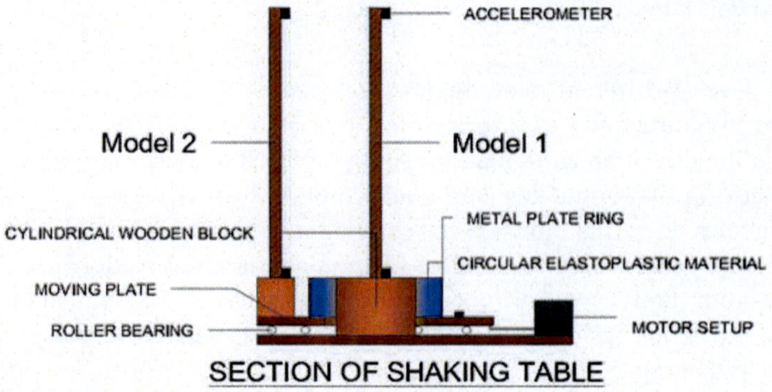


**Fig. 4** Section of shaking table

## 3.1 Components

**Motor**: A DC motor of 200 RPM if 12 V is used to vibrate the table.

**Electronic circuit**: A electronic circuit is used to increase and decrease the speed with frequency range 1–5 Hz and frequency control of ± 5%

**Arduino**: The Arduino nanoboard based on the AT mega 328 is used to connect accelerometer and display readings on the serial monitor.

**Speed regulator**: DC motor PWM speed regulator, 5V-2A switch function.

**Three-axis accelerometer**: The MPU6050 consists three-axis accelerometer with micro-electro mechanical (MEMs) technology. It used to detect angle of tilt or inclination along the *X*, *Y* and *Z*-axes.

## 3.2 Testing

The testing was done on a frequency of 3 Hz and readings were obtained along the three axes for time period of 35 s.

## 3.3 Result and Analysis

The earthquake wave was simulated by shaking table method to analyse the effect of acceleration against time. Acceleration time graph for a fixed frequency range was plotted for three axes. It can be clearly observed from Figs. 5, 6 and 7 that the proposed technique could absorb the vibrations and remained stable at the beginning of the time stamp, whereas the conventional structure deflected instantly. The vibrations were reduced to 30–40%. Hence, it is evident that these vibrations were absorbed by elastoplastic materials.

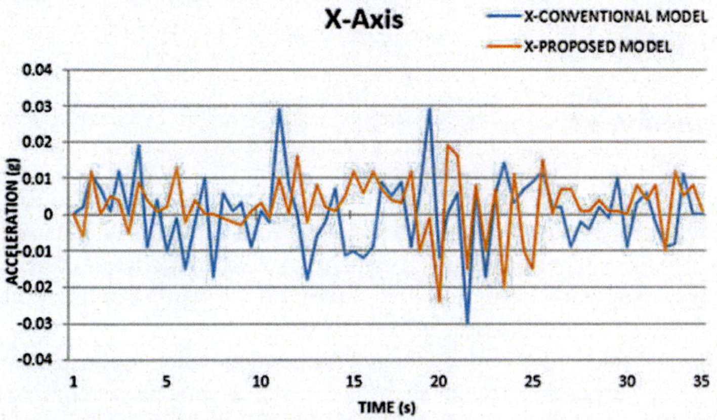


**Fig. 5** Accleration time history of prototype in *X* direction

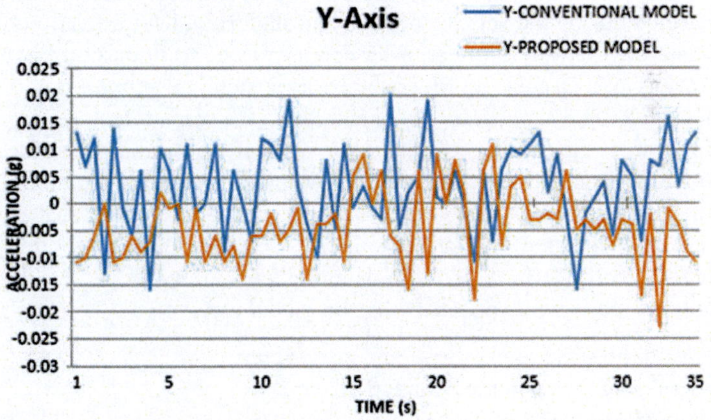


**Fig. 6** Accleration time history of prototype in *Y* direction

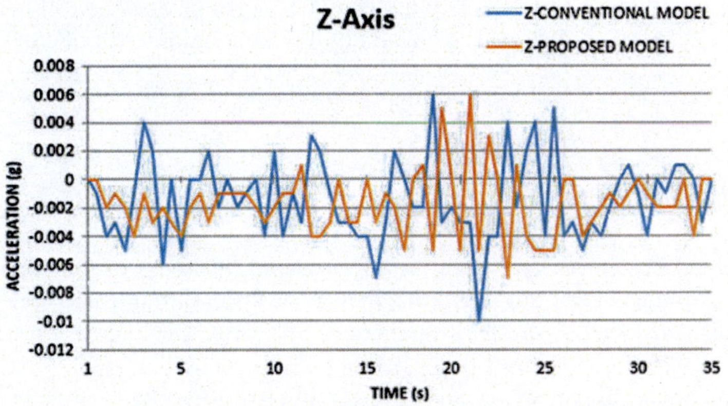


**Fig. 7** Accleration time history prototype in *Z* direction

# 4 Field Test

## *4.1 Methodology*

Various methods have been introduced to make structures resistant to earthquakes. However, methods can be employed to make the foundation itself resistant to earthquakes. A structure experiences the seismic impact when the lateral forces, i.e., the surface waves (Rayleigh waves and love waves) hit the footing of the foundation. It sways away the structure and may also collapse it.

The setup consists of a circular concrete foundation which replicates circular raft foundation. The purpose of circular shape is to resist seismic waves coming from any direction and to distribute the load of the superstructure over large area, especially in soft soils where the amplification of waves is greater. These foundations are suitable for a wide range of soil types. It resists the differential settlement of soil which generally occurs during the seismic event by strong ground amplifications. It can be applied where the shallow foundation is necessary but soil condition is poor. The bearing capacity of a circular footing is more because if soil support of a continuous footing yields, all the soil particles move parallel to the plane which is perpendicular to the centre line of the footing. On the other hand, if the soil support of a square or circular footing yields, the soil particles move in radial and not in parallel, so bearing capacity increases. The failure will occur in accordance with the narrow side of the footing in case of isolated and strip footings. Therefore, circular footing, without preference for specific failure direction, is more "cost-effective".

Assumptions made while constructing the model are:

- Soil beneath the structure is dense and compacted.
- Only lateral forces act on the foundation.
- The footing rests on the hard strata.

To observe and analyse the results and behaviour of structure, a whole unit consisting of structure and foundation techniques is set up and the following setup is then subjected to surface waves created by the Jack hammer throughout the soil structure.

## *4.2 Construction*

The earthquake's effects vary with the softness and thickness of the sediment. As seismic waves travel through the ground, they move faster through hard rock than soft soil. When waves are transmitted from hard to soft earth, they increase in amplitude (or size). A bigger wave causes more shaking.

Soft soil means bigger waves and stronger amplification. As seismic waves have strong amplifications in soft soil, the model has been tested for the same and properties of soil are mentioned in Table 1. For testing, the excavation was done up to 800 mm

**Table 1** Material properties

| Sr. No | Material | Properties | |
|---|---|---|---|
| 1 | Jack hammer | Name—11 kg demolition hammer<br>Type—hammer drill<br>Speed range—1–6<br>BPM range—900–1890 | |
| 2 | Concrete<br>As per<br>IS 456-2000 | Grade of concrete<br>Type of cement<br>Cement<br>Water<br>Fine aggregate<br>Coarse aggregate<br>Water–cement ratio | M 20<br>OPC 53 grade<br>399 kg/m$^3$<br>191.58 kg/m$^3$<br>650.62 kg/m$^3$<br>1222.14 kg/m$^3$<br>0.4 |
| 3 | Steel<br>As per<br>IS 1786-2008 | Yield strength<br>Diameter | 415 N/mm$^2$<br>10 mm |
| 4 | Rubber | Natural rubber<br>Polychloroprene (neoprene)<br>Polyisobutylene (butyl) | |
| 5 | Soil<br>As per<br>IS 2720-1986 | Optimum moisture content<br>Maximum dry density<br>Cohesion $C_u$ (kg/cm$^2$)<br>Angle of friction $\phi$ (degrees) | 22.00%<br>1.594 gm/cm$^3$<br>0.22<br>21.31 |
| 6 | Sensor | Accelerometer<br>Arduino | Three axes ($X$, $Y$, $Z$)<br>MPU 6050 |

deep in soil as shown in Fig. 8. A thin waterproofing membrane of thin layer was placed on it as shown in Fig. 9. Then, the formation of hard strata included rubble and hard rocks were done as shown in Fig. 10. Thereafter, a 120 mm layer of plain cement concrete of ratio 1:6 is layer as shown in Fig. 11. Thereafter, a circular footing of 400 mm diameter and 300 mm depth of M20 grade concrete was placed on PCC as shown in Fig. 12. Tamping rod was used to achieve the desired compaction. The curing of the footing was done till 28 days. A steel bar of grade Fe415 of 1000 mm length was embedded in a concrete footing as shown in Fig. 12. Sensors were placed on soil and bar to measure the accelerations as shown in Fig. 13.

The sheets of natural rubber, polychloroprene (neoprene), polyisobutylene (butyl) which of commonly used in seismic bearings having properties mentioned in Table 1 were layer in a concentric manner around the footing as shown in Fig. 14. The mechanical (tear strength, high strain fatigue resistance and creep resistance) and low temperature (cold crystallization) properties of natural rubber are superior to those of most synthetic rubbers. Natural rubber, followed by neoprene, is the most frequently specified material for use in rubber bearings. Butyl rubbers have the lowest temperature properties among the hydrocarbon rubbers and they are often substituted for neoprene for low temperature applications. The sheet of thickness 8, 6, 2 mm has been used. The layers were overlapped on joints as shown in Fig. 15 to distribute the impact forces radially on it.

**Fig. 8** Excavation

**Fig. 9** Thin waterproof membrane

**Fig. 10** Hard strata

**Fig. 11** PCC

**Fig. 12** Footing

**Fig. 13** Placing of sensor

**Fig. 14** Laying of sheets

**Fig. 15** Concentric layers

**Fig. 16** Field test of cloaked model

## 4.3 *Procedure*

The test has been conducted on Model 1 and Model 2 as mentioned above.

Model 1 consists of placing of rubber sheets in concentric layers around the footing and vibration of uniform intensity were generated by Jack hammer without changing the position as was during testing of Model 1, i.e. 1 m. The sensors were placed on soil and on bar and readings were noted along the three axes as shown in Fig. 16.

Model 2 consists of conventional circular footing. The vibrations of uniform intensity were generated by Jack hammer placed at same position, i.e. 1 m away from the centre of footing under the same soil conditions as in Model 1. The sensors were placed on soil and on bar and readings were noted along the three axes as shown in Fig. 17.

# 5 Result and Analysis

The results were obtained by using accelerometers to calculate acceleration in *X*, *Y*, *Z* direction of both models. An accelerometer was placed at the top of the bar and another on the soil beneath. The result in *X* direction shows more variation in acceleration in soil and less deflection on bar due to the presence of rubber sheets

**Fig. 17** Field test of conventional model

as shown in Fig. 18, whereas the absence of rubber sheets showed more variation in acceleration bar than soil. Overall the rubber sheets reduced the accelerations by 35%–45%, respectively (Figs. 19, 20, 21, 22 and 23).

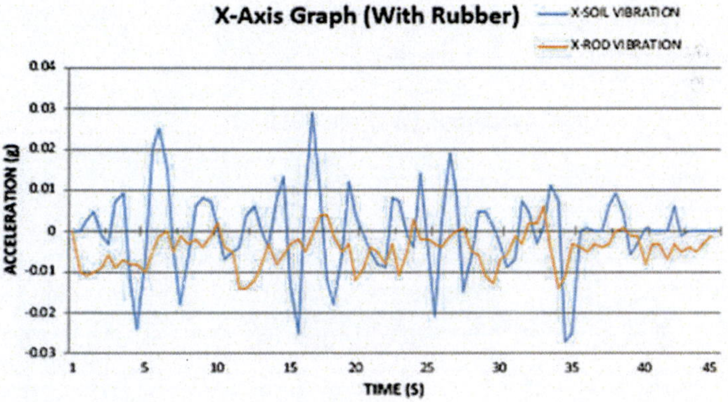


**Fig. 18** Acceleration time history of *X* direction

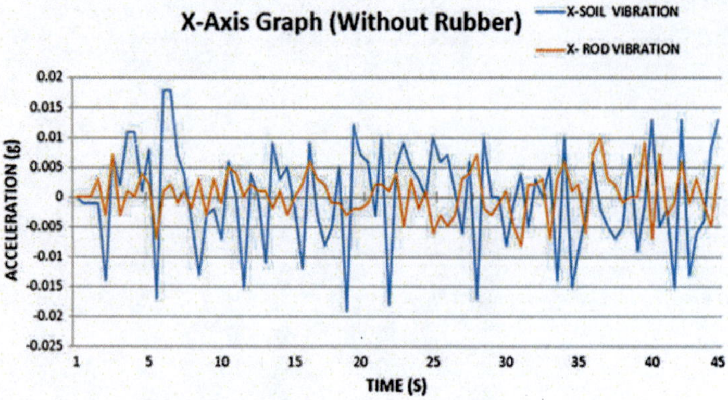


**Fig. 19** Acceleration time history of *X* direction

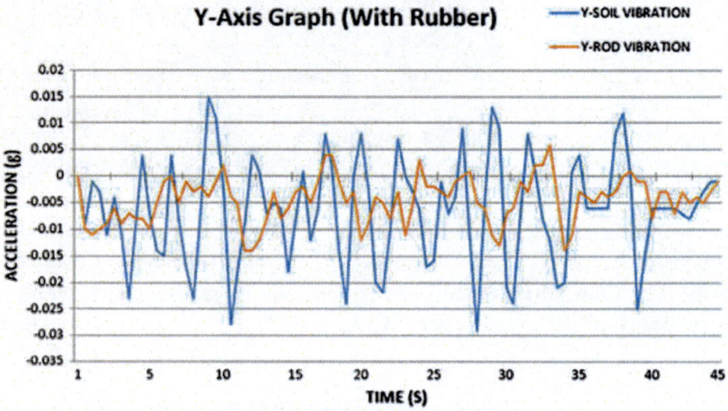


**Fig. 20** Acceleration time history of *Y* direction

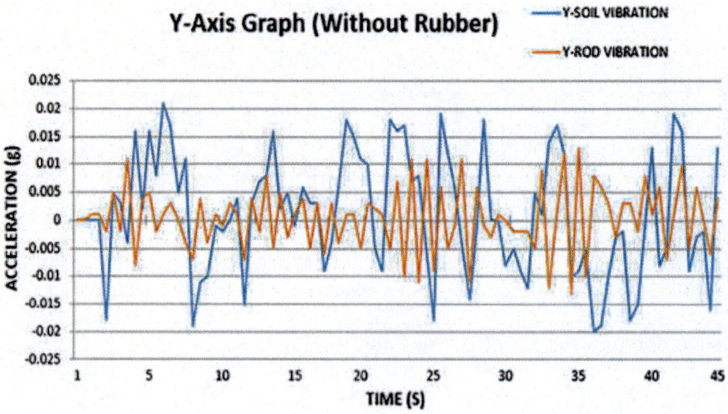


**Fig. 21** Acceleration time history of *Y* direction

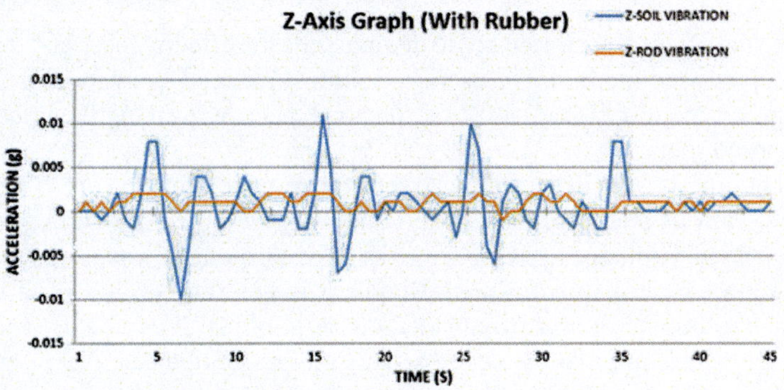


**Fig. 22** Acceleration time history of Z direction

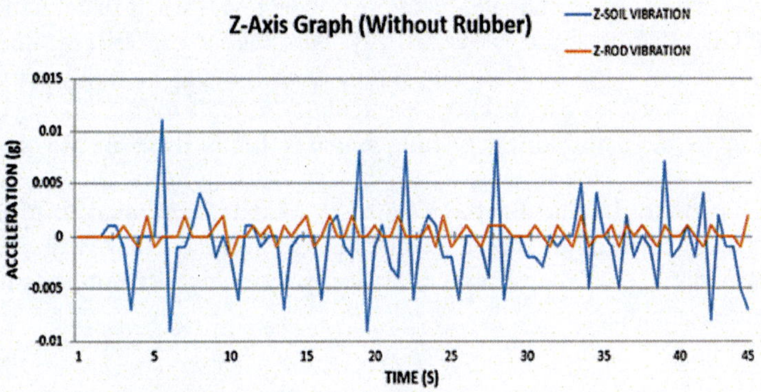


**Fig. 23** Acceleration time history time of Z direction

## 6 Conclusion

The current research work attempts to address and understand the seismic mitigation measure of using an elastoplastic material consisting of concentric layers of rubber sheets placed in concentric manner around the foundation. The setup acted as an underground barrier to the vibrations reaching the foundation. The proposed technique resisted the incoming vibrations generated by harmonic action of seismic waves, thereby minimizing the hampering of the stability of the structure. The cloaked foundation technique was helped by redirecting the wave energy elsewhere away from the structure. As the earthquake response of building is greatly influenced by contact condition between foundation and supporting soil, the application of the rubber layers around the foundation resulted in decrease in resonance phenomenon of building to a large extent. A series of simulations on field test and prototype test and parametric study has been carried out to demonstrate the effectiveness and the robustness of the proposed method. On average field test reduced accelerations by 35%–45% and prototype test as 30%–40%, respectively, hence, proving the effectiveness of the proposed method.

The technique is still in its infancy, yet government agencies are beginning to take notice. Various mathematical models also suggest a pattern that might provide a "zero stop band", which can stop a wide range of earthquake waves and that would be really a key to get it to work well. Maybe it is possible, on the scale of a city, to diminish the impact of an earthquake considerably [8]. It can be used to protect expensive infrastructure that cannot be seismically retrofitted, such as oil pipelines; electric pylons in vulnerable locations. Also, the nuclear power plants, and perhaps even whole cities, could be protected from earthquakes using "seismic invisibility cloaks [8]. It can reduce the effects of magnitude 7 earthquake down to magnitude 5 or even lower and the earthquake-cloaking concept on a small scale in the ground using machines that generate vibrations [9]. This rubber can reduce the effect of seismic vibrations on the buildings on top of them it could be easily retrofitted to existing buildings at low cost, making it particularly suitable for developing countries [10]. This is the beginning of the technique and in future there are ways of testing it through large-scale experiments, which involve creating full-size model buildings and shaking them to simulate the force from recorded real earthquakes. If the theory gets scaled up to larger objects, then it could be used to create cloaks to protect buildings and structures, or perhaps more realistically to protect very important specific parts of those structures.

## References


1. Barras C (2009) An invisibility cloaks could hide buildings from earthquake. https://www.newscientist.com/article/dn17378-invisibility-cloak-could-hide-buildings-from-quakes/amp/.html. Accessed 8 Jan 2019
2. Padua D (2016) An invisibility cloak to spare from earthquakes. http://www.itechpost.com/articles/68555/20161226/invisibility-cloak-spare-structures-earthquakes.html. Accessed 8 Jan 2019
3. Zitzman L (2019) How earthquake proof buildings are designed. https://www.bigrentz.com/blog/earthquake-proof-buildings?amp/html. Accessed 5th Mar 2019
4. Noon D (2018) A novel approach in the construction of an Earthquake, Tsunami and Radiation resistant system. Google science fair
5. Mohamed F, Sebastien G, Stefan E (2009) Ultrabroadband elastic cloaking in thin plates. https://doi.org/10.1103/PhysRevLett.103.024301
6. Tsang HH, Sheikh MN, Lam N (2007) Rubber-soil CUSHION for earthquake protection. In: Lam N, Wilson J, Gibson G, Anderson S (eds) Australian earthquake engineering society conference. Australian Earthquake Engineering Society, Australia, pp 1–8
7. Shimamura A, Kashiwa H, Miyamoto Y (2012) Study on earthquake response reduction of direct foundation using composite geomaterial. In: 15th world conference on earthquake engineering, vol 1, ISBN: 978-1-63439-651-6
8. Cosmos 57 by Philip Dooley (2014) Earthquake cloaking could protect our cities. https://cosmosmagazine.com/geoscience/earthquake-cloaking-could-protect-cities-temblors.html. Accessed 2 April 2019

9. Ghose T (2016) Could an earthquake invisibility cloak shield building from damage? https://www.livescience.com/57246-earthquake-cloaks-could-protect-buildings.html. Accessed 8 April 2019
10. Bernal J (2018) Could rubber tyre foundations help protect buildings from earthquake? https://www.citymetric.com/fabric/could-rubber-tyre-foundations-help-protect-buildings-during-earthquake-4139.html. Accessed 11 Jan 2020

# Effect of Near-Fault Vertical Ground Motions on the Seismic Response of Bridges with Rocking Foundations

**Omar El-Hawat, Behzad Fatahi, and Mohammadreza Mostafa**

**Abstract** Rocking foundations can be used in bridges to reduce the seismic demand of the structure and prevent inelastic behaviour from developing at the piers during large earthquake events. Several studies have proven that rocking is an effective seismic isolation technique under lateral earthquake loading. However, limited research has been conducted on the effect of the vertical component of earthquakes on the rocking behaviour of bridge piers. This paper aims to numerically investigate the effect of the vertical component of near-fault earthquakes on the seismic performance of bridges with rocking pile foundations. Two identical bridge configurations with different foundation systems (conventional fixed base foundation and rocking foundation) are subjected to two loading cases: (1) horizontal ground motions only, (2) combined horizontal and vertical ground motions. Three-dimensional models of the bridges are developed with the appropriate material models to capture possible inelastic behaviour, as well as to model the soil–structure interaction. Four near-fault ground motions with three components are selected and scaled to the appropriate seismic hazard and applied to the bridges using nonlinear dynamic time history analyses. The dynamic responses of the bridges are compared in terms of deck displacements, deck bending moments, and pier axial and bending moments. The results show that the vertical component of ground motions can considerably increase the dynamic response of the bridge with the rocking foundation when compared to the fixed base foundation, leading to increased deck displacements and inertial actions on the bridge structure.

**Keywords** Near-fault earthquakes · Vertical ground motions · Rocking pile foundation · Soil–structure interaction

---

O. El-Hawat (✉) · B. Fatahi
School of Civil and Environmental Engineering, University of Technology, Sydney (UTS), Sydney, NSW, Australia
e-mail: Omar.el-hawat@uts.edu.au

B. Fatahi
e-mail: Behzad.fatahi@uts.edu.au

M. Mostafa
Stantec, Pasadena, CA, USA

 
T. G. Sitharam et al. (eds.), *Seismic Design and Performance*, Lecture Notes in Civil Engineering 120, https://doi.org/10.1007/978-981-33-4005-3_8

## 1 Introduction

Conventional bridges with fixed base foundations are designed to withstand severe seismic loading through the development of plastic hinges at their piers. However, the damage caused by the inelastic behaviour of the hinges can lead to lengthy and costly repairs, traffic disturbances and possible demolition of the entire bridge [1].

Rocking isolation is an alternative design concept that can be implemented in the design of bridges, as a form of seismic base isolation. This can not only reduce the seismic demand on the structure, but also prevent substantial plastic damage from occurring [2]. Rocking shallow foundations and rocking pile foundations are common methods of implementing rocking isolation in bridges. Several numerical, analytical and experimental studies were conducted to investigate the seismic performance of rocking shallow foundations [3, 4]. It was concluded that the rocking shallow foundations would significantly reduce the seismic inertial forces on the bridge structures; however, they would cause considerable soil settlement and residual rotations. The rocking pile foundation was introduced to address the issues associated with settlement by allowing the pile cap to rock about the piles. Although the rocking piles system performed nominally elastically, it lacked energy dissipation that would otherwise be present from the soil/plastic hinge damage, and as a result, experienced increased seismic displacements [5]. Consequently, observations from previous large earthquakes have indicated that large deck displacements could lead to bearing failure, pounding damage between the decks and abutments, and in some cases, unseating of the deck, and ultimately, collapse of the bridge [6].

Most existing researches on seismic performance of rocking foundations in bridges have been conducted using only the horizontal ground motions, while neglecting the vertical components. This is because it is believed that in general, the vertical component of the earthquake has a relatively smaller peak ground acceleration (PGA) when compared to its horizontal counterparts ($V/H$ ratio assumed to be 2/3) [7]. Furthermore, the structure is believed to be very strong in the vertical direction [7]. Nevertheless, several studies have confirmed that this is not the case for structures subjected to near-fault ground motions as peak $V/H$ exceeded the 2/3 rule, as observed in the 1994 Northridge and 2011 Christchurch events [8]. In addition, several numerical and experimental studies have confirmed that bridge structures are vulnerable to vertical ground motions [9]. The fluctuating axial force in the piers caused by the vertical excitations and can increase the shear demand on the pier while decreasing its shear capacity [10]. Additionally, the vertical excitations increase the bending moments of the deck causing yielding at midspan and at the pier supports.

Several studies were conducted on the arrival time of the propagating $P$ waves (vertical component) with the $S$ waves (horizontal component). It was concluded that for distances of 5 km or less to the earthquake source, and the time interval between the arrival of the vertical and horizontal component should be taken as zero [8, 11]. This makes it possible for the peak horizontal acceleration to coincide with the peak vertical acceleration. This can be critical for bridges with rocking foundations, as the vertical excitations can increase the $P$-delta effects, which intern increases the horizontal displacements and pier bending moments. Limited research has been conducted on the effect of the vertical component of ground motions on the performance of rocking foundations. Hence, it is of importance to investigate the effect of vertical earthquake excitations on the horizontal deck displacements and pier bending moments for bridges with rocking pile foundations.

## 2 Bridge Description

The bridges adopted for this study represent a typical highway bridge located in a high seismic region of New Zealand. A total of two bridges are investigated in this study: (1) Bridge with a rocking pile foundation (RP) designed to uplift and rock about the piles, (2) bridge with a conventional fixed base foundation (FB) designed to develop plastic hinging at the pier base. The bridges are designed to comply with NZS 3101 and NZS 1170.5 [12]. Additionally, the bridges are configured with geometries and dimensions similar to existing bridges in the same seismic region. The configuration of the adopted bridges is presented in Fig. 1.

The concrete bridges are symmetrical with four equal spans of 45 m, and a total length of 180 m. The bridge deck comprises of a concrete box girder with an overall depth of 1.7 m and maximum width of 14 m. The concrete used in the bridge has a compressive strength of 50 MPa and a unit weight of 25 kN/m$^3$. Furthermore, the reinforcing steel has a yield strength of 500 MPa and a unit weight of 77 kN/m$^3$. The bridge decks are supported by three intermediate single pier bents with equal heights of 8.0 m. The concrete piers comprise of a rectangular hollow section with dimensions of 2.2 × 4.2 m with wall thicknesses of 0.5 m and longitudinal reinforcing steel ratio of 2%.

The piers are supported by a deep pile foundation consisting of four piles connected to a pile cap. The dimensions of the pile cap are 7.2 × 7.2 × 1.9 m. The piles have a diameter of 1.2 m and longitudinal reinforcing steel of 1.5%. For the RP bridge, these piles are unconnected to prevent the transfer of tensile forces and to allow uplift during strong earthquake shaking. Additionally, the piles protrude 0.4 m into the pile cap. This is to ensure that full contact of the piles is maintained, and the adequate transfer of shear forces is achieved for the duration of the earthquake.

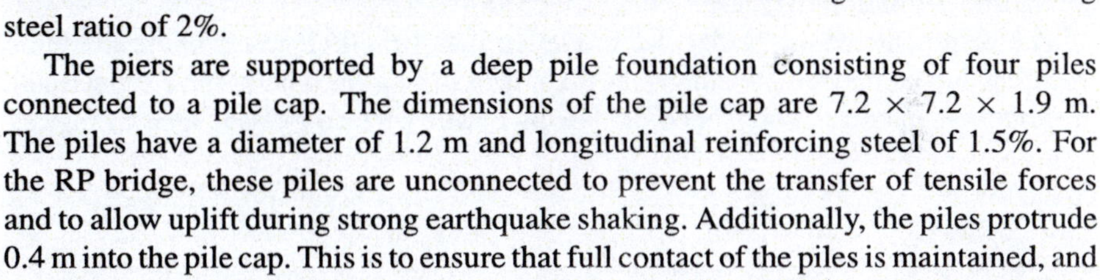


**Fig. 1** Typical longitudinal section of the bridges adopted in this study

The bridge deck is supported by four elastomeric bearings at the abutments with a deck seat width of 650 mm. The elastomeric bearings have a diameter of 600 mm and thickness of 200 mm. The piles at the abutment have a diameter of 0.8 m and longitudinal reinforcement ratio of 1.5%. All the piles are end bearing with a 1 m socket into class A rock. In addition, the piles are founded in soft clay with properties classifying it into class D soil as defined in the NZS 1170.5.

## 3 Numerical Model

The finite element software SAP2000 V20.1 was used to develop the three-dimensional models of the bridges and to perform the nonlinear time history analyses. Figure 2 presents the numerical model of the adopted bridges. The bridge deck was modelled as a spine through the use of multiple frames elements and lumped masses [13]. The bridge deck is modelled using frame elements that are characterized to capture biaxial bending, shear, axial and torsional actions on the member.

The intermediate piers are modelled using frame elements which are connected to the underside of the deck using hinge connections to allow rotation while restricting lateral displacements. As displayed in Fig. 3a, flexural plastic hinges (Fibre P-M2-M3) were assigned to the base of the piers to capture possible inelastic behaviour that can develop during the seismic loading. All assigned flexural hinges adopt a Takeda type hysteresis for the concrete and a kinematic hardening type hysteresis for the reinforcing steel. Furthermore, the nonlinear material relationship of the plastic hinge accounts for changes in the axial-movement interaction surface, axial strain, and moment-rotation response.

The abutments of the bridge are modelled using shell elements while the piles are modelled using frame elements as displayed in Fig. 3b. The passive capacity of the abutment backwall soil is modelled using elastic perfectly plastic area springs as

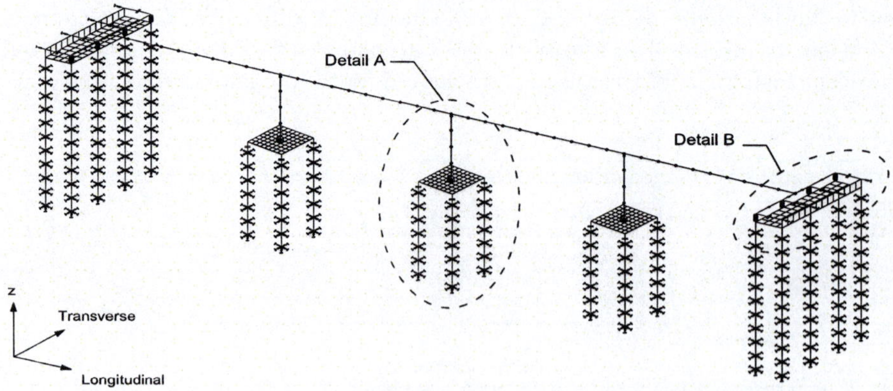


**Fig. 2** Schematic diagram of numerical model of the bridges

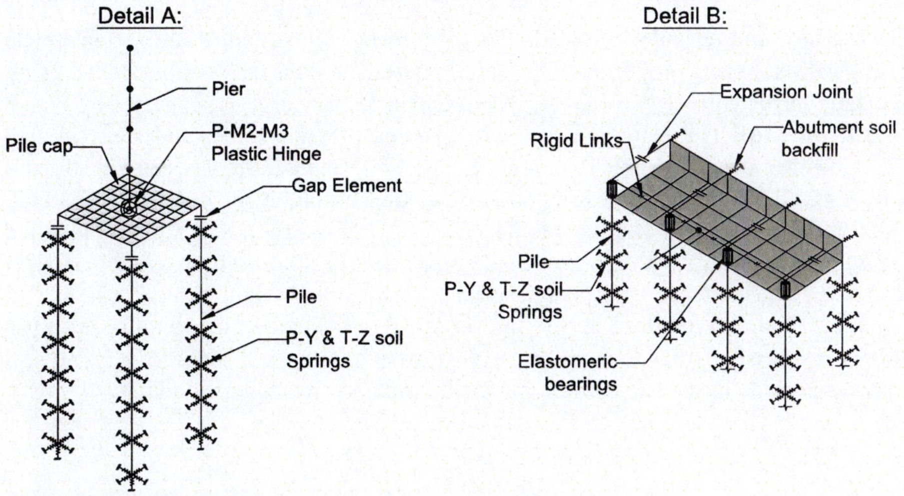


**Fig. 3** **a** Numerical model of the pier and rocking foundation interface, **b** numerical model of the abutment

per Caltrans recommendations [14]. The area springs have compressive stiffnesses of 28.7 kN/mm/m with an upper bound limit of 3.8 MN. The abutment backwall soil springs are only activated once the expansion gap is closed and the deck makes contact with the abutment backwall. The expansion gap is modelled using a nonlinear gap element with compressive stiffness that is equal to the stiffness of the deck and an opening of 300 mm [15].

Furthermore, the elastomeric bearings are modelled using multi-linear springs with a horizontal stiffness given by Eq. (1) [5]:

$$k_h = \frac{G_r A}{t_r} \tag{1}$$

where $G_r$ is the shear modulus of the rubber assumed to be 0.9 MPa, A is the plan area of the bearing and $t_r$ is the total height of the bearing (200 mm). Eurocode-8 specifies that the maximum shear strain for an elastomeric bearing is 200%. Therefore, if the bearings experience a shear displacement that is twice its height (400 mm), the bearing will rupture, and its stiffness reduces to zero.

The pile cap foundation is modelled using a shell element while the piles are modelled using frame elements as presented in Fig. 3a. The piles for the FB bridge are fixed to the foundation. However, for the RP foundation, a nonlinear gap element is modelled between the pile cap and pile head to allow the pile cap to uplift under earthquake loading. The gap element is assigned with a compressive stiffness that is equal to the stiffness of the pile and a zero opening. Pile group effects were not considered since the spacing between piles were larger than $3 \times D$ (pile diameter $D = 1.2$ m at the piers and $D = 0.8$ m at abutments).

The surrounding soil was modelled to capture the soil–structure interaction effects using various spring properties. The lateral resistance of the clay was modelled using a plastic spring with a P-y force–displacement backbone curve as suggested by Welch and Reese [16]. These springs can only take compressive forces and are modelled on opposite sides of the pile to capture the possible separation between the soil and pile, as well as the lateral bearing strength of the soil under seismic lateral loading. Additionally, the skin friction of the pile from the surrounding soil is modelled using plastic springs with a T-z force–displacement backbone curve as suggested by the API [17]. Both the P-y springs and T-z springs are assigned to the pile at 1 m intervals along its depth. The pile-toe resistance is modelled using a linear spring with compressive stiffness equal to the stiffness of the rock. Additionally, all soil springs provide hysteretic damping using a kinematic hysteresis model.

# 4 Input Ground Motions and Seismic Hazard

The bridges investigated in this study are hypothetically located within 5 km from a fault in a high seismic region of New Zealand. Therefore, near-fault ground motions are selected for the analyses of this study. The chosen ground motions and their properties are displayed in Table 1.

## 4.1 *Scaling of Ground Motions*

The three components of ground motion are utilized for the time history analyses. The ground motion records are scaled according to NZS 1170.5 [12] for site class *D*, hazard factor of $Z = 0.4$ and return period factor of $R_s/R_u = 1.8$ (2% probability of exceedance in 50 years). Figure 4 presents the target response spectrum for both the horizontal and vertical component of the ground motions for a damping ratio of 5%. The ground motion records were scaled to the target response spectrum for periods between $T = 0.01$ s and $T = 3$ s. The NZS 1170.5 employs relatively new ground motion prediction equations for determining the vertical response spectrum

**Table 1** Summary of ground motion records

| Earthquake | Year | Station | M*a | $R_{rup}$ (km) | PGA (g) | | | V/H |
|---|---|---|---|---|---|---|---|---|
| | | | | | $H_1$ | $H_2$ | V | |
| Kobe | 1995 | Takarazuka | 6.9 | 0.27 | 0.34 | 0.47 | 0.427 | 0.91 |
| Christchurch | 2011 | Pumping station | 6.2 | 1.98 | 0.60 | 0.67 | 1.90 | 2.83 |
| Imperial Valley | 1979 | El Centro | 6.53 | 0.56 | 0.70 | 0.61 | 0.58 | 0.83 |
| Tabas Iran | 1978 | Tabas | 7.35 | 2.05 | 0.85 | 0.86 | 0.64 | 0.74 |

*Note* M*a represents moment magnitude and $R_{rup}$ represents closest distance to fault rupture

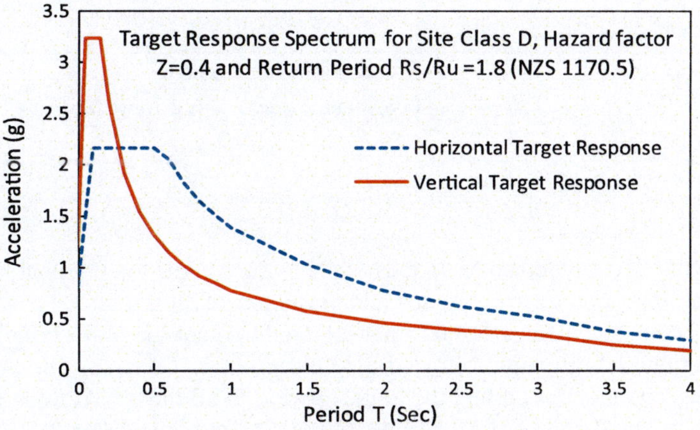


**Fig. 4** Target response spectrum for the horizontal and vertical components

for different soil classes. This is because recent researches have revealed that a factor of 2/3 of the horizontal response spectrum can be largely inaccurate for determining the vertical response of the earthquake [9].

## *4.2 Application of Ground Motions*

In order to investigate the effect of vertical excitations on the seismic response of the bridges, the ground motions were applied in two separate load cases: (1) application of both horizontal components ($H$) and (2) application of all three components of ground motion ($H + V$). The time difference between the arrival of the vertical component with the horizontal component was taken to be zero [10]. Furthermore, the vertical ground motion is adjusted such that the peak horizontal ground acceleration coincides with the peak vertical ground acceleration [18]. Rayleigh's mass and proportional stiffness damping were used with a 5% damping ratio for the first and second fundamental periods of the bridges [19].

# 5 Results and Discussion

## *5.1 Modal Analysis*

A modal analysis was conducted for the bridges using Ritz vectors with a 98% mass participation ratio. Furthermore, a sine sweep analysis was conducted to approximate the fundamental period of for the RP bridges due to the nonlinearity present with the

**Table 2** Fundamental periods of the bridges

| Bridge types | Natural period of the bridge | | |
|---|---|---|---|
| | Longitudinal (First) | Transverse (Second) | Vertical |
| Fixed base (FB) bridge | 0.48 | 0.42 | 0.11 |
| Rocking pile foundation (RP) bridge | 1.85 | 1.42 | 1.12 |

gap elements. The fundamental periods of the bridges in the longitudinal, transverse and vertical direction were computed and presented in Table 2.

It can be observed from Table 2 that the predominant mode for both bridges is in the longitudinal direction followed by the transverse direction. The bridge with the rocking pile foundation had the largest period in all three directions due to the reduction of lateral and vertical stiffness of the pier and foundation. With reference to the shape of the response spectrum in Fig. 4, an increase in the natural period is beneficial for reducing the inertial forces on the structure. However, a larger structural period will result in larger deformations which can cause other failures to occur in the bridge.

The increased period of the RP bridge in the vertical direction is due to the connection release between the pile and pile cap. Again, with reference to the vertical response spectrum in Fig. 4, the reduction in vertical stiffness can reduce the inertial forces in the piers, but also increase the vertical displacements of the deck. The vertical deck displacements caused by the vertical excitations cannot only increase the inertial and secondary forces (P-delta effect) on the pier, but also increase the horizontal deck displacements.

## 5.2 *Nonlinear Time History Analyses*

Nonlinear dynamic time history analyses were performed on the bridges using four ground motion records for two separate load cases indicated. The earthquakes were applied to the longitudinal direction of the bridges (refer to Fig. 1), and their dynamic response is reported in Table 3.

**Pier Bending Moments**

All three piers experienced identical results due to the symmetrical configuration of the bridge. Hence, only the seismic response of the central pier (Pier 2, refer to Fig. 1) is reported in Table 3 and Fig. 5. It can be seen from Table 3 and Fig. 5a, the RP bridge experienced much smaller pier bending moments when compared to the FB bridge for both load cases. This is because the RP bridge had a larger structural period and therefore attracts smaller inertial forces to the pier. The piers in the FB bridge exceeded their ultimate capacity ($M_{\text{ult}} = 100$ MN.m) and developed plastic hinges. Furthermore, the results from the flexural plastic hinges confirm that the piers belonging to the FB bridge experienced concrete crushing and significant yielding in

**Table 3** Dynamic response of the bridges for the longitudinal direction

| Earthquake | Bridge and loading type | | Maximum pier bending moment | | Maximum pier axial force | | Maximum deck displacement | | Maximum deck bending moment | |
|---|---|---|---|---|---|---|---|---|---|---|
| | | | MN.m | Δ (%) | MN | Δ (%) | mm | Δ (%) | MN.m | Δ (%) |
| 1995 Kobe | FB | H | 277 | | 9.45 | | 139 | | 8.5 | |
| | | H + V | 277.2 | 0 | 31.4 | 232 | 142 | 2 | 50.3 | 492 |
| | RP | H | 37.8 | | 22.4 | | 839 | | 67.4 | |
| | | H + V | 60 | 59 | 28.2 | 26 | 1005 | 20 | 94.1 | 40 |
| 2011 Christchurch | FB | H | 278 | | 9.45 | | 119 | | 8.5 | |
| | | H + V | 280 | 1 | 28.6 | 203 | 122 | 3 | 56.9 | 569 |
| | RP | H | 36.7 | | 18.83 | | 523 | | 45.2 | |
| | | H + V | 46.2 | 26 | 28.2 | 50 | 675 | 29 | 63.8 | 41 |
| 1979 Imperial Valley | FB | H | 240 | | 9.45 | | 150 | | 8.5 | |
| | | H + V | 240.4 | 0 | 33.6 | 256 | 141 | −6 | 53 | 524 |
| | RP | H | 48.7 | | 22.5 | | 605 | | 71.7 | |
| | | H + V | 73.5 | 51 | 24.3 | 8 | 901 | 49 | 80.2 | 12 |
| 1978 Tabas | FB | H | 245 | | 9.45 | | 102 | | 8.5 | |
| | | H + V | 246.2 | 0 | 33.4 | 253 | 101 | −1 | 58.7 | 591 |
| | RP | H | 37.7 | | 20.1 | | 572 | | 54.5 | |
| | | H + V | 61.5 | 63 | 35.6 | 77 | 700 | 22 | 117.1 | 115 |

Δ is the percentage difference between the ($H$) and ($H + V$) load cases

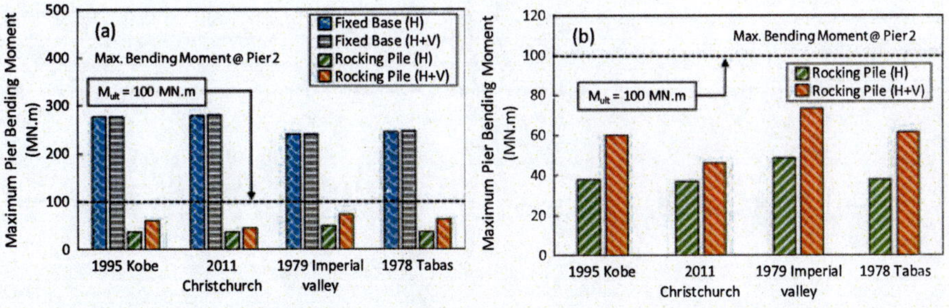


**Fig. 5** **a** Maximum bending moments experienced in the piers for both bridges, **b** maximum bending moments in the piers of the RP bridge for both load cases

the reinforcing steel. When comparing the two load cases for the FB bridge in Fig. 5a and Table 3, it can be seen that the vertical component of the ground motion had an insignificant effect on the bending moment in the piers with a maximum increase of 1% for all the earthquakes. In contrast, the RP bridge experienced a significant increase in the pier bending moments when the vertical excitations are considered. Table 3 reports that for the 1978 Tabas earthquake, the bending moment in the piers of the RP system increased by a maximum of 63%. This increase is due to the large period of the RP bridge in the vertical direction. As can be seen in Table 2 and Fig. 4, the increased period in the vertical direction also increased the vertical displacements of the deck and consequently increased the secondary forces on the pier during lateral displacement of the deck. The RP bridge is more susceptible to the vertical ground motions when compared to the FB bridge, as it experienced significant increases to the bending moment of the piers.

## Deck Displacements

Figure 6 presents the deck displacement time histories of the bridges subjected to the four earthquakes. Referring to Table 3 and Fig. 6 for the FB bridge, the vertical excitations had a minor impact on the deck displacements. Table 3 shows a maximum of 3% increase in the deck displacement for the 2011 Christchurch earthquake, but a 6% decrease for the 1979 Imperial Valley earthquake. This is because the small period of the FB bridge for both the longitudinal and vertical direction prevents the bridge from developing large displacements. Furthermore, the accelerations from the vertical component of the ground motion can either help stabilize or further

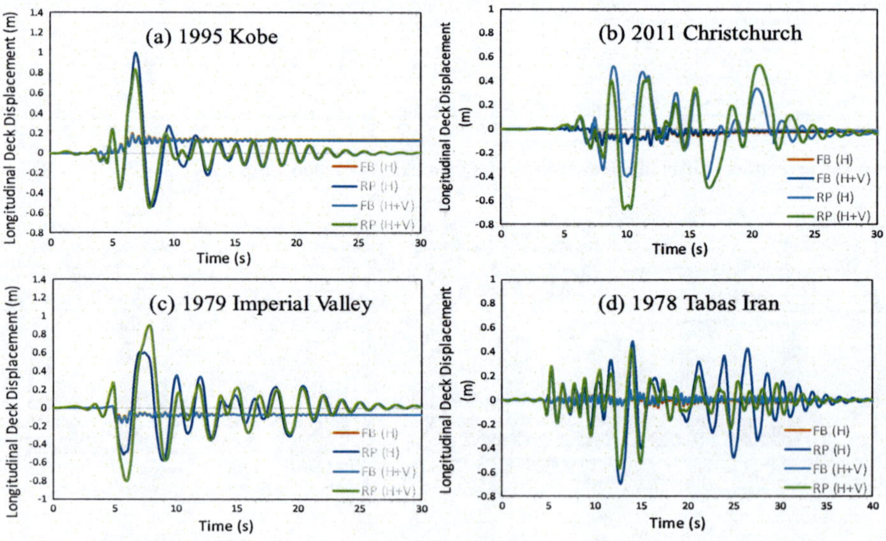


Fig. 6 Deck displacement time histories for **a** 1995 Kobe, **b** 2011 Christchurch, **c** 1979 Imperial Valley, **d** 1978 Tabas Iran

excite the structure during lateral seismic loading by reducing or increasing the secondary effects on the bridge. Furthermore, the high frequency content of the vertical excitation prevented large pier and bridge deck deformations from occurring.

Referring to Table 3 and Fig. 6 for the RP bridge, the inclusion of the vertical excitations significantly increased the deck displacement of the bridge. This is also due to the structure's larger period in both the horizontal and in the vertical component. Due to the reduced lateral stiffness in the RP bridge and consequently increased natural period, the structure experienced larger horizontal displacements. Additionally, the RP bridge also had a larger period in the vertical direction, which led to larger vertical displacements. When both the vertical and horizontal ground motions coincide, the secondary forces on the piers, increased the horizontal displacements of the bridge deck. The deck seat width at the abutments is 650 mm. For the load case without the vertical ground motion, the RP bridge survived three of the four earthquakes. However, when the vertical component of the earthquake is included in the analyses, the horizontal bridge deck displacements increased by 20–49% which resulted in the bridge decks to pound and unseat of the supports, and the subsequent collapse of the RP bridge for all the earthquakes. As such, it is crucial to consider the effect of the vertical component of the earthquake in the design of bridges with rocking pile foundations. Larger deck displacements can result in shear failure in the elastomeric bearings, pounding damage between the deck and abutments, and unseating of the bridge deck.

## Pier Axial Forces and Deck bending Moments

Figure 7a displays the maximum seismic axial forces experienced by the bridge piers and Fig. 7b displays the maximum deck bending moments experienced by the bridge. It is evident from Table 3 and Fig. 7 that for the case with horizontal only loading, the RP bridge had significantly higher axial forces in the piers and deck bending moments than the FB bridge. This is mainly due to the pounding forces generated from the rocking behaviour and further due to the lateral seismic loading acting axially on the piers during uplift. In addition, the uplift rocking motion of the RP bridge generates larger axial forces in the vertical direction due to its pounding effect.

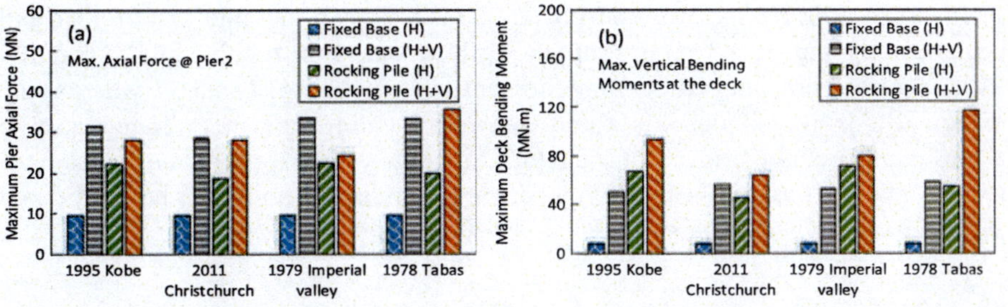


**Fig. 7** **a** Maximum experienced axial forces in the pier, **b** maximum experienced bending moments at the decks

For the load case with combined horizontal and vertical ground motions, Fig. 7 and Table 3 indicate that the axial forces in the piers become similar and almost equal for both the FB and RP bridges. Referring to Table 3, the axial load in the FB bridge pier increased up to 256% for all the earthquakes. This is because the vertical accelerations have a high frequency nature and correlate well with the low natural period of the bridge in the vertical direction, increasing the seismic response of the pier. Furthermore, a similar effect is observed for the maximum deck bending moments. As shown in Table 3 for the FB bridge, the deck bending moment increased by a maximum of 591% as a result of low vertical period of the FB bridge and high frequency excitation of the earthquakes. It is important to note that the length of the bridge deck is a significant factor in the magnitude of these forces.

Referring to Table 3 for the RP bridge, the maximum increase in the axial force in the piers is 77% for all four earthquakes. This difference between the two load cases is substantially smaller for the RP bridge when compared to the FB bridge. This is because the induced inertial forces on the pier are smaller due to the prolonged period and reduced vertical stiffness of the vertical direction of the bridge. Similarly, the maximum bending moment experienced in the deck of the RP bridge increased by 115% for all the earthquakes due to the increased period of the structure. Furthermore, the combined forces in the RP bridge resulting from the inertia and forces generated from pounding during rocking exceed the forces experienced in the FB bridge. This issue requires caution during the analysis and design of the RP bridges.

## 6 Conclusion

This paper investigated the effect of near-fault ground motions on the seismic performance of bridges with rocking pile foundations. Three-dimensional numerical models of two RP and FB bridges were developed, and nonlinear time history analyses were performed using four ground motions. The analyses were performed using two separate load cases to exclude and include the vertical component of the earthquake. The results showed that the fixed base bridge experienced insignificant increases to the pier bending moment and horizontal deck displacements. However, the fixed base bridge experienced substantial increases to the pier axial forces and deck bending moments. On the contrary, the bridge with the rocking pile foundation experienced significant pier bending moments and deck displacements, whereas the experienced increased pier axial forces and deck bending moments were not as high as those in fixed base bridge. It is concluded that the vertical ground motion component is more critical for bridges with rocking foundations and must be considered for the design in near-fault sites.

## References


1. Mehrsoroush A, Saiidi MS (2016) Cyclic response of precast bridge piers with novel column-base pipe pins and pocket cap beam connections. J Bridge Eng 21(4):04015080
2. Xu R, Fatahi B (2018) Geosynthetic-reinforced cushioned piles with controlled rocking for seismic safeguarding. Geosynth Int 25(6):561–581
3. Anastasopoulos I et al (2010) Soil failure can be used for seismic protection of structures. Bull Earthq Eng 8(2):309–326
4. Antonellis G et al (2015) Shake table test of large-scale bridge columns supported on rocking shallow foundations. J Geotech Geoenviron Eng 141(5):04015009
5. Antonellis G, Panagiotou M (2013) Seismic response of bridges with rocking foundations compared to fixed-base bridges at a near-fault site. J Bridge Eng 19(5):04014007
6. Raheem SEA (2009) Pounding mitigation and unseating prevention at expansion joints of isolated multi-span bridges. Eng Struct 31(10):2345–2356
7. Warn GP, Whittaker AS (2008) Vertical earthquake loads on seismic isolation systems in bridges. J Struct Eng 134(11):1696–1704
8. Dehghanpoor A et al (2019) Soil-pile-superstructure interaction effects in seismically isolated bridges under combined vertical and horizontal strong ground motions. Soil Dyn Earthq Eng 126:105753
9. Kunnath SK et al (2008) Effect of near-fault vertical ground motions on seismic response of highway overcrossings. J Bridge Eng 13(3):282–290
10. Kim SJ, Holub CJ, Elnashai AS (2010) Analytical assessment of the effect of vertical earthquake motion on RC bridge piers. J Struct Eng 137(2):252–260
11. Chen Y et al (2016) Impact of vertical ground excitation on a bridge with footing uplift. J Earthq Eng 20(7):1035–1053
12. Committee SNZt (2004) Structural design actions (NZS 1170.5), Wellington, New Zealand
13. Hassoun M, Fatahi B (2019) Novel integrated ground anchor technology for the seismic protection of isolated segmented cantilever bridges. Soil Dyn Earthq Eng 125:105709
14. Caltrans S (2010) Caltrans seismic design criteria version 1.6. California Department of Transportation, Sacramento
15. Fatahi B et al (2018) Three-dimensional response of neighboring buildings sitting on pile foundations to seismic pounding. Int J Geomech 18(4):04018007
16. Welch RC, Reese LC (1972) Lateral load behavior of drilled shafts. University of Texas, Austin
17. API Recommended practice for planning, designing and constructing fixed offshore platforms–working stress design–in Twenty. Conference location 2000
18. Shrestha B (2015) Seismic response of long span cable-stayed bridge to near-fault vertical ground motions. KSCE J Civil Eng 19(1):180–187
19. Xu R, Fatahi B (2018) Influence of geotextile arrangement on seismic performance of mid-rise buildings subjected to MCE shaking. Geotext Geomembr 46(4):511–528

# Effect of Composite Pile Foundation System on the Behavior of Soft Ground

**Dimpa Moni Kalita and S. D. Anitha Kumari**

**Abstract** Construction of bridge approach embankment in soft ground experiences soil settlement due to different support condition of the approach embankment and the abutment. At the bridge-embankment transition zone, this difference will result in formation of bump at the bridge end. Hence, the present study focuses on developing and efficient foundation system that can minimize the bump at the bridge end. Ground improvement techniques using composite piles can be adopted to improve the soft ground. A series of numerical analysis is carried out to compare embankment system without and with composite pile made of cement-fly ash-gravel (CFG) and lime in reducing settlement. The analysis results show that settlement at the transition zone is reduced by 45% when composite piles are used as ground improvement. Performance analysis of CFG and lime pile shows that the stress carrying capacity of CFG piles is higher than lime piles. Meanwhile, the parametric study results show that pile spacing of 6 times diameter, pile length of 25 times diameter and sand as cushion material are the most optimum design parameters that can effectively smooth the bridge-embankment transition zone. Further, the effect of constant and varying pile length in the foundation system is studied. It is observed that varying length piles reduce soil settlement by approximately 25% compared to constant length piles.

**Keywords** Soft ground · Bridge-embankment transition zone · Composite piles

## 1 Introduction

An effective and well-developed transportation system has a significant role in the economic development of a country. The rapid development in the transport facilities is forcing the construction of this system on the available land which may

D. M. Kalita (✉) · S. D. Anitha Kumari (✉)
M. S. Ramaiah University of Applied Sciences, Bangalore, India
e-mail: dimpakalita912@gmail.com

S. D. Anitha Kumari
e-mail: anitha.vipin@gmail.com

 
T. G. Sitharam et al. (eds.), *Seismic Design and Performance*, Lecture Notes in Civil Engineering 120, https://doi.org/10.1007/978-981-33-4005-3_9

be sandy, silt, clay, loamy or in general soft and weak soil. Soft ground has problems like insufficient bearing capacity, high post construction settlement, instability on excavation and embankment construction. Geotechnical engineering faces major challenge in designing and construction of highways and railways on this ground. In the bridge approach embankment system, the approach embankment rests on natural soil, whereas the bridge abutment is supported by deep piles. Therefore, the stiffness of the abutment and the embankment is different. At the bridge-embankment transition zone, this difference in stiffness will result in the formation of bump at the bridge end. Therefore, ground improvement technique should be adopted underneath the embankment to reduce the bump at the bridge end. In recent times, ground improvement technique is concerned in delivering built facilities that are environment friendly as well as financially viable. In the construction cycle, sustainable geotechnical practice can significantly improve the overall sustainability of the project. One of the sustainable techniques is to use fly ash in construction. In this study, fly ash is used in piles to form a composite pile foundation for ground improvement. The composite pile formed by the mixture of cement, fly ash and gravel will be a rigid pile which is known as cement-fly ash-gravel (CFG) pile.

So far, many researchers have investigated the behavior of composite piles as ground improvement in soft soil. Zhang et al. [1, 2] reported a case study on the performance of geosynthetic-reinforced and pile-supported embankments (GRPS) over silty clay and found that the total postconstruction settlement is reduced by 21–38.6% where 8.8–17.3% was reduced in the reinforced area and 27.5–44% in the substratum soil. An analytical solution for estimating the stress concentration ratio of GRPS embankment for high speed railways was proposed in the paper. It was reported that the combination of geosynthetics in the platform and CFG piles in the subsoil reduced the lateral displacement in the soil. Moreover, parametric study on pile length showed that change in pile length affected the stress distribution ratio. Jiang et al. [3] conducted a numerical analysis of well-monitored pile slab supports (PSS) embankment with CFG piles in the foundation. They observed that the cushion reduced shear force in slab by 28% and maximum bending moment by 17%. They also mentioned that increase in the pile stiffness by 10 times reduced the maximum settlement, differential settlement and lateral displacement of the embankment by 50%, 67% and 84%, respectively. Xiao et al. [4] studied the effect of CFG piles on the abutment piles for different ground replacement ratio and reported that increase in ground replacement ratio reduced the axial force, normal contact pressure, bending moment and horizontal displacement of abutment piles. Wei and Bian [5] analyzed the dynamic performance of CFG pile-supported bridge embankment with constant length and varying length pile and observed that multistage improvement strategy using varying length CFG piles gave better result in reducing settlement compared to constant length pile.

As per the study conducted on existing literature, researchers have mainly focused on reducing soft ground settlement by using geosynthetic-reinforcement and pile-supported embankment. Limited researches were carried out to compare the effect of lime and CFG pile-supported embankment with various cushion material. The present study is focused on the effects of lime and CFG pile as ground reinforcement for reducing the soft ground settlement. Moreover, the effect of design parameters on the embankment performance is also studied to obtain the most feasible foundation system and compared its performance with varying length pile foundation system.

As per the study conducted on existing literature, researchers have mainly focused on reducing soft ground settlement by using geosynthetic-reinforcement and pile-supported embankment. Limited researches were carried out to compare the effect of lime and CFG pile-supported embankment with various cushion material. The present study is focused on the effects of lime and CFG pile as ground reinforcement for reducing the soft ground settlement. Moreover, the effect of design parameters on the embankment performance is also studied to obtain the most feasible foundation system and compared its performance with varying length pile foundation system.

## 2 Methodology

Three-dimensional numerical model of the bridge approach embankment is analyzed using numerical methods. FEM software ABAQUS was used for modeling the system. The model consists of embankment, abutment, sand cushion, composite piles and abutment piles. The modeled embankment is resting on subsoil consisting of two layers: silty clay and clay. Mohr–Coulomb failure criterion is adopted for the soil layers and material parameters used are tabulated in Table 1. The parameters and material properties used in modeling the piles are mentioned in Table 2. For analysis, a length of 10 m for CFG and lime piles and 17 m for the abutment piles are used. The position of abutment piles and composite piles with spacing in longitudinal and transverse direction is shown in Fig. 1. The model is analyzed for self-weight of the embankment, i.e., 120 KN/m$^2$.

**Table 1** Material properties of geomaterial [4]

| Material | Height H (m) | Density $\gamma$ (kN/m$^3$) | Young's modulus, E (kN/m$^2$) | Poisson ratio, $\mu$ | Friction angle $^\circ$ | Cohesion c (kN/m$^2$) | Dilation angle, $\psi^\circ$ |
|---|---|---|---|---|---|---|---|
| Silty clay | 7 | 19.1 | $9 \times 10^3$ | 0.33 | 12 | 35 | 0.1 |
| Clay | 13 | 19.5 | $15.3 \times 10^3$ | 0.33 | 30 | 8.3 | 1 |
| Embankment | 6 | 20 | $80 \times 10^3$ | 0.3 | 30 | 15 | 4 |
| Sand cushion | 0.5 | 20 | $100 \times 10^3$ | 0.3 | 35 | 0 | 7 |

**Table 2** Material properties of elastic piles [4, 6]

| Material | Diameter, D (m) | Density $\gamma$ (kN/m$^3$) | Young's modulus, E (kN/m$^2$) | Poisson ratio, $\mu$ |
|---|---|---|---|---|
| Abutment pile | 0.96 | 24.52 | $70 \times 10^6$ | 0.2 |
| CFG pile | 0.48 | 14.72 | $35 \times 10^6$ | 0.25 |
| Lime pile | 0.48 | 19.62 | $15 \times 10^6$ | 0.25 |

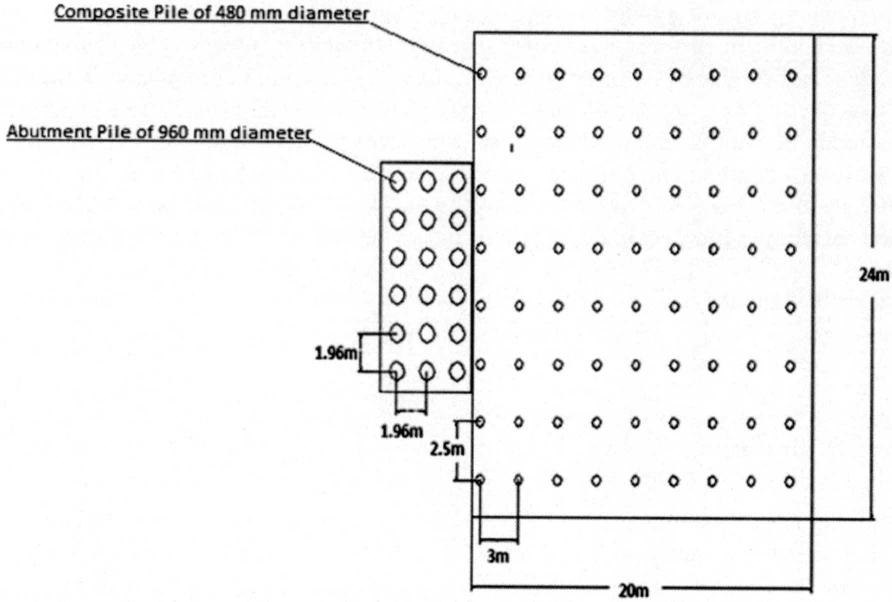


**Fig. 1** Top view of the composite pile foundation

# 3 Analysis of Bridge Approach Embankment System

The bridge approach embankment system is analyzed for two different cases. In the first case, the bridge approach embankment system is analyzed for the embankment without and with composite pile foundation underneath the embankment. In the second case, the bridge approach embankment system is analyzed for CFG pile composite foundation and lime pile composite foundation to obtain the most suitable pile material.

## *3.1 Bridge Approach Embankment with and Without Composite Piles*

The numerical model of the bridge approach embankment is analyzed for two models; (i) without composite pile foundation and (ii) with CFG pile composite foundation. It is observed that the soil without any composite piles underneath the embankment undergoes settlement in the cushion as well as in the soil. Figure 2a and b compares the magnitude of settlement in the embankment system without and with composite pile foundation. The stress in the cushion is 75.54 kPa and settlement is 79.64 mm for the embankment system without composite pile foundation at the bridge-embankment transition zone. In the case of CFG pile foundation, the stress in the cushion is

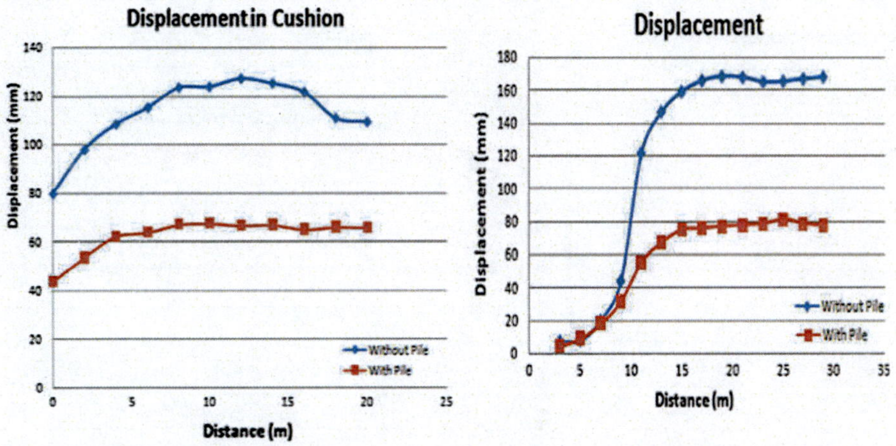


(a) Sand cushion (b) Soil

**Fig. 2** Comparison of displacements for embankment with and without composite piles

69.68 kPa and settlement is 43.67 mm. It is also observed that as the distance from the transition zone increases, the stress and displacement in the cushion also increase gradually. Comparing the stress for both the cases, it is observed that the stress in embankment system without pile is 8% more in the bridge-embankment transition zone and 33% more at a distance of 10 m from the transition zone compared to embankment with composite pile. In Fig. 2a, the displacement increases up to a certain length for the embankment system with reinforced CFG piles and then it becomes almost constant. The increase in the settlement is due to the load from the embankment which is stabilized when CFG piles are used. Thus, CFG piles have effectively stabilizes the subsoil. In Fig. 2b, settlement at the extreme end of the soil in longitudinal and transverse directions is zero as no load is acting at that point. whereas near the bridge-embankment transition zone, the dead load of the embankment produces an impact on the underlying soil that results in settlement. Comparing the settlement for both the cases in cushion and soil, it is observed that CFG pile foundation underneath the embankment can effectively reduce the cushion and soil settlements by approximately 45% and soil settlement by 54%, respectively.

### *3.2 Comparison of CFG and Lime Piles in Composite Pile Foundation*

The three-dimensional model of the bridge approach embankment is analyzed for two different pile foundation system; (i) CFG pile foundation and (ii) lime pile foundation under static loading. The embankment system with CFG pile foundation, the maximum stress obtained from the numerical analysis at the pile top is 29.86 kPa,

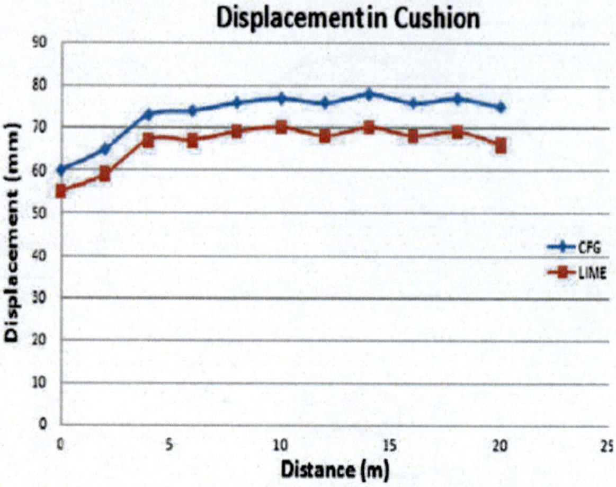


**Fig. 3** Settlement comparisons for CFG and lime pile foundation

and for lime pile foundation system, it is 23.47 kPa. Thus, the initial stress generated on the CFG piles is 21% more compared to lime piles. But it is observed that the stress throughout the pile length is reducing gradually along the depth of the pile. The total percentage of stress transferred to the lower depth of CFG piles is 45% and to the lower depth of lime piles is 65%. This indicates that smaller depth of CFG for pile can carry more stress compared to lime pile. However, the displacement at the bridge-embankment transition zone is 60 mm for CFG pile foundation and 55 mm for lime pile foundation. Figure 3 shows the displacement comparison for CFG and lime pile in cushion. Therefore, further analysis on the design parameters needs to be studied to obtain the most effective pile material.

## 4 Parametric Studies

In this study, the three-dimensional numerical model of the bridge approach embankment is analyzed for various influencing parameters. In order to obtain the most suitable pile material, analysis of CFG and lime pile foundation is done for different pile spacing in each model. With the most suitable pile material and pile spacing, further analysis on the effect of pile length and the behavior of cushion materials is studied.

### *4.1 Effect of Pile Spacing*

The effect of pile spacing on the overall performance is studied by considering 14 different configurations. The 14 different configuration includes seven CFG piles and seven lime piles and are compared for pile spacing of 2D, 3D, 4D, 5D, 6D,

7D and 8D, where D is the diameter of the pile, i.e., 480 mm. Figure 4a and b shows the displacement comparisons for different spacing with CFG piles and lime pile respectively. It has been observed that pile spacing between 2D–4D performs very well in reducing settlement as the piles are very compactly placed and the performance of 8D spacing is not satisfactory. Therefore, the performance of 5D, 6D and 7D is compared. At the bridge-embankment transition zone, soil settlements are 34 mm, 37 mm and 41 mm for 5D, 6D and 7D spacing, respectively, with CFG pile foundation. The lime pile foundation soil settlements obtained are 52 mm, 58 mm and 65 mm for 5D, 6D and 7D spacing. This indicated that CFG piles give better results in reducing settlement compared to lime pile at the transition zone. Moreover, it has been observed that in 7D pile spacingm the displacement is varying alternately which indicates that this pile spacing is not able to improve the soft soil much. The settlement for both CFG and lime foundation in 5D and 6D shows uniform variation and is approximately same. Therefore, 6D spacing is considered as the optimum pile spacing for this foundation system. However, in the transition zone, settlement of 37 mm with 6D spacing will not minimize the bump at the bridge end completely. So, further investigation on the effect of pile length needs to be studied.

### 4.2 *Effect of Pile Length*

Figure 5a and b shows the displacement for different pile depth in sand cushion and soil respectively. The influence of CFG pile composite foundation is analyzed with pile lengths of 8 m, 10 m, 12 m and 14 m. At the bridge-embankment transition zone, cushion displacement observed for 8 m, 10 m, 12 m and 14 m pile length is 45.93 mm, 42.34 mm, 34.66 mm and 32.3 mm, respectively. The soil displacement observed for 8 m, 10 m, 12 m and 14 m pile length is 41 mm, 37 mm, 34 mm and 32 mm, respectively. It is observed that higher the length of pile, lower is the displacement at the bridge-embankment transition zone. Comparing the displacements for different pile length, CFG pile of length 12 m and 14 m shows approximately the same settlement at the bride end. Therefore, an effective pile foundation system is modeled with a pile depth of 12 m.

### 4.3 *Effect of Cushion Material*

The three-dimensional model of the bridge approach embankment system is analyzed for two different load transfer medium; sand cushion and concrete slab. Figure 6a and b shows displacement in sand cushion and soil respectively for different cushion materials. At the bridge-embankment transition zone, the settlement in sand cushion is 34 mm and the settlement in concrete slab is 27 mm. It indicates that when a concrete slab is used as load transfer medium, the cushion settlement gets reduced by 23%. Result shows that the maximum settlement in sand cushion is 52 mm and in

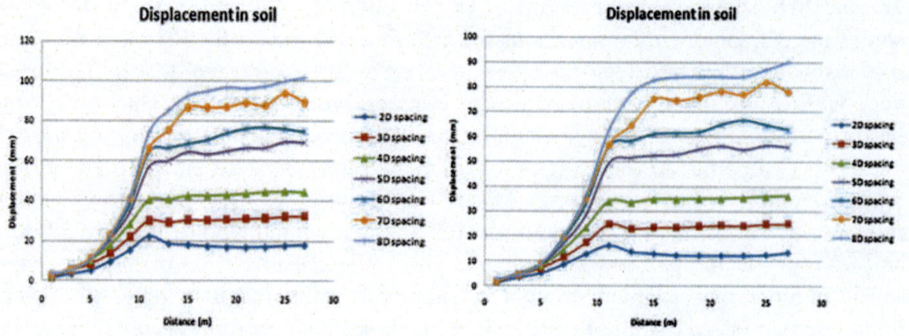


(a) CFG Piles (b) Lime Piles

**Fig. 4** Displacement comparison for different pile spacing

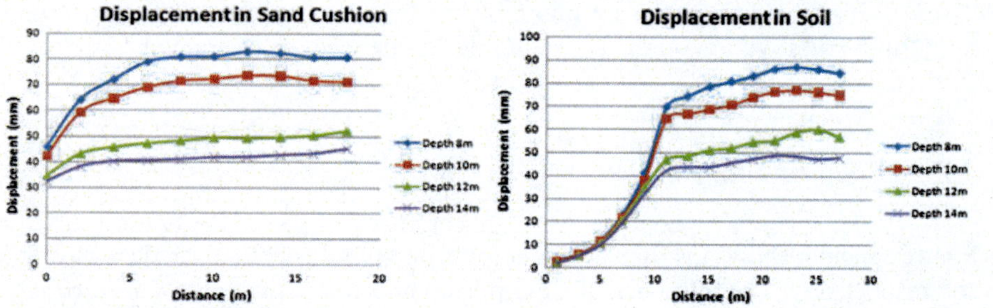


(a) Sand Cushion (b) Soil

**Fig. 5** Displacement for different pile length

concrete slab is 42 mm. But settlement of a concrete slab by 27 mm and 42 mm will result in cracking at the concrete surface. Thus to prevent cracking in the concrete slab, it is required to provide proper reinforcement design at the transition zone. The present study is not focused on designing an optimum-reinforced concrete slab therefore; sand is taken as the most suitable cushion material to carry the load transfer mechanism in the embankment system.

### *4.4 Comparison of Constant Length and Varying Length Pile in Foundation*

Based on the results obtained from the parametric studies, a foundation system with constant length pile is designed. The study focuses on the performance of soft ground reinforced by constant and varying lengths CFG pile foundation system. Figure 7a and b shows the 2D view of constant length and varying length pile foundation system

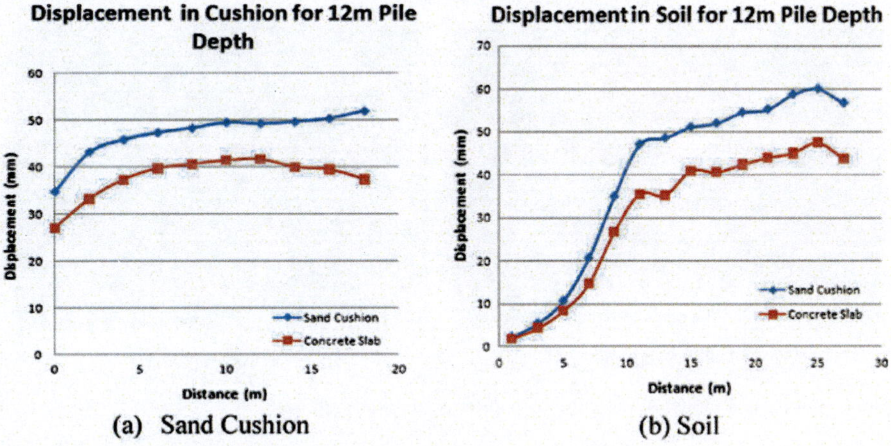


**Fig. 6** Displacement for different cushion material

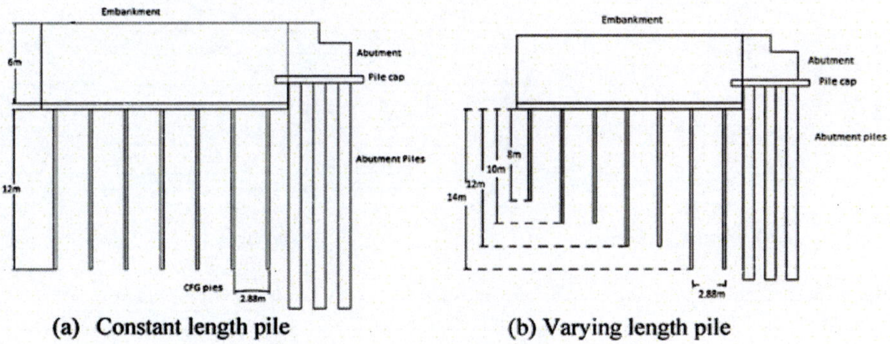


**Fig. 7** Two-dimensional view of the composite pile foundation system

used for analysis. The numerical model with constant length and varying length pile is shown in Fig. 8a and b respectively. Figure 9a and b compares the displacement for constant length and varying length pile in cushion and in soil respectively. Results show that at the transition zone, displacement in sand cushion is 34 mm for constant length piles and 22 mm for varying length piles and in the soil is 35 mm for constant length piles and 16 mm for varying length piles. Varying length pile foundation reduces settlement at the bridge-embankment transition zone by 25% compared to constant length pile foundation.

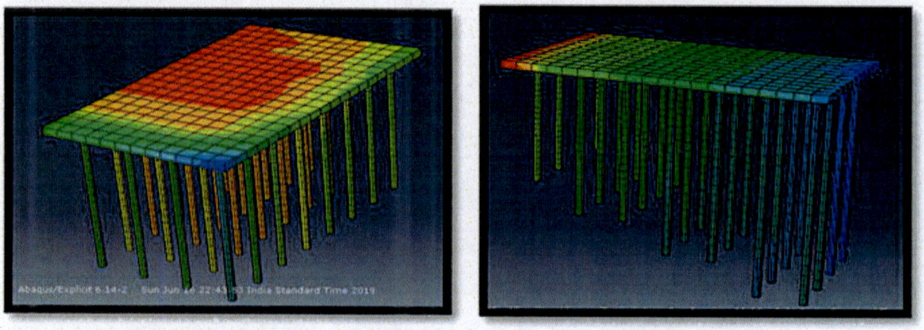

(a) Constant length pile (b) Varying length pile

**Fig. 8** Numerical model of the composite pile foundation system

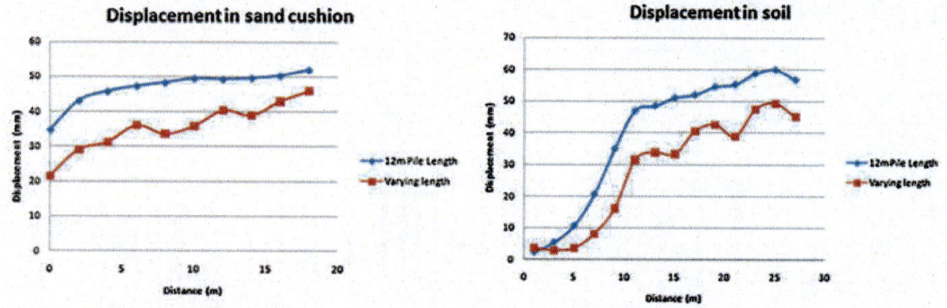


(a) Sand Cushion (b) Soil

**Fig. 9** Settlement comparison of constant and varying length pile foundation system

## 5 Conclusions

The behavior of soft ground reinforced by CFG and lime pile in a bridge approach embankment system is compared. Results obtained from the analysis indicate that shorter span of CFG pile can carry more stress therefore; the stress carrying capacity of CFG pile is more compared to lime piles. The displacement at the bridge transition zone is minimized when CFG piles are used in the foundation compared to lime piles. In order to get the most optimum design parameters, the influence of pile spacing, pile length and cushion material is analyzed. It has been observed that pile spacing of 6 times diameter, pile length of 25 times diameter and sand as cushion are the most feasible parameters in a foundation system. Further, the performance of constant length and varying length piles is compared to understand the effect on the behavior of the foundation system. The analysis indicates that for a smooth bridge-embankment transition zone, a foundation system with varying pile length shows better response against settlement compared to constant length pile in a foundation resting on soft ground. Thus, it is recommended to adopt a pile length of 14 m for a distance of

4.5 m from the bridge-embankment transition zone, i.e., three times its distance in the varying length pile foundation system.

**Acknowledgements** The authors would like to extend their gratitude to M.S. Ramaiah University of Applied Sciences for providing all the facilities for conducting the research work.

## References

1. Zhang CL, Jiang GL, Liu XF, Wang ZM (2015) Lateral displacement of silty clay under cement-fly ash-gravel pile-supported embankments: analytical consideration and field evidence. J Cent South Univ 22(4):1477–1489
2. Zhang C et al (2014) Deformation performance of cement-fly ash-gravel pile-supported embankments over silty clay of medium compressibility: a case study. Arab J Geosci. https://doi.org/10.1007/s12517-014-1559-8
3. Jiang Y, Han J, Zheng G (2013) Numerical analysis of a pile–slab-supported railway embankment. Acta Geotech 9(3):499–511. https://doi.org/10.1007/s11440-013-0285-9
4. Xiao D, Jiang GL, Liao D, Hu YF, Liu XF (2018) Influence of cement-fly ash-gravel pile-supported approach on abutment piles in soft ground. J Rock Mech Geotech Eng 10(5):977–985
5. Li W, Bian X (2016) Dynamic performance of pile-supported bridge-embankment transition zones under high-speed train moving loads 143:1059–1067
6. Zheng JJ, Abusharar SW, Wang XZ (2007) Three-dimensional nonlinear finite element modeling of composite foundation formed by CFG-lime piles. Comput Geotech 637–643

# Study on Pull-Out Resistance of Single-Belled Anchor in Layered Sand and Soft Silty-Clay Deposits: Model Tests and Numerical Modelling

**Tanaya Deb and Sujit Kumar Pal**

**Abstract** The piles that assist to counter any pull-out forces and overturning moments enforced on foundation are idealized as anchor piles. Belled anchors offer passive form of resistance against pull-out forces by the dead-weight and shear resistance of breakout region of foundation media. Experimental investigations are conducted in two different series, for determining pull-out resistance of single-belled anchor models in layered sand and soft silty-clay deposits, having different anchor characteristics and embedment depths. Pull-out resistances are evaluated in layered soil deposits where the thickness of underlying soft soil is constant for all models. The consistency indices are 0.23 and 0.46 in soft soil. Pull-out resistance of belled anchor piles is primarily dependent on the consistency of soft soil and thus on the cohesive strength in undrained condition and secondarily, dependent on overlying sand deposit of different thickness, embedment ratios and diameter ratios. But the effect of bell angle is found to be negligible on pull-out resistance in layered sand and soft soil deposit. All the models are found to be much effective to attain higher pull-out resistance when consistency index is higher in underlying soft soil, as these values are increased by approximately 60–83%. Few numerical analysis for comparatively larger models (scale factor = 1/10) through finite element method in Plaxis 3D Foundation is performed. The reason of variation in pull-out resistance is explored by displacement contour from the Plaxis 3D analysis. So, the Plaxis results are correct and can be used for parametric study.

**Keywords** Belled anchor · Layered deposit · Pull-out resistance · Plaxis 3D Foundation · Displacement contour

T. Deb (✉)
TTAADC Polytechnic Institute, Khumulwng, Tripura, India
e-mail: debtanaya88@gmail.com

S. K. Pal
Civil Engineering Department, NIT Agartala, Agartala, Tripura, India
e-mail: skpal1963@gmail.com

 
T. G. Sitharam et al. (eds.), *Seismic Design and Performance*, Lecture Notes in Civil Engineering 120, https://doi.org/10.1007/978-981-33-4005-3_10

# 1 Introduction

The anchors, possessing enlarged bases, are recommended in foundation engineering to resist the pull-out forces as well as overturning moments imposed on the foundation. The prediction of anchor behaviour and the failure pattern, based on foundation media under tensile loading, is a significant issue in foundation design. Although the tensile strength of the soil can be safely ignored for safe design, whereas pull-out loads induce tensile stresses in soil, in such situations, "*anchors*" itself act as reinforcement. The expected tensile load and overturning moments are basic design considerations to ascertain anchor's type, shape and size, for off-shore structures and on-land structures.

Adams and Radhakrishna [1] and Radhakrishna and Adams [15] studied the in situ pull-out resistance of full-scale and intermediate-scale belled piles in clay deposit. Ismael and Kylm [7] performed field test on full-scale belled anchor pile in layered cohesive deposits. The previous studies on model plate anchors buried at different depths of undrained clay are also available in experimental models [4, 5], Das and Puri [6], Rao et al. [16], Bhattacharya et al. [3]. The pull-out resistance of screw anchors in soft clay had been studied by Rao et al. [18] and Ashni and Janani [2]. The satisfactory performance of pull-out resistance of plate anchor in clay was well studied by Sahoo and Kumar [19], Meyerhof and Adams [11], Khatri and Kumar [8] and O'Kolly et al. [13]. Few studies were carried out on pull-out resistance of screw anchors in clay [23], Merifield et al. [8, 10]. McClelland [9] presented the detailed construction process of under-reamed footing in the North Sea, where dense sand was overlying hard clay. Model test on circular plate anchors was performed to observe the pull-out resistance of circular plate in layered cohesive-frictional soil by Stewart [22], where artificially created frictionless clay (named as "blyben") was underlying dense sand. Peck et al. [14] noticed that bells should not be attempted unless the ground is cohesive enough to allow the roof of a bell to stand without collapse during the time between excavations, clean up of the bottom and placement of the concrete in the bell.

# 2 Aim of the Present Study

This study attempts to investigate the (a) behaviour of pull-out resistance versus model displacement curves and (b) variation in pull-out resistance of belled anchors on accounts of variation in embedment ratios, diameter ratios and bell angles in different layered sand and soft silty-clay deposits. It is also aimed to outline reason behind the variations in experimental data by finite element analysis in Plaxis 3D Foundation (numerical model having, $\lambda = 1/10$) in both the layered soil deposits and correlated with respective failure wedges as obtained from software analysis and parametric study.

## 3 Experimental Set-up and Methodology

The schematic view of the experimental set-up in the laboratory is shown in Fig. 1. The experimental set-up consists of loading frame, loading frame, loading arrangement, load and displacement measuring devices, testing tank and model anchors.

The steel channels are used to fabricate the loading frame. The base of the loading frame is bolted in the ground by concreting for stability. The reaction frame, which is top horizontal beams, is carrying the loading arrangement. The loading arrangement basically consists of a screw and nut system. The nut arrangement along with the ball-bearing system is seated on the centre of top horizontal beams by welding. The screw is long vertical pulling shaft which can freely move in upward and downward direction through the nut arrangement based on the ball-bearing action. At the bottom of the pulling shaft, the model is connected via proving ring. A rotating circular wheel is connected with nut arrangement and is operated manually in clockwise direction to control the vertical movement of the shaft and thus the upward displacements of the anchors. Two magnetic bases are seated on two steel bars laid on top edges of the testing tank, and a couple of dial gauges are vertically held by these magnetic bases (0.01 mm accuracy). The response of (1.0 kN capacity, in tension) proving ring demonstrates the gross pull-out resistance of model anchors, and the corresponding upward displacement (at the ground surface) of the anchors is recorded from the average reading of both the dial gauges.

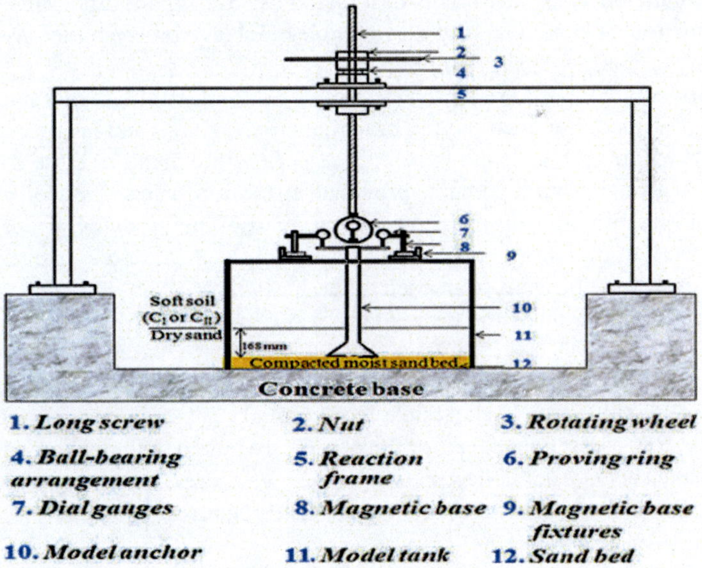


**Fig. 1** Schematic diagram of experimental set-up in laboratory

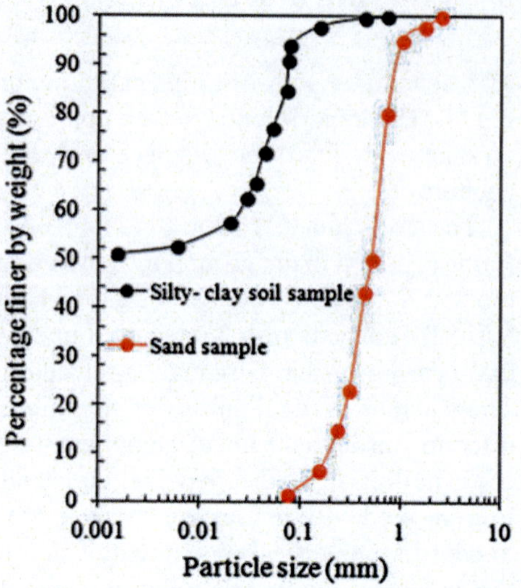


**Fig. 2** Grain size distribution of silty-clay soil and sand samples

The locally available silty-clay soil is collected from Howrah river bank, near Agartala, Tripura, India. The soil in the present test condition belongs to water content at 90% and 80% of the liquid limit. Based on the water contents, consistency indices ($I_c$) calculated are 0.23 and 0.46, and according to the state of consistency, they are recognized as $C_I$ and $C_{II}$, respectively. A dry sand is used to characterize the layered foundation beds, and this is designated as "$S$". Figure 2 presents the grain size distribution of sand and silty-clay samples. Table 1 presents the physical and engineering properties of soil samples.

The bell and shaft parts are fabricated from steel rounded solid bars, and the parts are internal welded. For each model, base diameter (i.e. $D_b$) and height are fixed at 26 mm and 650 mm. The detailing of $\beta$, $D_s/D_b$, $L/D_b$ are listed in Table 2 (columns [i]–[iii]). A grooving arrangement, provided at the top of each model, internally consists of threading to grip the proving ring gently. The grooving arrangement is carrying two flat steel strips externally by welding, so that the dial gauges can rest on them gently at 180° apart from each other.

The inside dimension of the testing tank is 600 mm (length) × 600 mm (width) × 700 mm (height). A steel frame is fabricated to shape the model tank, and the bottom is finished with steel plate. The four sides of the steel frame are choked-up by pieces of Plaxi glass of 12 mm thick. Few vertical steel stiffeners are welded with the frame in four sides. Four sheets of Plaxi glass are bolted with stiffeners and frame to prevent the sheets from deflecting outwards. The dimension of the model tank is large enough, so the boundary effects can be safely ignored.

**Table 1** Physical and engineering properties of soil samples

| Physical and engineering properties | Test results | |
|---|---|---|
| | Sand sample | Silty-clay sample |
| Medium sand, 2 to 0.425 mm (%) | 77.00 | – |
| Fine sand, 0.425 to 0.075 mm (%) | 23.00 | 6.00 |
| Silt 0.075 to 0.002 mm (%) | 1.50 | 43.00 |
| Clay $\leq$ 0.002 mm(%) | 0.00 | 51.00 |
| Effective grain size, $D_{10}$ (mm) | 0.23 | – |
| Coefficient of curvature, $C_c$ | 1.33 | – |
| Coefficient of uniformity, $C_u$ | 3.26 | – |
| Name of soil (USCS) | SW | CH |
| Specific gravity, $G_s$ | 2.69 | 2.62 |
| Plasticity index (%) | – | 24.50 |
| Minimum void ratio, $e_{min.}$ | 0.49 | – |
| Maximum void ratio, $e_{max.}$ | 0.79 | – |
| Void ratio, at placement density, $e_{exptl.}$ | 0.58 | 1.28($C_I$),1.02($C_{II}$) |
| Placement density (kN/m$^3$) | 16.90 | 16.97($C_I$), 17.73($C_{II}$) |
| Water content (%) at placement density | 0.00 | 50.40 ($C_I$), 44.80($C_{II}$) |
| Relative density $D_r$ (%) | 63.94 | – |
| Angle of shearing resistance, $\phi$ (°) | 39.50 | – |
| Undrained cohesion, $c$ (kN/m$^2$) | 0.00 | 2.7($C_I$), 4.8($C_{II}$) |

## 3.1 *Layered Bed Preparation for Models*

In the test programme, for the model, the layered sand and soft silty-clay foundation media is prepared with the help of sand and one type of soft soil. The placement density of dry sand is 16.90 kN/m$^3$ and is achieved by free rainfall technique using soil tray. The height of free fall of sand from soil tray is set at 700 mm after calibration. The density of dry sand can be controlled easily within the range from loose to dense through the rainfall technique. Local densities of sand are ensured with the help of three wooden boxes of capacity 80 cc.

The silty-clay soil is brought from the site and pulverized in the laboratory. The soil is mixed with a requisite amount of water (90 and 80% of LL) and transferred in a water-tight polythene bag for seven days. The bed of soft silty-clay soil bed is formed in the model tank by layers of 50 mm thickness with hand packing and patted uniformly to achieve exact density by observing the lift of each layer.

A compacted moist sand bed of 100 mm thickness is made inside the model tank prior to preparation. The bed of soft soil is formed as explained earlier. The lift of each layer is observed from the outer surface of Plaxi glass wall. The similar method is followed to prepare the bed of both the consistencies, $C_I$ and $C_{II}$. The soft soil bed is prepared as suggested by Shin et al. [20] and Singh and Ramaswamy [21].

**Table 2** Detailing of 3D model-belled anchor characteristics, embedment ratios, thickness (%) of silty-clay ($C_I$ and $C_{II}$) and sand ($S_{II}$) layers and their designations in $S_{II}/C_I$ and $S_{II}/C_{II}$ deposits

| $\beta$ (°) (i) | $D_s/D_b$ (ii) | $L/D_b$ (iii) | $[L(S)/L] \times 100$ (iv) | $[L(C_I)/L] \times 100$ (v) | Model designations based on $\beta$, $D_s/D_b$, $L/D_b$, $L(S)/L$, $L(C_I)/L$ and $C_I$ (vi) | $(L(S)/L) \times 100$ (vii) | $(L(C_{II})/L) \times 100$ (viii) | Model designations based on $\beta$, $D_s/D_b$, $L/D_b$, $L(S)/L$, $L(C_{II})/L$ and $C_{II}$ (ix) |
|---|---|---|---|---|---|---|---|---|
| | | 3 | 39 | 61 | M:45-0.28-3-(0.39S/0.61CI) | 39 | 61 | M:45-0.28-3-(0.39S/0.61CII) |
| 45 | 0.28 | 4 | 54 | 46 | M:45-0.28-4-(0.54S/0.46CI) | 54 | 46 | M:45-0.28-4-(0.54S/0.46CII) |
| | | 5 | 63 | 37 | M:45-0.28-5-(0.63S/0.37CI) | 63 | 37 | M:45-0.28-5-(0.63S/0.37CII) |
| | | 3 | 30 | 70 | M:45-0.33-3-(0.30S/0.70CI) | 30 | 70 | M:45-0.33-3-(0.30S/0.70CII) |
| 45 | 0.33 | 4 | 47 | 53 | M:45-0.33-4-(0.47S/0.53CI) | 47 | 53 | M:45-0.33-4-(0.47S/0.53CII) |
| | | 5 | 58 | 42 | M:45-0.33-5-(0.58S/0.42CI) | 58 | 42 | M:45-0.33-5-(0.58S/0.42CII) |
| | | 3 | 18 | 82 | M:45-0.38-3-(0.18S/0.82CI) | 18 | 82 | M:45-0.38-3-(0.18S/0.82CII) |
| 45 | 0.38 | 4 | 38 | 62 | M:45-0.38-4-(0.38S/0.62CI) | 38 | 62 | M:45-0.38-4-(0.38S/0.62CII) |
| | | 5 | 51 | 49 | M:45-0.38-5-(0.51S/0.49CI) | 51 | 49 | M:45-0.38-5-(0.51S/0.49CII) |
| | | 3 | 100 | 100 | M:45-0.46-3-(CI) | 100 | 100 | M:45-0.46-3-(CII) |
| | 0.46 | 4 | 25 | 75 | M:45-0.46-4-(0.25S/0.75CI) | 25 | 75 | M:45-0.46-4-(0.25S/0.75CII) |
| | | 5 | 40 | 60 | M:45-0.46-5-(0.40S/0.60CI) | 40 | 60 | M:45-0.46-5-(0.40S/0.60CII) |
| | | 3 | 39 | 61 | M:63-0.28-3-(0.39S/0.61CI) | 39 | 61 | M:63-0.28-3-(0.39S/0.61CII) |
| 63 | 0.28 | 4 | 54 | 46 | M:63-0.28-4-(0.54S/0.46CI) | 54 | 46 | M:63-0.28-4-(0.54S/0.46CII) |
| | | 5 | 63 | 37 | M:63-0.28-5-(0.63S/0.37CI) | 63 | 37 | M:63-0.28-5-(0.63S/0.37CII) |
| | | 3 | 30 | 70 | M:63-0.33-3-(0.30S/0.70CI) | 30 | 70 | M:63-0.33-3-(0.30S/0.70CII) |
| 63 | 0.33 | 4 | 47 | 53 | M:63-0.33-4-(0.47S/0.53CI) | 47 | 53 | M:63-0.33-4-(0.47S/0.53CII) |
| | | 5 | 58 | 42 | M:63-0.33-5-(0.58S/0.42CI) | 58 | 42 | M:63-0.33-5-(0.58S/0.42CII) |
| | | 3 | 18 | 82 | M:63-0.38-3-(0.18S/0.82CI) | 18 | 82 | M:63-0.38-3-(0.18S/0.82CII) |

(continued)

**Table 2** (continued)

| $\beta$ (°) (i) | $D_s/D_b$ (ii) | $L/D_b$ (iii) | $[L(S)/L]$ ×100 (iv) | $[L(C_I)/L]$ ×100 (v) | Model designations based on $\beta$, $D_s/D_b$, $L/D_b$, $L(S)/L$, $L(C_I)/L$ and $C_I$ (vi) | $(L(S)/L)$ ×100 (vii) | $(L(C_{II})/L)$ ×100 (viii) | Model designations based on $\beta$, $D_s/D_b$, $L/D_b$, $L(S)/L$, $L(C_{II})/L$ and $C_{II}$ (ix) |
|---|---|---|---|---|---|---|---|---|
| 63 | 0.38 | 4 | 38 | 62 | M:63-0.38-4-(0.38S/0.62CI) | 38 | 62 | M:63-0.38-4-(0.38S/0.62CII) |
| | | 5 | 51 | 49 | M:63-0.38-5-(0.51S/0.49CI) | 51 | 49 | M:63-0.38-5-(0.51S/0.49CII) |
| | | 3 | 100 | 100 | M:63-0.46-3-(CI) | 100 | 100 | M:63-0.46-3-(S) |
| 63 | 0.46 | 4 | 25 | 75 | M:63-0.46-4-(0.25S/0.75CI) | 25 | 75 | M:63-0.46-4-(0.25S/0.75CII) |
| | | 5 | 40 | 60 | M:63-0.46-5-(0.40S/0.60CI) | 40 | 60 | M:63-0.46-5-(0.40S/0.60CII) |
| | | 3 | 39 | 61 | M:72-0.28-3-(0.39S/0.61CI) | 39 | 61 | M:72-0.28-3-(0.39S/0.61CII) |
| 72 | 0.28 | 4 | 54 | 46 | M:72-0.28-4-(0.54S/0.46CI) | 54 | 46 | M:72-0.28-4-(0.54S/0.46CII) |
| | | 5 | 63 | 37 | M:72-0.28-5-(0.63S/0.37CI) | 63 | 37 | M:72-0.28-5-(0.63S/0.37CII) |
| | | 3 | 30 | 70 | M:72-0.33-3-(0.30S/0.70CI) | 30 | 70 | M:72-0.33-3-(0.30S/0.70CII) |
| 72 | 0.33 | 4 | 47 | 53 | M:72-0.33-4-(0.47S/0.53CI) | 47 | 53 | M:72-0.33-4-(0.47S/0.53CII) |
| | | 5 | 58 | 42 | M:72-0.33-5-(0.58S/0.42CI) | 58 | 42 | M:72-0.33-5-(0.58S/0.42CII) |
| | | 3 | 18 | 82 | M:72-0.38-3-(0.18S/0.82CI) | 18 | 82 | M:72-0.38-3-(0.18S/0.82CII) |
| 72 | 0.38 | 4 | 38 | 62 | M:72-0.38-4-(0.38S/0.62CI) | 38 | 62 | M:72-0.38-4-(0.38S/0.62CII) |
| | | 5 | 51 | 49 | M:72-0.38-5-(0.51S/0.49CI) | 51 | 49 | M:72-0.38-5-(0.51S/0.49CII) |
| | | 3 | 100 | 100 | M:72-0.46-3-(CI) | 100 | 100 | M:72-0.46-3-(CII) |
| 72 | 0.46 | 4 | 25 | 75 | M:72-0.46-4-(0.25S/0.75CI) | 25 | 75 | M:72-0.46-4-(0.25S/0.75CII) |
| | | 5 | 40 | 60 | M:72-0.46-5-(0.40S/0.60CI) | 40 | 060 | M:72-0.46-5-(0.40S/0.60CII) |

This bed possesses a constant thickness of 168 mm (i.e. $L(C_I)$ and $L(C_{II})$) of soft soil. Three wooden boxes (80 cc capacity) are placed in the corners of model tank to check the regularities in local density of soft soil. After finishing off the soft bed, the surface is properly covered with a polythene sheet. After 24 h, the polythene is removed, and sand is poured up to required height, i.e. $L(S)$, to maintain the total desired embedment depth. The thickness of overlying sand bed is varying based on different models to meet the total embedment depth. The height of free fall and regularity of local densities are maintained cautiously. All the tests are conducted after 24 h of completion of layered bed preparation, i.e. $S/C_I$ and $S/C_{II}$ deposits.

The details of the thickness of layers are explained in Table 2 [Columns (iv) and (v)] for $S/C_I$ deposit, where $L(S)/L$ represents the percentage of depth covered by overlying dry sand and $L(C_I)/L$ symbolizes the percentage of depth covered by underlying soft silty-clay, i.e. $C_I$. Similarly, the details of the thickness of layers are reported in the same table [Columns (vii) and (viii)] for layered soil deposit, where $L(C_{II})/L$ symbolizes the percentage of depth covered by underlying soft silty-clay, i.e. $C_{II}$ in $S/C_{II}$ deposit.

## 4 Test Programme

In the present experimental investigation, the following parameters are considered as variables:

- Embed ment ratios $(L/D_b) = 3$, 4 and 5
- Diameter ratios $(D_s/D_b) = 0.28$, 0.33, 0.38 and 0.46,
- Bell angles $(\beta) = 45$, 63 and 72°,
- Stages of consistency indices of soft silty-clay deposits $(I_c) = 0.23$ and 0.46.

In total ($3 \times 4 \times 2 \times 3 =$), 72 numbers of tests are performed in the investigation.

## 5 Model Designations

In Table 2 (columns [i], [ii], [iii], [vi], [vii], [ix] and [x]), all the models are presented for both layered sand and soft silty-clay soil ($S/C_I$ and $S/C_{II}$) deposits. By using a common coding system, each model is symbolized. The first, second, third and fourth parts demonstrate the type—M (model), $\beta$, $D_s/D_b$, $L/D_b$, respectively, as presented in Table 2 (columns [iv] and [v]). The symbol M:72-0.28-4 implies that a model of $\beta = 72°$ and $D_s/D_b = 0.28$ is embedded at $L/D_b = 4$. In the detail designation of a model, the fifth part indicates the percentages of thickness of the sand in the upper layer (in numerator) and soft silty-clay soil ($C_I$ and $C_{II}$) in bottom layer (in denominator). When a model is possessing $\beta$ of 45°, $D_s/D_b$ of 0.33 at $L/D_b$ of 4, its 53% of total $L$ is in upper layer ($S$) and 47% of total $L$ is in bottom layer $C_I$ respectively, and then, it is denoted by M:45-0.33-4-($0.53S/0.47C_I$). The symbol M:63-0.28-3-($0.61S/0.39C_{II}$)

implies that a model having $\beta$ of 63°, $D_s/D_b$ of 0.28 and embedded at $L/D_b$ of 3, and its 61% of total $L$ is in upper layer ($S$) and 39% of total $L$ is in the bottom layer ($C_{II}$).

## 6 Numerical Modelling

In order to investigate the reason behind the variations pull-out resistance data involved in the present numerical study, finite element method of analysis, based on Plaxis 3D Foundation software modelling has been performed. Large-scale experimental aspects (with scale factor, $\lambda = 1/10$) are considered in the finite element package, and physical properties of sand and soft soil are believed to obey the Mohr–Coulomb failure criteria. Drained strength parameters $\phi'$, $\psi'$ and $c'$ are used to model the layered sand deposits. In order to ensure the rough interface between the anchor and the sand, the strength reduction factor ($R_{\text{inter}}$) in the anchor–soil interface is considered as 0.1 and is already provided in finite element programming of Plaxis 3D Foundation software.

$$\tan \phi_i = R_{\text{inter}} \times \tan \phi$$

In the modelling technique, the belled anchor is modelled as an embedded pile element. In the 3D finite element model, the pile is constituted of beam element with three degree of freedom per node, i.e. two transitional degrees of freedom and one rotational degree of freedom. The discretization of entire layered soil block is shown in typical Fig. 3, and the domain size in the numerical model is assumed to be similar to the foundation model in the laboratory. The mesh refinement has been carried out from the belled anchor to the boundary with smaller and clustered elements to bigger and scattered elements. Figs. 4, 5 and 6 present the displacement contour for few models.

**Fig. 3** 3D mesh generation

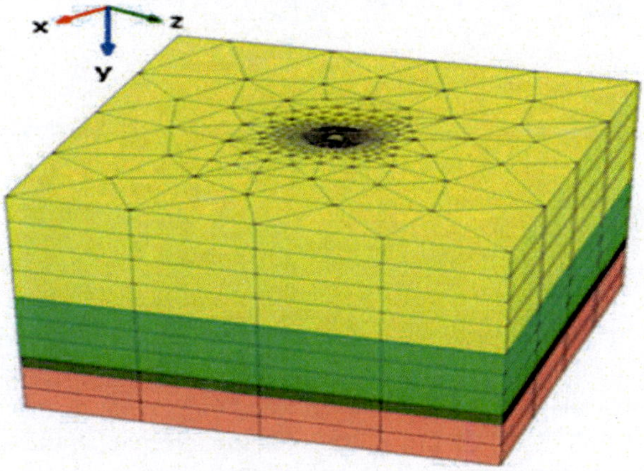

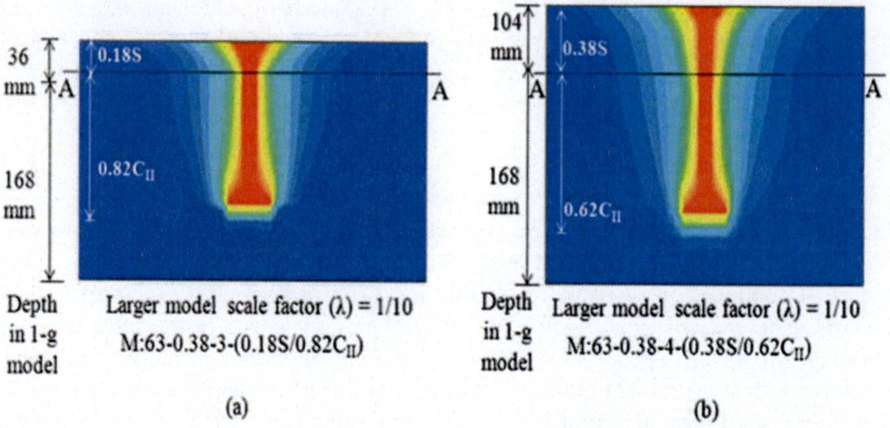


**Fig. 4** Displacement contour for model M:63-0.38 at the embedment ratios of **a** 3 and **b** 4

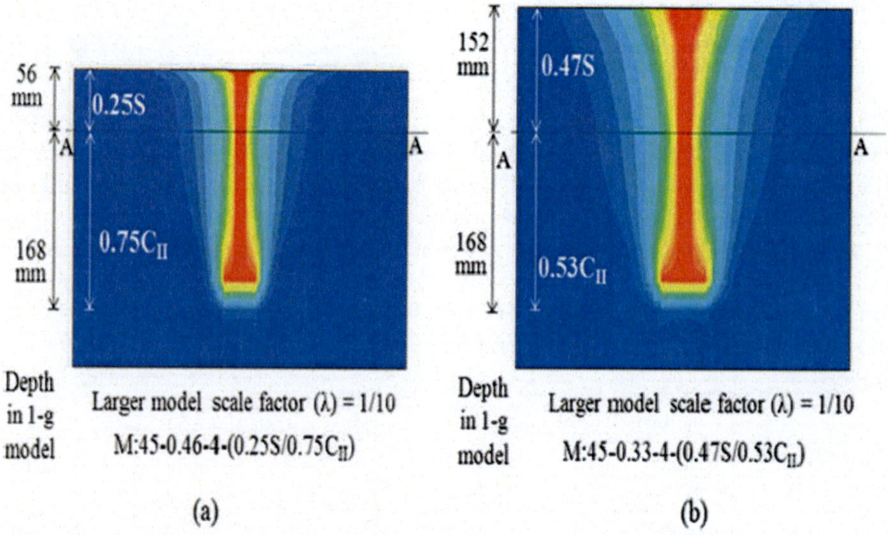


**Fig. 5** Displacement contour for models, both at the $L/D_b = 4$ and $\beta = 45°$, having $D_s/D_b =$ **a** 0.46 and **b** 0.33

## 7 Results and Discussions

On the basis of experimental observations, the response of load-displacement curves, effects of $L/D_b$, $D_s/D_b$ and $\beta$ in two types of layered sand and soft silty-clay soil deposits, i.e. $S/C_I$ and $S/C_{II}$ and comparison of net ultimate uplift capacities, i.e. $Q_u(S/C_I)$ and $Q_u(S/C_{II})$ based on the referred parameters are discussed with the help of related figures.

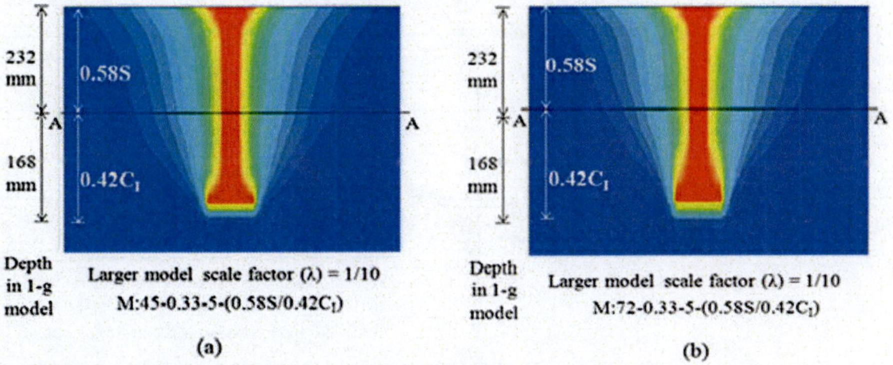


**Fig. 6** Displacement contour for model both at the $L/D_b = 5$ and $D_s/D_b = 0.33$, having $\beta =$ **a** 45° and **b** 72°

## *7.1 Behaviour of Load-Displacement Curves in Different Layered Sand And Soft Silty-Clay Deposits*

The typical net pull-out resistance versus model displacement behaviour for models M:45-0.33 and M:63-0.28 buried in $S/C_I$ and $S/C_{II}$ deposit are presented in Fig. 7a and b, respectively, both installed at $L/D_b$ of 3, 4 and 5. The figures reveal that when the models are embedded in $S/C_I$ and $S/C_{II}$ deposits, the pull-out resistance versus model displacement curves constitutes of cohesion in underlying soft layer and friction in overlying sand. In the initial linear part, the maximum strength mobilization of soft silty-clay soil occurs in a very small displacement of the anchor; the true elastic response is limited in this part and is very rapid. A continuous pulling of the models endorses the pull-out resistance versus model displacement curve to take a curvilinear shape, where anchor undergoes large displacement offering more or less constant tensile resistance. At this stage, tension cracks and upheaval of soft soil may appear in the soil layer interface. This phenomenon was also discussed by Thorn et al. [23] and Singh and Ramaswamy [21]. This pattern of pull-out resistance versus model displacement curves resemble full strength mobilization (collapse) within the soft soil layer; there is a gradual increase in upward displacement of anchor with corresponding approximately constant pull-out resistance. In the secondary stage, pull-out resistance versus model displacement curve shows a sudden upward jump as uplift capacity is fully achieved with a small displacement in the overlying dry sand layer. This is due to the fact that dry sand fully collapses at a very small upward displacement. The similar nature in pull-out resistance versus model displacement curve is observed in both the layered sand and soft silty-clay deposits. The similar pattern of the load-displacement curve was explained by Stewart [22] also.

**Fig. 7** Net ultimate pull-out resistance versus model displacement relationships for model anchors **a** M:45-0.33 and **b** M:63-0.28 both installed in $S/C_I$ deposit at $L/D_b = 3$, 4 and 5

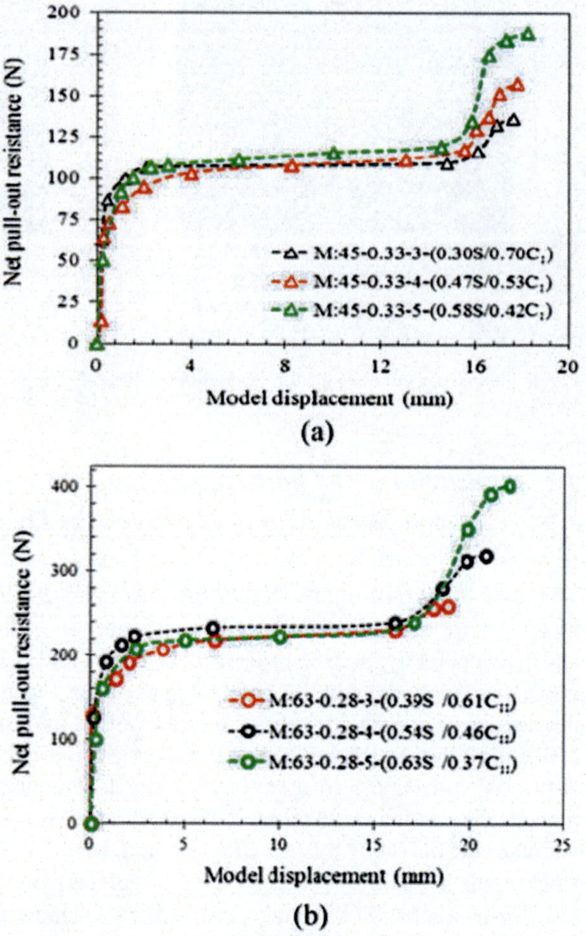


## 7.2 *Comparison of Net Ultimate Pull-Out Resistances ($Q_u(S/C_I)$ and $Q_u(S/C_{II})$) Due to Variation in Embedment Ratios ($L/D_b$)*

The typical Fig. 8 reveals that the primary (at $L/D_b = 3$ to 4) and secondary (at $L/D_b =$ 4 to 5) increments in $Q_u(S/C_I)$ are 18.08 and 29.08%, respectively, for M:45-0.28. In $S/C_{II}$ deposit, the same model for similar increase in embedment ratios, the primary and secondary increments in $Q_u(S/C_{II})$ are 14.60 and 22.53%, respectively. In the same figure, a comparison study illustrates that at $L/D_b = 3$, 4 and 5, the values of $Q_u(S/C_{II})$ are higher than $Q_u(S/C_I)$ in the order of 59.56%, 65.56% and 46.99%, respectively.

In this present study, for all the models, the underlying layer of soft silty-clay has a constant thickness, so with the increase in the value of embedment ratio,

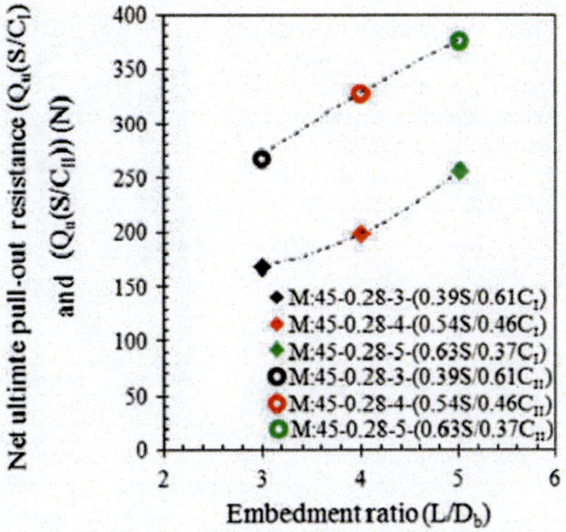


**Fig. 8** Net ultimate pull-out resistance ($Q_u(S/C_I)$ and $Q_u(S/C_{II})$) versus embedment ratio ($L/D_b$) relationships for model anchors M:45-0.28 installed in $S/C_I$ and $S/C_{II}$ deposit at $L/D_b = 3$, 4 and 5

only the thickness of the overlying sand layer increases. So, during uplift at higher embedment depth, when the failure starts to progress through the sand layer, more volume of sand is uplifted in contact with the belled shaft. Hence, for the cumulative resistances from dead-weight of soil layers, cohesion and frictional shear, pull-out resistances, is gradually increased; a phenomenon is independent of diameter ratios, bell angles and consistency of soft underlying soil. In Fig. 4a and b, the similar phenomenon is explained in the displacement contour for model M:63-0.38 at the embedment ratios of 3 and 4, respectively. Stewart [22] also presented a similar trend in results of pull-out resistance of circular plate anchor in layered cohesive-frictional soil.

### 7.3 Comparison on Net Ultimate Pull-Out Resistances ($Q_u(S/C_I)$ and $Q_u(S/C_{II})$) Due to Variation in Diameter Ratios ($D_s/D_b$)

In $S/C_I$ deposit, the typical Fig. 9 illustrates the decreasing trend in both net pull-out resistances (i.e. $Q_u(S/C_I)$ and $Q_u(S/C_{II})$) with gradually higher $D_s/D_b$ for all the models of $\beta = 72°$ and 45°, respectively, each installed at $L/D_b = 3$, 4 and 5. Fig. 9 shows that for 72° models due to increase in $D_s/D_b$ from 0.28 to 0.33, 0.33 to 0.38 and 0.38 to 0.46, decrease in $Q_u(S/C_I)$ values are subsequently 0.65, 0.84 and 0.61 times for $L/D_b = 4$. In the same plots of Fig. 8, for all models of $\beta = 72°$, having $D_s/D_b = 0.28$, 0.33, 0.38 and 0.46, the increment in the value of $Q_u(S/C_{II})$ than $Q_u(S/C_I)$ is 59.56, 59.80, 60.59 and 66.67% at $L/D_b = 4$.

In this present study, the increase in diameter ratio, i.e. decrease in base area of failure cylinder by the size (though the thickness of soft soil layer is constant),

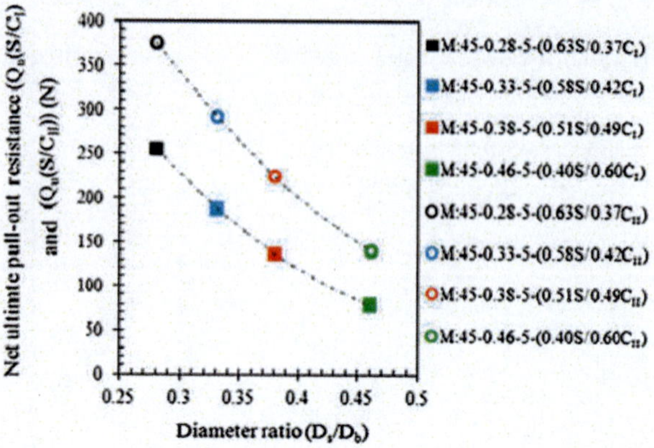


**Fig. 9** Net ultimate pull-out resistance ($Q_u(S/C_I)$ and $Q_u(S/C_{II})$) versus diameter ratio ($L/D_b$) relationships for model anchors having $\beta = 45°$, $L/D_b = 5$ and $D_s/D_b = 0.28, 0.33, 0.38$, and 0.46

results in lower value of dead-weight as well as cohesion in the failure cylinder of soft soil at the collapse stage. Again, for the same value of embedment ratio, total embedment depth decreases with the decrease in bell size (though the thickness of soft silty-clay layer is constant), and lesser thickness of sand overlay results in a lesser volume of breakout sand; less amount of dead-weight of dry sand as well as frictional shear is developed within failure zone. Hence, for the cumulative effect of dead-weight of soil layers, cohesion and frictional shear pull-out resistance are gradually decreased for increasing values of diameter ratios, irrespective of the value of embedment ratios, bell angles and consistency of soft silty-clay in layered soil. In Fig. 5a and b, the similar phenomenon is explained in the displacement contour for model M:45-0.46 and M:45-0.33, respectively, at the embedment ratios of 4. Rao et al. [17] and Bhattacharya et al. [3] made a experimental observation for plate anchors, where the lower sizes of anchors at the same embedment ratios in undrained clay can achieve lesser pull-out resistance in reinforced and unreinforced clay. Rao et al. [18], and Mistri and Singh [12] stated that the variation in the size of cylindrical failure zone results in differences in pull-out resistances of anchor plates of various sizes in soft soils.

## 7.4 Effect on Net Ultimate Uplift Pull-Out Resistances ($Q_u(S/C_I)$ and $Q_u(S/C_{II})$) Due to Variation in Bell Angles ($\beta$)

The typical Fig. 10 illustrates the relationship between net ultimate pull-out resistance (i.e. $Q_u(S/C_I)$ and $Q_u(S/C_{II})$) and $\beta$ for all the models, irrespective of the values of $L/D_b$ and $D_s/D_b$. The figure presents the models having $D_s/D_b$= 0.33 due to increase in β from 45 to 63° and 45 to 72°; $Q_u(S/C_I)$ is changed to −7.97 and −9.87% at $L/D_b$ = 4. Here, it is observed that the variation in $Q_u(S/C_I)$ and $Q_u(S/C_{II})$ due to change

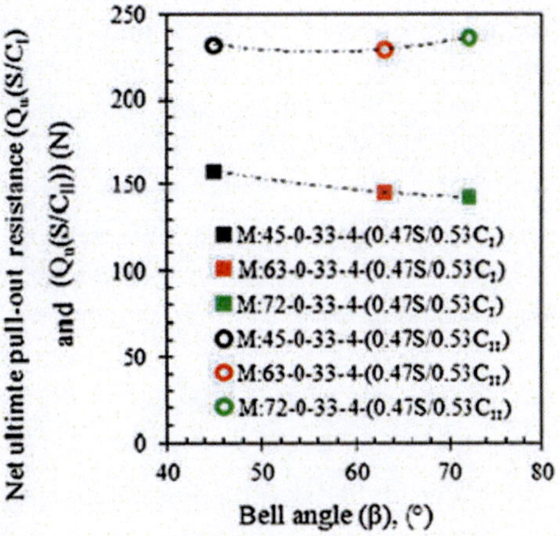


**Fig. 10** Net ultimate pull-out resistance ($Q_u(S/C_I)$ and $Q_u(S/C_{II})$) versus bell angles ($\beta$) relationships for model anchors $D_s/D_b =$ 0.33, $L/D_b = 4$ and $\beta = 45°$, 63° and 72°

in β is not worth mentioning, as because almost same values of pull-out resistance are obtained for the values of $\beta = 45$, 63 and 72°, when other parameters are constant.

In this present study, in spite of increase in bell angles (having particular embedment ratio and diameter ratio), almost same cohesive resistance is developed in breakout zone of similar extension in underlying layer, i.e. there may be the formation of almost cylindrical failure zones of approximately similar extents in soft soil of low consistency; hence, the variation in bell angles may not have any effect on the cylindrical failure zones in silty-clay soil. So, same volume of breakout dry sand wedge may develop, as the thickness of sand overlay is constant. Hence, for the almost equal cumulative effect of dead-weight of layers in breakout soils, cohesion and frictional shear, pull-out resistances are almost same for increasing values of bell angles regardless the values of embedment ratios and diameter ratios. This trend is found consistent irrespective of the consistencies of the underlying soft soil layer. In Fig. 6a and b, the similar phenomenon is explained in the displacement contour for model, both at the $L/D_b = 5$ and $D_s/D_b = 0.33$, having $\beta = 45°$ and 72°, respectively.

## 8 Significance of the Study

On the basis of the data recorded, all the models are found to be effective to attain higher $Q_u(S/C_{II})$ values than $Q_u(S/C_I)$ based on embedment ratios ($L/D_b$), diameter ratios ($D_s/D_b$) and bell angles ($\beta$), and these are within the range of approximately 60–83%.

Anchors are suggested when foundations have to bear extensive compressible loads and to carry tension loads in both on-land and off-shore structures. In few unavoidable circumstances, foundations of transmission line towers and power poles are to be erected on low lying land, where inter-bedded silty-clay horizon possessing low bearing capacity overlying hard strata may often be encountered in the field. Very old ponds or low lying lands may contain a thick bed of soft plastic silt and most probably clay. At such cases, the construction of the bell portion in compressible layer and to rest on underlying firm soil strata make it economically feasible than the stripping out of weak soil. The grade of depressed areas may be raised by following controlled filling procedures in layers with a cohesionless deposit, and thus, the anchor installed in shallow clay layer can be transformed into a deep anchor. The blanket of sand acts as surcharge loading and accelerates the stabilization process.

## 9 Findings of the Study

The major findings of the present study of this chapter are highlighted herein:

- For the same model, the pull-out resistance versus model displacement response is similar in layered sand and soft silty-clay deposits, irrespective the values of soft consistency indices of underlying soft soil.
- The pattern of pull-out resistance versus model displacement curve is dominated by the cohesion of the underlying soft soil and the frictional shear of overlying dry sand.
- As embedment ratios increase gradually, the net ultimate pull-out resistances of the same belled anchor always follow an ascending trend, irrespective of diameter ratios, bell angles and consistency indices of underlying soft soil.
- For higher diameter ratios, progressively declining net ultimate pull-out resistances are attained irrespective of bell angles and embedment ratios. This trend is followed irrespective of consistency indices of the underlying soft soil.
- With a gradual increase in bell angles irrespective of diameter ratios and embedment ratios, net ultimate pull-out resistance data show almost more or less same value. These findings are consistently independent of the consistency indices of the underlying soft soil.
- A comparative study reveals that belled anchors can achieve higher $Q_u(S/C_{II})$ than $Q_u(S/C_I)$ based on $L/D_b$, $D_s/D_b$ and $\alpha$. As the same model is showing different uplift capacities based on undrained shear strength of the underlying soft silty-clay layer, hence it may be remarked that pull-out resistance is essentially dependent on the consistency indices and undrained shear strength of underlying soft layer, in layered sand and soft silty-clay layer deposits.
- The reason behind the variation in pull-out resistances can be satisfactorily analysed from the displacement contour of numerical models (comparatively larger models, $\lambda = 1/10$), as revealed from the data of Plaxis 3D Foundation (FE analysis), for the same embedment ratios, diameter ratios, bell angles and foundation properties.
- All the models are found much effective to attain higher $Q_u(S/C_{II})$ values than $Q_u(S/C_I)$ within the range of approximately 60–83%.

## References


1. Adams JI, Radhakrishna HS (1971) Uplift resistance of augered footings in fissured clay. Can Geotech J, 8:452–462
2. Ashni MA, Janani V (2017) Experimental study on pull-out capacity of helical pile in clayey soil. Int J of Civil Eng Technol 8:1514–1521
3. Bhattacharya P, Bhowmik D, Mukherjee SP (2008) Pullout behaviour of square anchors in reinforced clay. In: 12th International Conference of International Association for computer Methods and Advances in Geomechanics, pp 1–6 October, Goa, India
4. Das BM (1978) Model test for uplift capacity of foundation in clay. Soils Found 18
5. Das BM (1980) A procedure for estimation of ultimate uplift capacity of foundations in clay. Soils Found 20
6. Das BM, Puri VK (1989) Holding capacity of inclined square plate anchors in clay. Soils Found 29:138–144
7. Ismael NF, Klym TW (1978) Behavior of rigid piers in layered cohesive soils. J Geotech Eng 104:1061–1074
8. Khatri VN, Kumar J (2011) Uplift capacity of axially loaded piles in clays. Int J of Geomech 11:23–28
9. McClelland B (1974) Design of deep penetration piles for ocean structures. J of Geotech Eng 100:1–37
10. Merifield RS, Sloan SW, Yu HS (2001) Stability of plate anchors in undrained clay. Geotech 51:141–153
11. Meyerhof GG, Adams JI (1968) The ultimate uplift capacity of foundations. Can Geotech J 5:225–244
12. Mistri B, Singh B (2011) Pullout behavior of plate anchors in cohesive soils. Elec J Geoteh Eng 16:1173–1184
13. O'Kelly BC, Brinkgreve RBJ, Sivakumar V (2014) Pullout resistance of granular anchors in clay or undrained condition. Soils and Foundations https://doi.org/10.1016/j.sandf.2014.11.009
14. Peck RB, Hanson WE, Thornburn TH (1974) Foundation Engineering. Second Edition, John Wiley and Sons, Inc., New York
15. Radhakrishna HS, Adams JI (1973) Long-term uplift capacity of augered footings in fissured clay. Can Geotech J, 10:647–652
16. Rao AS, Phanikumar BR, Babu RD, Suresh K (2007) Pullout behavior of granular pile-anchors in expansive clay beds in situ. J Geotech Geoenviron Eng 133:531–538
17. Rao SN, Latha KH, Pallavi B, Surenndran S (2006) Studies on pullout capacity of anchors in marine clays for mooring systems. Appl Ocean Res 28:103– 111
18. Rao SN, Prasad YVSN, Veeresh C, (1993) Behaviour of embedded model screw anchors in soft clays. Geotech 43:605–614
19. Sahoo JP, Kumar J (2012) Vertical uplift resistance of two horizontal strip anchors with common vertical axis. Int J of Geotech Eng https://doi.org/10.3328/IJGE.2012.06.04 485–495
20. Shin EC, Das BM, Puri VK, Yen SC, Cook EE (1993) Ultimate uplift capacity of model metal piles in clay. Geotech and Geol Eng 11:203–215
21. Singh SP, Ramaswamy SV (2008) Effect of shape on holding capacity of plate anchors buried in soft soil. Geomech Goeng: An Int J 3:157–166
22. Stewart W (1985) Uplift capacity of circular plate anchors in layered soil. Can Geotech J 22:589–592
23. Thorne CP, Wang CX, Carter JP (2004) Uplift capacity of rapidly loaded strip anchors in uniform strength clay. Geotech 54:507–517

# Behavior of Pile Group in Liquefied Soil Deposits Under Earthquake Loadings

**G. M. Basavanagowda, P. Gowthami, S. V. Dinesh, L. Govindaraju, and S. Muttana Balareddy**

**Abstract** Pile foundation is extensively used for the support of bridges and other structures to safely transfer structural loads to the bearing ground through the pile with skin friction and end bearing mechanism and to avoid lateral movement and excessive settlement. Lateral loads come from a variety of sources, including earthquake, ship impact, waves, and wind. Design of foundation in seismic regions typically if the soil strata are likely to get liquefied is a challenging task. Piles have failed due to large bending moments at the interface of liquefiable non-liquefiable layers. The behavior of single pile and group of piles in liquefied deposit is a complex issue and uncertainty exists regarding quantification of pile displacement, bending moment, shear force, and other parameters. In the current study, the numerical simulations were carried out using finite difference software FLAC3D. The soil model has three layers, slightly cemented sand at the top and bottom, Nevada sand at the middle with water table at the ground level. The Mohr-Coulomb model was assigned to the soil grid as the constitutive model, and the pile was modeled as a structural element. Analysis was carried out for the pile and pile group having floating pile head condition. The simulation was carried out by considering Cyclic, Bhuj, and El Centro acceleration time history as input ground motions. The pile analysis results were represented in terms of displacement, bending moment, shear force, and pore pressure ratio along the pile length. The numerical analysis results were compared with the results in the

G. M. Basavanagowda
Department of Civil Engineering, RIT, Bangalore 560054, India
e-mail: basavanagowdagm@gmail.com

P. Gowthami (✉) · S. V. Dinesh · S. M. Balareddy
Department of Civil Engineering, Siddaganga Institute of Technology, Tumkur 572103, India
e-mail: gowthamip2013@gmail.com

S. V. Dinesh
e-mail: dineshsv2004@gmail.com

S. M. Balareddy
e-mail: msb@sit.ac.in

L. Govindaraju
Department of Civil Engineering, UVCE, Bangalore 560056, India
e-mail: lgr_civil@yahoo.com

 
T. G. Sitharam et al. (eds.), *Seismic Design and Performance*, Lecture Notes in Civil Engineering 120, https://doi.org/10.1007/978-981-33-4005-3_11

literature for validation. It was observed that the pile displacement is maximum at the pile top, and maximum bending moment occurs at or near the interface of liquefiable and non-liquefiable layers. The pore pressure ratio indicates liquefaction of Nevada sand. The pile displacement and bending moment were dependent on peak ground acceleration, duration of strong ground motion.

**Keywords** Pile group · Earthquake · Liquefiable · Displacement · Bending moment

## 1 Introduction

In the recent past, buildings supported on pile foundations resting in liquefied soil deposits have failed during earthquake. The key factors causing failure are lateral spreading of the ground and the inertial effects. There are many reports which have been documented on the damage of pile foundation in liquefied soil deposits [1–6]. Piles are subjected to large passive earth pressure developed due to non-liquefiable layers located on either at top or bottom of liquefiable layer. Further, piles are subjected to large bending moment at the interface of liquefiable and non-liquefiable layers. The damage to pile foundation during 1995 Kobe earthquake in Japan indicated failure at the top interfaces and large deformation at the pile head [7]. Hazzar et al. [8] modeled a group of piles embedded in clayey and silty soil using FLAC3D to understand the interaction effects and the pile group behavior subjected to axial and lateral loads. The lateral loads show higher lateral deflection in sandy soil than the clayey soil. The presence of vertical load in clayey soil is reduced lateral stress resulting in reduced resistance to lateral deformation of the pile. Ruoshi et al. [9] investigated the seismic response of soil-pile structure simulations using FLAC3D software. The superstructure and foundation were founded on a soft clayey soil. The results indicated that by increasing the number of piles, the bending moment, shear force, and base shear increases resulting in the increase of story displacement.

The pile foundation behavior under vertical or lateral loads has been investigated for more than a century through full scale tests [10], centrifuge model tests [11, 12], and analytical [13] as well as numerical solutions [8]. The procedure used for modeling soil range from rigorous soil continuum discretization such as finite element (FE) or finite difference (FD) formulation to simplified interaction models such as the subgrade reaction approach. In the conventional subgrade approach, soil is modeled by spring elements attached to the pile at different depths. The springs generally have non-linear load displacement characteristics called t-z and p-y curves for vertical and lateral loadings, respectively. These two types of springs are generally uncoupled, and therefore, soil reactions along the corresponding degrees of freedom are also uncoupled.

Hussien et al. [14] used a simplified soil-pile interaction FE model and reported slight increase in the lateral capacity of free head piles installed in sandy soil due to the presence of vertical loads and attributed this increase to the increase in the confining pressures in the sand deposit surrounding the upper part of the pile. In fact, the scopes of previous attempts examining the behavior of piles have been limited to study the response of piles embedded in homogeneous sandy or clayey soils. Further, the mechanism of pile bending in non-homogeneous and/or layered soils is quite different from those piles resting in ideal homogeneous soil conditions.

In view of the above specified issues, this paper discusses the results of 3D FD analyses carried out using FLAC3D [15, 16] in order to evaluate the influence of liquefiable layer on the top and bottom non-liquefiable layer under group of pile condition.

## 2 Modeling

The 3D Finite Difference (FD) FLAC3D was employed to study the behavior of single pile under lateral load and pile group model under different seismic excitations.

### 2.1 Soil Model

The Mohr-Coulomb model was adopted in this study to simulate the non-linear behavior of soil. The parameters required to effectively define the soil behavior are elastic bulk modulus, $K$, elastic shear modulus, $G$, the mass density, $\rho$, the friction angle, $\Phi$, and the cohesion, $C$.

### 2.2 Pile Model

Pile element is modeled as linear elastic material. Parameters required to define the pile material behavior are elastic bulk modulus, elastic shear modulus, and the mass density.

### 2.3 Validation of FLAC3D Model for Single Pile (Static Analysis)

Before analyzing the influence of pile group on the layered soil deposit, the applicability of the adopted model was verified by evaluating the single pile response under lateral loading from FD half model of Hazzar et al. [8]. The entire soil block consists of a single layer of sandy soil. The pile length considered is 10 m, and diameter of pile is 1 m. Thickness of soil stratum is taken as 16 m.

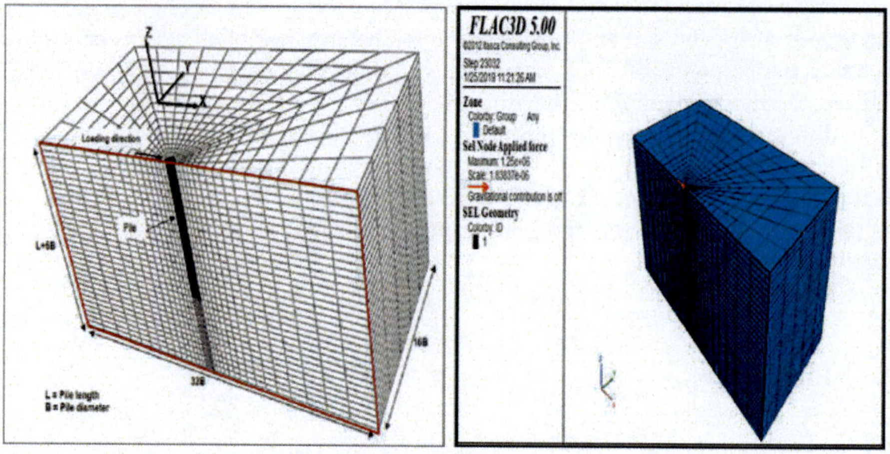

**Fig. 1** General layout and meshing of FLAC3D model

Geotechnical properties of soil layer [8] (Dense Sand) are density 1800 kg/m$^3$; friction angle [$\Phi$] is 36, elastic bulk modulus [$K$] 41.7 Mpa, elastic shear modulus [$G$] 19.2 Mpa, Poisson's ratio [$\mu$] 0.2, and the properties of pile are density 2300 kg/m$^3$, elastic modulus [$E$] 25 GPa, and Poisson's ratio [$\mu$] 0.15 were used in analysis.

Figure 1 shows the general layout and meshing of FLAC3D model used for the analysis of soil-pile system.

The lateral load was applied gradually at the pile head without vertical load till the lateral deflection of pile reaches 100 mm. From the result, it is observed that as lateral load increases, the deflection also increase as shown in Fig. 2. The lateral deflection results (Hazzar et al. [8]) are compared with the present investigation of FLAC3D. The present study results show good agreement with the results from Hazzard et al. [8].

## 2.4 Parametric Studies

**Pile group soil model subjected to different earthquake excitations.**

FLAC3D was employed to study the behavior of group of pile under different earthquake excitations.

A soil grid of 15 m x 15 m x 10 m dimension consisting of three layers in which top and bottom 2 m thickness is Slightly Cemented Sand (SCS), and in between 6 m, Nevada Sand (Nsand) layer is considered.

A four pile group was installed and attached to the pile cap, the pile length and diameter are 10 m and 0.6 m, respectively. The pile was connected to the floating pile cap having dimensions of 2 m × 2.5 m × 0.5 m with 1.5 m center to center distance

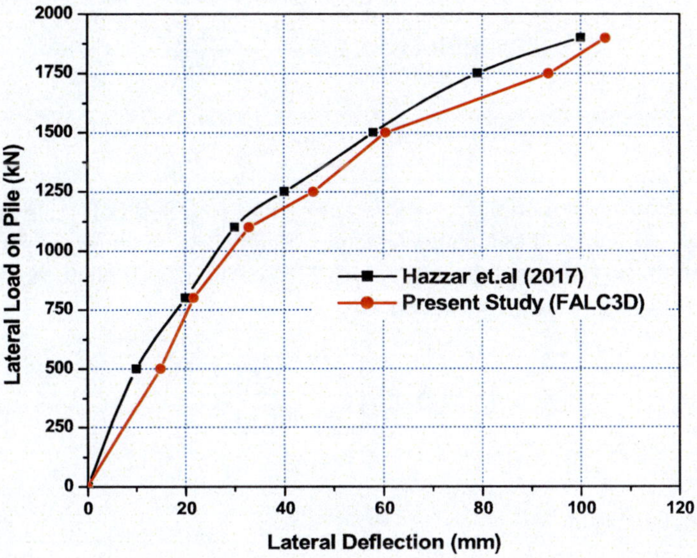


**Fig. 2** Lateral load versus lateral pile deflection

as shown in Fig. 3. In order to simulate the non-linear behavior and the possible shear failure of the soil model, a Mohr-Coulomb plasticity model was used.

Soil-pile model for the group of piles developed using FLAC3D is as shown in Fig. 3.

The pile was modeled as a two-noded finite element of six degrees of freedom per node. The pile structural element was divided into 32 sub-divisions. The pile was interacting with the surrounding soil through coupled shear and normal springs, to transfer motion and forces between the pile and soil zone. The shear and normal stiffnesses were computed using Eqs. (1) and (2).

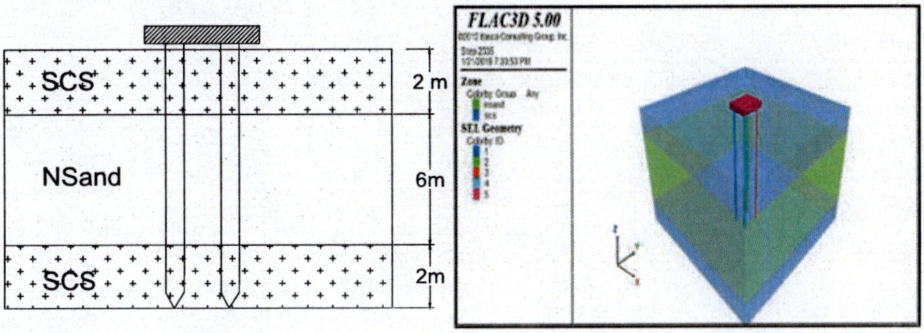


**Fig. 3** Group of pile installed in soil model

$$\text{Shear stiffness, } K_S = \frac{32(1-\mu)Gr}{7-8\mu} \tag{1}$$

$$\text{Normal stiffness, } K_n = \frac{4Gr}{1-\mu} \tag{2}$$

where $G$ = Soil shear modulus, $r$ = Pile radius, and $\mu$ = Soil Poisson's ratio.

Pile cap was modeled as a plate element, which resists bending.

Normal stiffness ($K_n$) and shear stiffness ($K_s$) were used to simulate the interaction between soil and pile cap given by Eq. (3).

$$K_n = K_s = \left[\frac{K + \frac{4}{3}G}{\Delta_{z\min}}\right] 4.5 \tag{3}$$

where

$K$ Soil bulk modulus
$G$ Soil shear modulus
$\Delta_{z\min}$ Smallest dimension of the adjacent grid zone.

The value of normal stiffness and shear stiffness obtained was $897.3 \times 10^6$ N/m using the above expression.

In the dynamic analysis, the waves lack the energy to propagate through the model. The free field condition was applied at the boundaries to facilitate wave propagation in one direction without any reflection.

The properties of Nsand and SCS, pile and pile cap material are tabulated in Table 1, and soil-pile interaction parameter is tabulated in Table 2.

Pile group response analyses were carried out for three-layered soil model, pile being resting at the bottom of SCS layer; soil model was subjected to cyclic loading having an 0.3 g uniform amplitude at 2 Hz frequency with 40 cycles, 2001 Bhuj

**Table 1** Properties of soil, pile, and pile cap (Abdoun et al. [1] and Bowles [17])

| Properties | Soil types | | Pile types | | Pile cap |
|---|---|---|---|---|---|
| | Nevada sand | Slightly cemented sand | Concrete | Polyetherimide | |
| Cohesion [Cu] (kPa) | – | 5.1 | – | – | |
| Density (kg/m$^3$) | 1800 | 1900 | 2400 | 1270 | 2830 |
| Friction angle [$\varphi$] (°) | 30 | 34.5 | – | – | |
| Poisson's ratio [$\mu$] | 0.25 | 0.3 | 0.2 | 0.36 | 0.35 |
| Relative density [%] | 40 | – | – | – | |
| Young's modulus [$E$] (MPa) | 10 | 60 | 25000 | 1260 | 69000 |

**Table 2** Soil–pile interaction values

| Parameters | Soil-pile interaction | Soil-pile cap interaction |
|---|---|---|
| Normal stiffness $[K_n]$ (N/m) | $6.4 \times 10^6$ | $89.73 \times 10^7$ |
| Shear stiffness $[K_s]$ (N/m) | $5.76 \times 10^6$ | $89.73 \times 10^7$ |

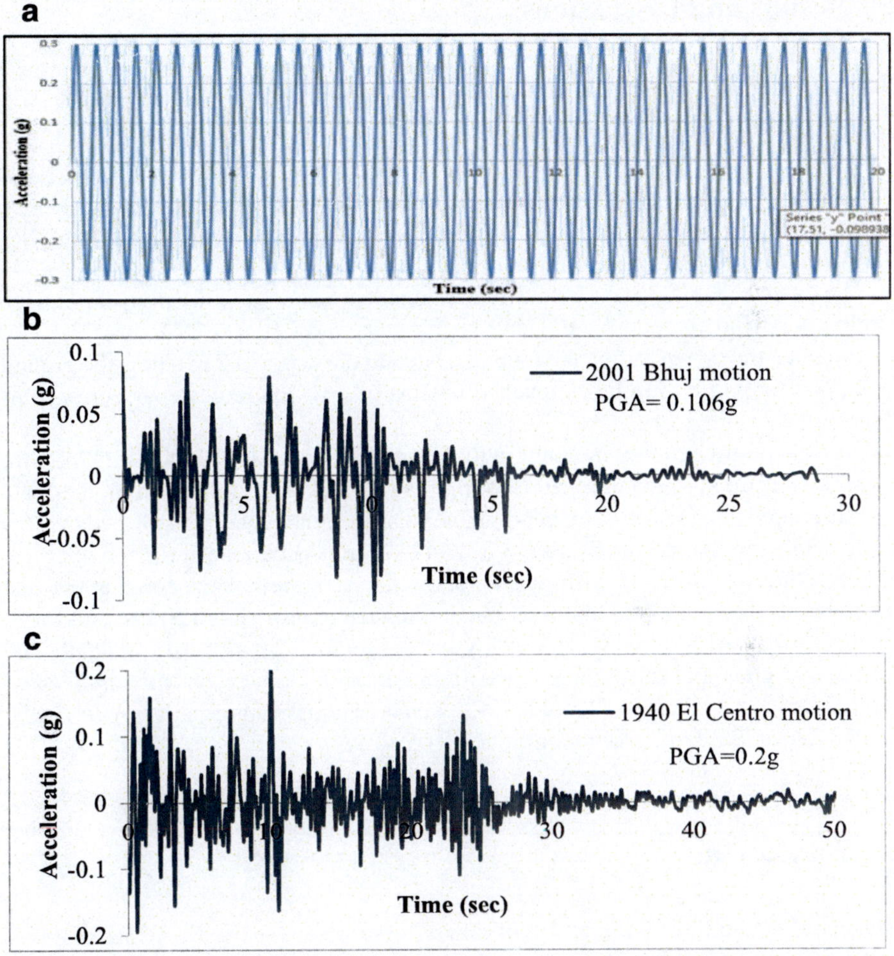


**Fig. 4** **a** Cyclic wave ground motion. **b** 2001 Bhuj ground motion. **c** 1940 El Centro ground motion

ground motion and El Centro ground motions as shown in Fig. 4a, b, c. The simulation was carried out for 2.5 and 5 s.

The response of the pile in terms of bending moment, shear force, pore pressure ratio, and displacement along the pile length was evaluated.

A range of duration, frequency, and intensities was covered to evaluate the soil-pile interaction in liquefied ground. The ground motion were applied along the *x*-direction at the base of the model.

## 3 Results and Discussions

### *3.1 Soil-Pile Group Model Subjected to Different Earthquake Excitations*

The analysis was carried out for a group of four piles, and the simulation was done up to 5 s. Figure 5a, b, c, d, e, f shows the comparison of the behavior of the piles during the above three ground motions.

The top pile displacement for cyclic loading is 61.6 cm. For 2001 Bhuj earthquake motion, it is 10.6 cm, and for El Centro earthquake motion, it is 28.25 cm.

Under all the three earthquake motions, beyond the depth of 2 m below the ground level, pore pressure ratio has a reached a value of 1 which indicates occurrence of liquefaction.

The maximum bending moment in the front pile for cyclic loading is 237 kN-m; for Bhuj, it is 62 kN-m; and for El Centro, it is 131.58 kN-m at the bottom interface. For the rear piles, the maximum bending moment for cyclic loading is 71.2 kN-m, for Bhuj is 35.9 kN-m, and for El Centro is 67.42 kN-m is occurring at the top interface between the two layers. The front piles show the maximum shear force at the pile top for cyclic loading of 80 kN; for Bhuj earthquake motion, it is 120 kN; and for El Centro earthquake motion, it is 93.48 kN and gradually decreases till 5 m depth, and for the rear piles, the shear force is maximum at the bottom layer. Maximum shear force for cyclic loading is 106.94 kN; for Bhuj earthquake motion, it is 117.37 kN; and for El Centro earthquake motion, it is 105.11 kN.

## 4 Conclusions

Numerical simulation of the group of pile was performed using FLAC3D in liquefiable soil deposits under different ground motions. The results were evaluated in terms of pile deflection, bending moment, shear force, and pore pressure ratio.

From the present study, the following conclusions are drawn:

1. The results bending moment and displacement of single pile show a great deal of consistency between present study and previous numerical investigations for earthquake response.

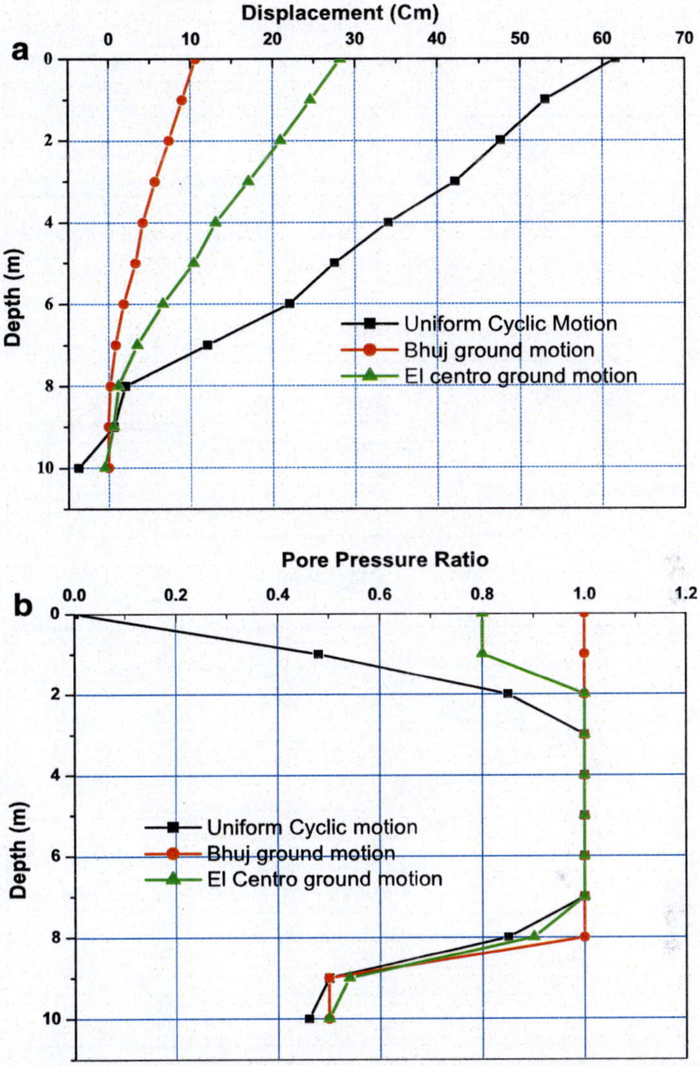


**Fig. 5** **a** Displacement of the pile group at 5 s. **b** Pore pressure ratio. **c** Bending moment along front pile. **d** Bending moment along rear pile. **e** Shear force along front pile. **f** Shear force along rear pile

2. The largest values of bending moments at any time during shaking have been observed at the interface between liquefied and non-liquefied layer, and these results are in accordance with the field observations during earthquakes. The pile group looses its confining pressure and begin to behave like a column. The front piles are subjected to maximum bending moment than rear piles.

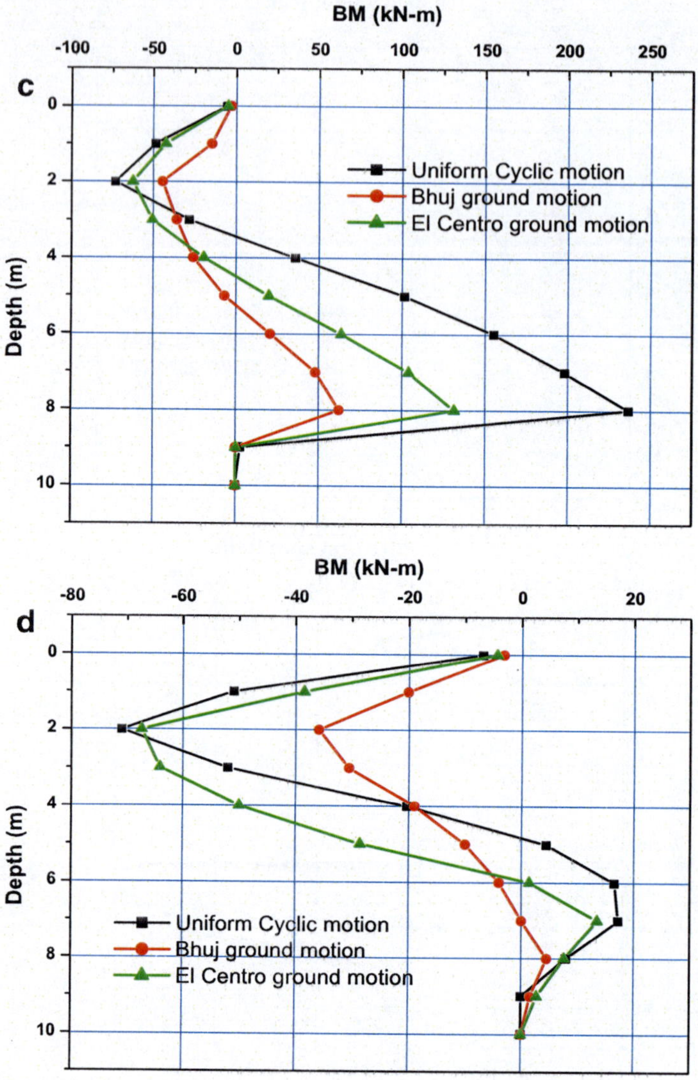


**Fig. 5** (continued)

3. The largest bending moments have occurred at some point during shaking, depending upon the intensity of PGA and also the occurrence of successive amplitudes and then decreased with increase in duration at the bottom interface which may lead to localized failure.
4. The pile deflection is maximum at the top and gradually decreases toward the bottom and tends to zero in the bottom non-liquefiable layer.

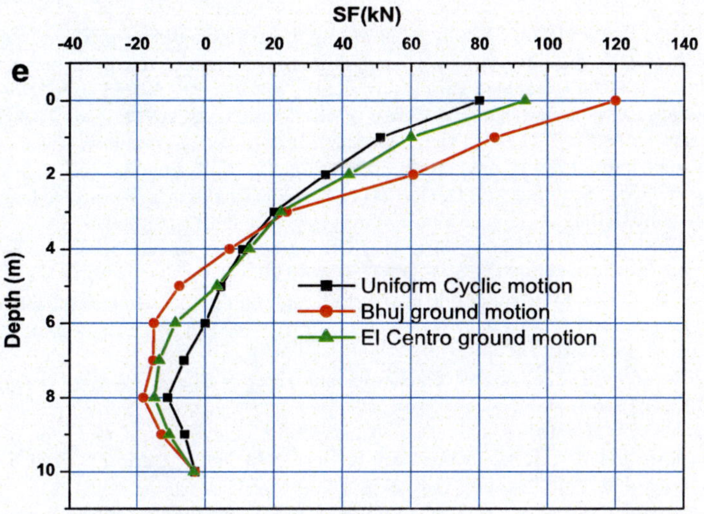


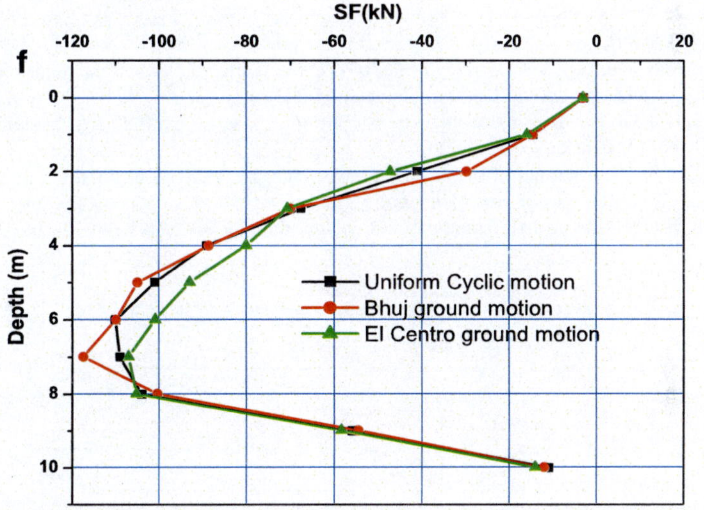


**Fig. 5** (continued)

# References

1. Abdoun T, Dobry R (2002) Evaluation of pile foundation response to lateral spreading. Soil Dyn Earthq Eng 22(9–12):1051–1058
2. Hamada M, O'Rourke TD (1992) Case studies of liquefaction and lifeline performance during past earthquakes, vol 1 Japanese case studies technical rep. NCEER-92-0001, National Center for Earthquake Engineering Research, Buffalo, NY
3. Liu L, Dobry R (1995) Effect of liquefaction on lateral response of piles by centrifuge model tests. NCEER Bull 9(1):7–11

4. Mizuno H, Liba M (1982) Shaking table testing of seismic building-pile-soil interaction. In: Proceedings of the 5th Japan earthquake engineering symposium, Tokyo, Japan, pp 1713–1720
5. Mizuno H (1987) Pile damage during earthquakes in Japan (1923–1983). In: Nogami T (ed) Proceedings, session on dynamic response of pile foundations. ASCE, New York, pp 53–77
6. Tokimatsu K, Asaka Y (1998) Effects of liquefaction-induced ground displacements on pile performance in the 1995 Hyogoken-Nambu Earthquake. Soils Found 2:163–178
7. Abdoun T, Dobry R, O'Rouke TD, Goh SH (2003) Pile foundation response to lateral spreads: Centrifuge modelling. J Geotech Geo-Environ Eng ASCE 129(10):869–878
8. Hazzar L, Hussien MN, Karray M (2017) On the behavior of pile groups under combined lateral and vertical loading. Ocean Eng 131:174–185
9. Xu R, Fatahi B (2018) Effect of pile group configuration on the seismic response of buildings considering the soil pile structure interaction. In: Advance in soil dynamics and foundation engineering, pp 279–287
10. Brown DA, Reese LC et al (1987) Cyclic lateral loading of a large scale pile group. J Geotech Eng 113(11):1326–1343
11. Abdoun T (1997) Centrifuge modeling of seismically induced lateral spreading of multilayer sand deposit and its effects on pile foundations. PhD thesis, Dept. Of Civil Engineering, Rensselaer Polytechnic Institute, Troy, N.Y
12. Tobita T et al. (2004) Group pile behavior under lateral loading in centrifuge model tests. Int J Phys Model Geotech 4(4):1–11
13. Matlock H, Reese LC (1960) Generalized solutions for laterally loaded piles. Soil Mech Found Eng 86(5):63–91
14. Hussien MN, Tobita T, lai S, Karray M (2014b) On the influence of vertical loads on the lateral response of pile foundation. Comput Geotech 55:392–403
15. FLAC3D (2012) Fast lagrangian analysis of continuum's version 5.0. Itasca Consulting Group, Minneapolis, Minnesota, U.S.A
16. Itasca (2009) User manual of FLAC 3D: Fast Lagrangian analysis of continua in 3D, version 5. Itasca Consulting Group inc, Minnapolis
17. Bowles JE (1997) Foundation analysis and design, 5th edition. McGraw-Hill, Inc, New York

# Dynamic Response of Soil Nailed Wall

Amrita, B. R. Jayalekshmi, and R. Shivashankar

**Abstract** Soil nailing is a technique used to stabilize steep cut slopes and to retain excavations. The method consists of inserting nail bars into the ground to stabilize the soil mass. The present study evaluates the dynamic response of a soil nailed wall supporting a vertical cut of height 6 m. The finite element analysis has been carried out to study the stability and performance of the soil nailed wall using PLAXIS 2D software. The seismic resistance and failure mechanism of the wall are analyzed under the El Centro ground motion. The seismic response variation for different parameters such as angle of inclination of nail and length of nail is studied. The results are observed in terms of maximum lateral displacement, development of maximum tensile force in nails and failure mechanism of soil nailed wall under static and dynamic conditions. The results of the study indicate that the soil nailed structure is an effective method which imparts stability to the retaining system under dynamic conditions.

**Keywords** Soil nailing · Finite element analysis · Seismic response

## 1 Introduction

Soil nailing can be used to stabilize vertical cuts. It is an effective ground improvement method commonly used to strengthen steep slopes and to retain deep excavation. The method consists of inserting nails into the pre-drilled holes made in the in situ soil mass and grouting them. The nails can be inserted horizontally or at small inclination to the horizontal. After placing the nails in position, facing is provided over the excavated part. The facing can be in the form of shotcrete, cast-in place reinforced shotcrete or reinforced concrete panels. The technique follows a top to bottom construction sequence. The method is widely used in railway and roadway embankments, tunnel construction, underground facilities, stabilizing steep cut slopes, etc.

Amrita (✉) · B. R. Jayalekshmi · R. Shivashankar
Department of Civil Engineering, National Institute of Technology Karnataka (NITK), Surathkal, Karnataka, India
e-mail: 23amritar@gmail.com

 
T. G. Sitharam et al. (eds.), *Seismic Design and Performance*, Lecture Notes in Civil Engineering 120, https://doi.org/10.1007/978-981-33-4005-3_12

Previous studies conducted on the soil nailed wall and reported in the literature show that it is an effective method for in situ ground improvement. The performance of the structure was studied under static and dynamic excitation. Various experimental and numerical studies were conducted by Stocker et al. [10], Shen et al. [8], Juran [5], Felio et al. [3], Vucetic et al. [11], Li et al. [7], Giri and Senguta [4], El-Emam [2], Babu and Singh [1], Singh and Babu [9] etc. to study the behaviour of soil nailed wall under dynamic conditions. The experimental studies reveal that the soil nailed wall is capable of resisting dynamic excitation as high as 0.7 g. However, detailed study on the dynamic performance of soil nailed wall has been conducted by a few. This can be accomplished by using an appropriate numerical simulation and analysis of a soil nailed wall. Thus, this study aims at evaluating the dynamic behaviour of soil nailed wall using finite element software. A soil nailed wall of height 6 m employed to retain sandy soil has been modelled, and its static and dynamic analysis are carried out using Plaxis software for various parameters.

## 2 Methodology

A soil nailed wall was designed to retain a vertical cut of height 6 m in conventional way using Federal Highway Administrative (FHWA) guidelines [6]. The wall was to retain sandy soil. The different material properties and other parameters which were used in the design of the soil nailed wall are given in Table 1.

**Table 1** Soil nailed wall properties and other parameters

| Parameter | Value |
|---|---|
| Vertical height of wall (m) | 6 |
| Backslope angle (degrees) | 0 |
| Soil type | Sandy soil |
| Soil cohesion (kPa) | 2 |
| Angle of internal friction (degrees) | 30 |
| Unit weight of soil ($kN/m^3$) | 18 |
| Modulus of elasticity of soil (MPa) | 30 |
| Nail type | Grouted nail |
| Diameter of nail (mm) | 20 |
| Drill hole diameter (mm) | 100 |
| Nail spacing $S_v \times S_h$ (m × m) | 1 × 1 |
| Yield strength of nail (MPa) | 500 |
| Modulus of elasticity of nail (GPa) | 200 |
| Compressive strength of grout (MPa) | 20 |
| Modulus of elasticity of grout (GPa) | 22 |
| Facing thickness (mm) | 200 |

The required length of the nail was determined to be 4.2 m from the conventional design, which was found to be sufficient to meet the minimum requirement of external and internal factor of safety values.

The soil nailed wall was modelled in the finite element software and the performance of the structure was determined under both static and dynamic condition. For the dynamic analysis, strong ground motion of El Centro earthquake was applied at the base of the model. The study was further carried out to determine the influence of nail length and nail inclination on the stability of the structure. For this, the nail length ratio (L/H) which is the ratio of length of the nail to the height of the wall was taken as 0.6, 0.7, 0.8 and 0.9. And the effect of nail inclination was studied for 0°, 10°, 15° and 20° inclination of the nail with respect to the horizontal.

## 3 Numerical Simulation

The response of the soil nailed wall was evaluated through numerical simulation using Plaxis software assuming a plane strain condition. The analysis was performed under both static load and dynamic excitation. The finite element discretization of the model was done using 15-noded triangular elements with medium mesh. The soil was modelled using hardening soil with small strain stiffness (HSS) model, which incorporates increased stiffness of soil at small strain values. The soil nails and facing of the wall were simulated as the linear elastic materials. The plate elements were used to model them.

For the analysis, the equivalent modulus of elasticity of the grouted nails was determined by considering the contribution of both grout and nail stiffness. Since the nail and the facing were modelled as the plate elements, their axial and bending stiffness were determined for the numerical analysis which is tabulated in Table 2.

The construction procedure of soil nailed wall was simulated using staged construction in which an excavation lift of 2 m height was performed in each stage. The static analysis of the model was carried out and the results were noted. For the dynamic analysis, the time history motion of El Centro earthquake scaled down to a peak ground acceleration of 0.3g was used which is shown in Fig. 1.

During the dynamic analysis, viscous boundaries were applied at the sides of the model, and the input data of the earthquake motion was applied at the base of the model. The finite element model of the soil nailed wall simulated in Plaxis software is shown in Fig. 2.

**Table 2** Material properties of grouted nail and facing of the wall

| Parameter | Axial stiffness (kN/m) | Bending stiffness ($kNm^2/m$) |
|---|---|---|
| Grouted nail | 22.87e4 | 142.94 |
| Facing | 4.4e6 | 14.67e3 |

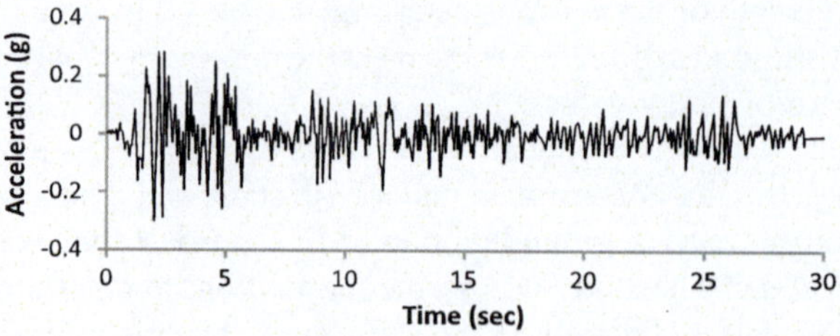


**Fig. 1** Time history input ground motion of El Centro earthquake

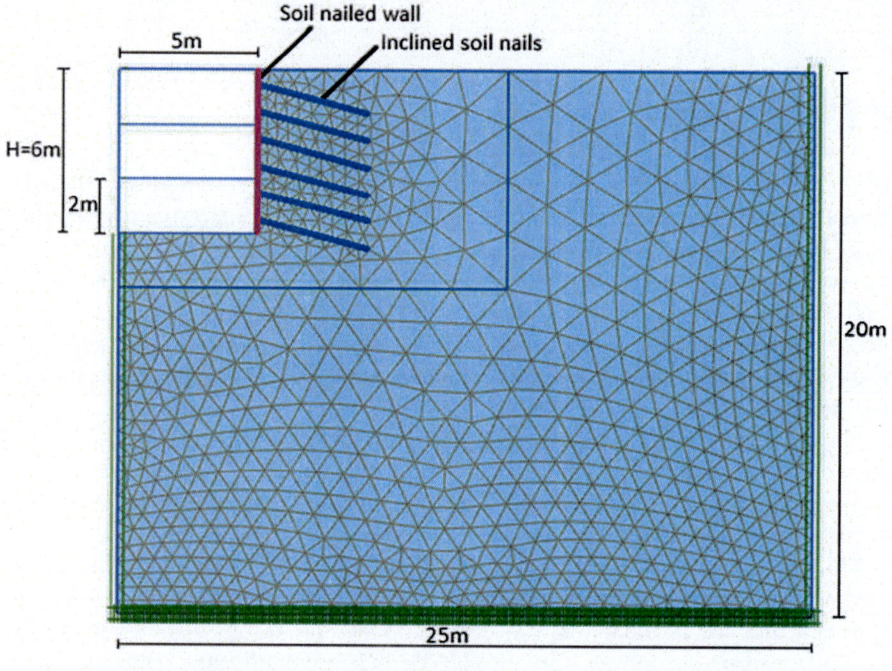


**Fig. 2** Finite element model of the soil nailed wall

## 4 Results and Discussion

The response of the soil nailed wall was evaluated under static and dynamic loading. The influence of length of the nail and inclination of the nail with horizontal on the performance of the wall was studied, and the results of the analysis are presented below.

## 4.1 Deformation of the Wall

The analysis results show that the displacement was maximum at the crest of the wall which decreased towards the bottom of the wall. The displacement of the soil nailed wall under seismic excitation for different nail length ratio is shown in Fig. 3.

It can be observed from the graph that the pattern of displacement of the wall was same for different nail length ratios. With the increase of nail length ratio, the displacement of the wall crest decreases. However, there was a significant reduction in the displacement when the nail length ratio was changed from 0.6 to 0.7. With further increase of the nail length ratio from 0.7 to 0.8 and then to 0.9, the reduction in displacement of the wall was in small magnitude. Thus, it can be concluded that the minimum nail length ratio of 0.7 was necessary for the soil nailed wall under dynamic excitation. The displacement response of the wall for different nail inclination is shown in Fig. 4. It can be seen that the displacement was minimum for the nail inclination of 15° and was maximum when the inclination was increased to 20°.

Figures 5 and 6 show the horizontal displacement of the wall plotted along the depth of the wall. It can be seen that the displacement was maximum when the length of the nail is 0.6 times the height of the wall and the displacement decreased as

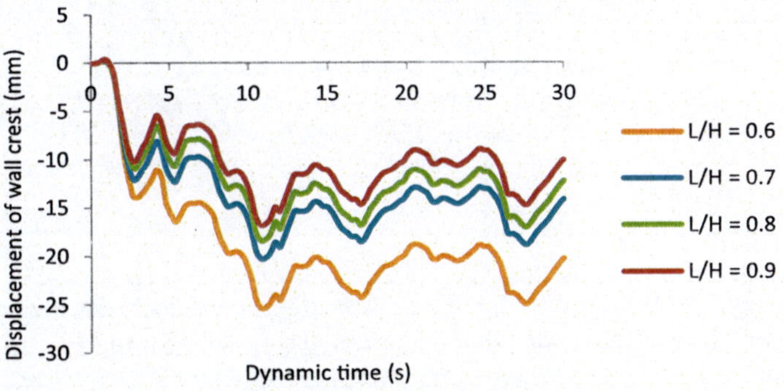


**Fig. 3** Displacement of the wall crest for different nail length ratio under dynamic loading

**Fig. 4** Displacement of the wall crest for different nail inclination under dynamic loading

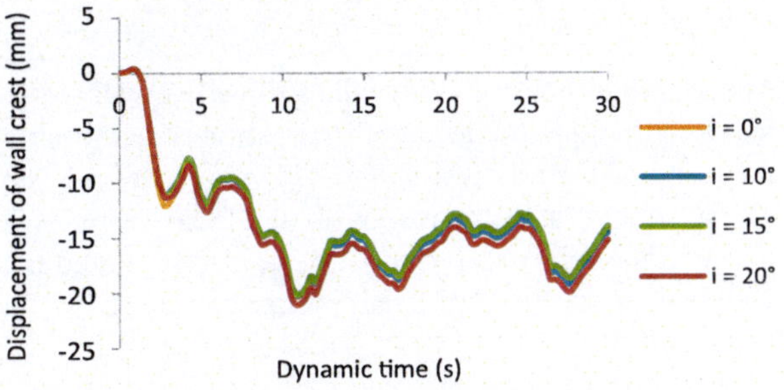

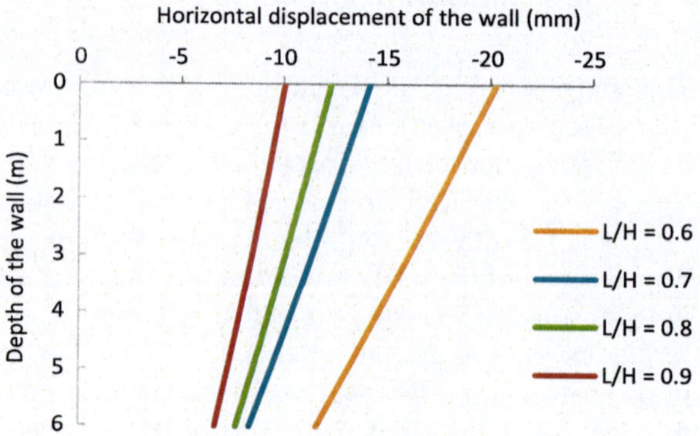


**Fig. 5** Horizontal displacement of the wall along the depth of the wall for different nail length ratio under dynamic loading

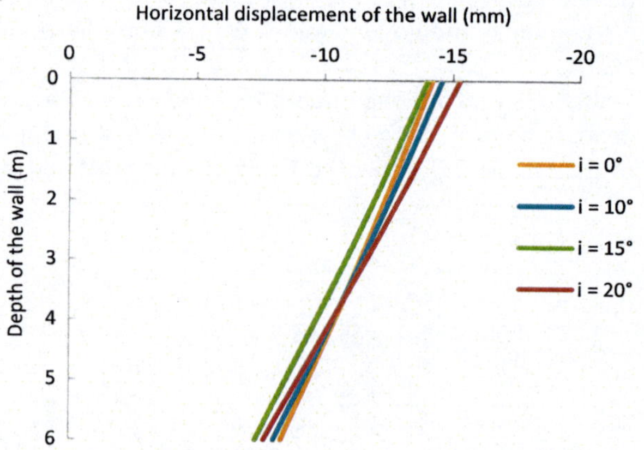


**Fig. 6** Horizontal displacement of the wall along the depth of the wall for different nail inclination under dynamic loading

this ratio increased from 0.6 to 0.9. There was significant decrease in displacement when the nail length ratio was changed from 0.6 to 0.7. Similarly, the minimum displacement was observed when the nail inclination was 15° to the horizontal.

## 4.2 *Acceleration of the Wall*

The acceleration response of the soil nailed wall under dynamic loading for different nail length ratio and different inclination of nail is shown in Figs. 7 and 8. The acceleration of the wall crest shows different pattern in comparison to the deformation. The acceleration of the wall with nail length ratio of 0.7 was higher when compared to the other nail length ratios. It can also be noted that acceleration response of the wall was highest with the nail inclination of 15°.

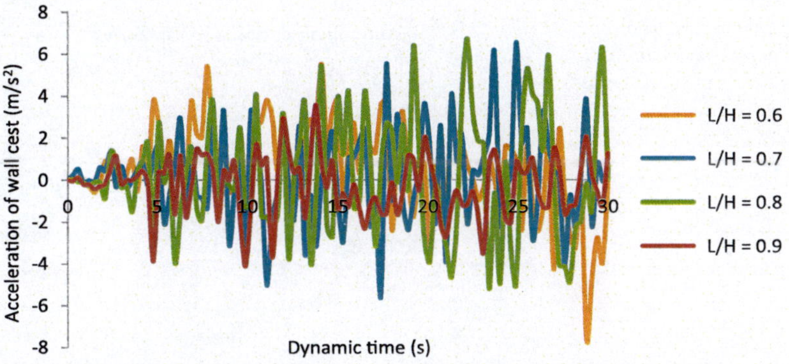


**Fig. 7** Acceleration of the wall crest for different nail length ratio under dynamic loading

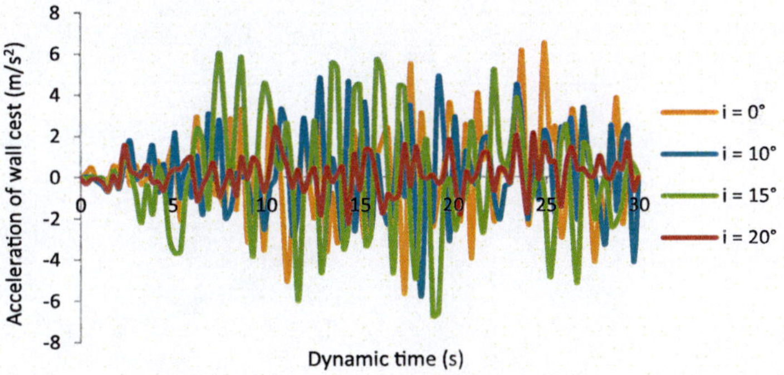


**Fig. 8** Acceleration of the wall crest for different nail inclination under dynamic loading

## *4.3 Axial Force in Nails*

Figures 9 and 10 show the maximum tensile nail force developed in the nails embedded at various depths. With the nails placed horizontally, irrespective of the nail length, the magnitude of the axial tensile force was similar for different nail length ratio which increased with the depth of the nail. This is because as the excavation proceeds, the size of soil mass to be retained increases resulting in larger stresses which give rise to higher axial force in nails.

For different nail inclination, the axial force developed in the nail also increases along the depth of the nail. The magnitude of the force was slightly higher for the inclination of 10° and 15° when compared to other inclination values.

Fig. 9 Maximum axial tensile nail force along the depth of the nail for different nail length ratio

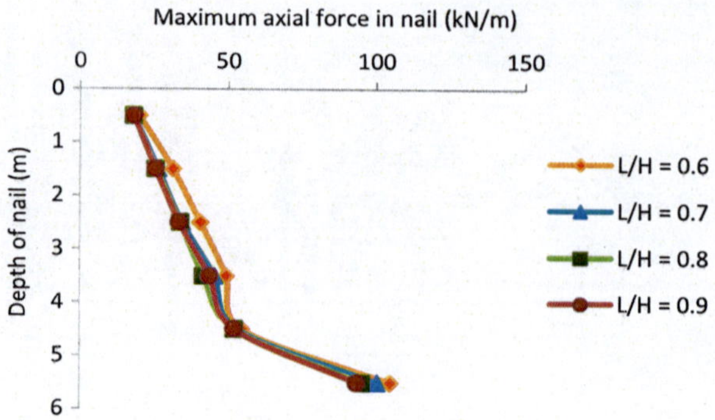


Fig. 10 Maximum axial tensile nail force along the depth of the nail for different nail inclination

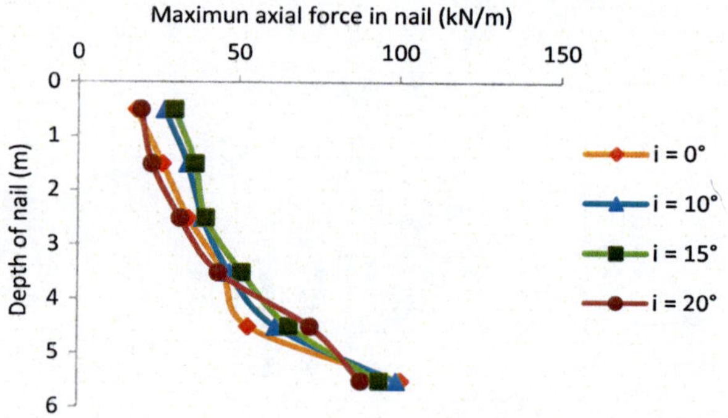


## 4.4 *Factor of Safety*

As the excavation proceeds, the factor of safety for stability of the soil mass decreases. The factor of safety was calculated for each stage of excavation and is plotted as a function of depth of construction of the wall in Figs. 11 and 12. It can be observed that the factor of safety decreases with the construction of the wall. The factor of safety was higher for the nail length ratio of 0.9, and it decreases with the decreasing nail length ratio.

The factor of safety value was higher when the nails were placed at an inclination to the horizontal as compared to horizontally placed nails. For inclined nails, the factor of safety was almost similar. However, the magnitude was slightly higher when the inclination was 15° to the horizontal.

## 4.5 *Effect of Nail Length Ratio*

The displacement of the wall is shown as a function of nail length ratio in Fig. 13. With the increase in nail length ratio, the deformation of the wall decreases. However,

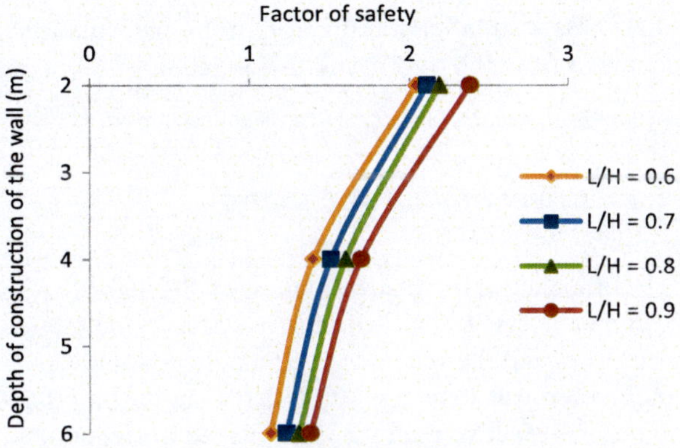


**Fig. 11** Factor of safety for construction depth of the wall for different nail length ratio

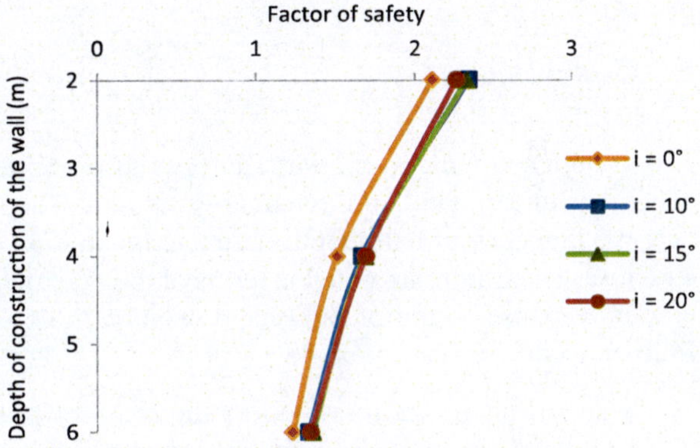


**Fig. 12** Factor of safety for construction depth of the wall for different nail inclination

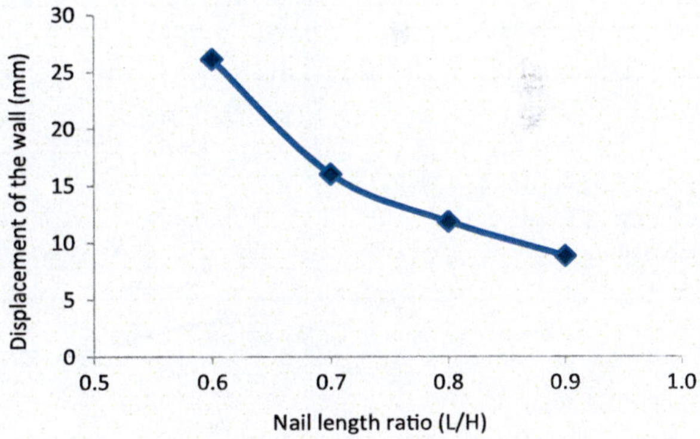


**Fig. 13** Displacement of the wall for different nail length ratio

there was a significant reduction in the deformation when the nail length ratio was changed from 0.6 to 0.7, and this percentage reduction is 38%.

### 4.6 *Effect of Nail Inclination*

The influence of inclination of nails on the deformation of the wall is shown in Fig. 14. It can be observed that with the increase in inclination of nails, the deformation first decreases and then increases. The deformation was found to be minimum when the inclination was between 10° and 15° suggesting that the inclined nails in this range are more effective than the nails placed horizontally.

## 5 Conclusions

A 6 m high soil nailed wall was analyzed using finite element method to study its behaviour under static and dynamic excitation. The study was conducted to evaluate the influence of nail length ratio and inclination of nails on the stability of the structure. The results are noted in terms of the maximum lateral displacement, acceleration response, factor of safety and axial nail force. Based on the evaluation, the following conclusions are made:

i. The nail length ratio influences the stability of the soil nailed wall. The nail length of 0.7 times the height of the wall was effective for the dynamic analysis.
ii. The displacement of the wall was minimum when the inclination of the nail was between 10° and 15° to the horizontal.
iii. The axial tensile force in nails increase with the depth of embedment of the nail.

**Fig. 14** Displacement of the wall for different nail inclination

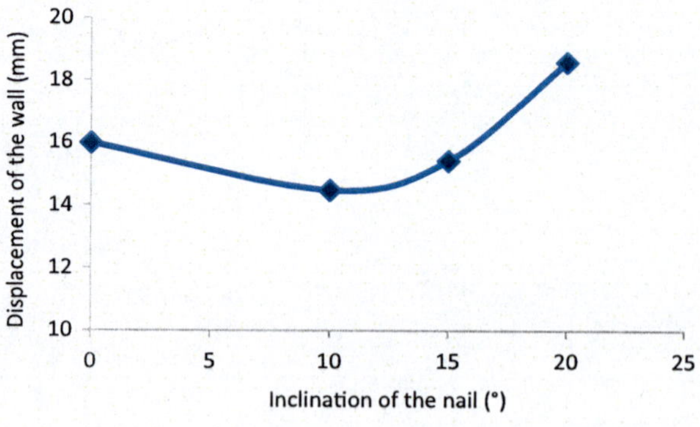

## References


1. Babu GS, Singh VP (2008) Numerical analysis of performance of soil nail walls in seismic conditions. ISET J Earthq Technol 45(1–2):31–40
2. El-Emam MM (2018) Experimental verification of current seismic analysis methods of reinforced soil walls. Soil Dyn Earthq Eng 113:241–255
3. Felio GY, Vucetic M, Hudson M, Barar O, Chapman R (1990) Performance of soil nailed walls during the October 17, 1989 Loma Prieta earthquake. In: Proceedings of the 43rd Canadian geotechnical conference, Quebec, Canada, vol 1, pp 165–173
4. Giri D, Sengupta A (2009) Dynamic behavior of small scale nailed soil slopes. Geotech Geolog Eng 27(6):687 (2009)
5. Juran I (1985) Reinforced soil systems-application in retaining structures. Geotech Eng 16(1):39–82
6. Lazarte CA, Robinson H, Gomez JE, Baxter A, Cadden A, Berg RR (2015) Geotechnical engineering circular no. 7 soil nail walls-reference manual. National Highway Institute (US)
7. Li J, Tham LG, Junaideen SM, Yue ZQ, Lee CF (2008) Loose fill slope stabilization with soil nails: full-scale test. J Geotech Geoenviron Eng 134(3):277–288
8. Shen CK, Herrmann LR, Romstad KM, Bang S, Kim YS, Denatale JS (1981) An in situ earth reinforcement lateral support system. NASA STI/Recon Technical Report
9. Singh VP, Babu GS (2010) 2D numerical simulations of soil nail walls. Geotech Geol Eng 28(4):299–309
10. Stocker MF, Korber GW, Gassler G, Gudehus G (1979) Soil nailing. In: Proceedings of the international conference on soil reinforcement, vol 2, pp 469–474
11. Vucetic M, Tufenkjian MR, Doroudian M (1993) Dynamic centrifuge testing of soil-nailed excavations. Geotech Test J 16(2):172–187

# Seismic Response of Buildings Resting on Pile-Raft Foundation in Soft Soil

**S. Vinoda Krishna and B. R. Jayalekshmi**

**Abstract** The combined pile-raft foundation is used for high-rise buildings on soft soil since it is an efficient and economical foundation system. The present study deals with evaluation of performance of 15 storey reinforced concrete structures built on piled-raft foundation resting on soft soil under earthquake loads. Finite element analysis of the three dimensional soil-pile-raft-building systems consisting of a combination of long and short piles have been carried out for transient response. The length ratio between the central piles and corner piles is varied from 0.33 to 3 such that the combinations consist of long piles placed at corners or at centre region while short piles are placed at centre or at corners. These systems are subjected to ground motions corresponding to El-Centro and Loma prieta earthquake. The seismic responses of the integrated systems vary with the change in the length of pile.

**Keywords** Pile-raft foundation · Finite element analysis · Pile length ratio

## 1 Introduction

Foundation systems are used to transfer the load safely into the soil ensuring overall stability and serviceability of the superstructures. A Combined Piled Raft Foundation (CPRF) is a composite structure which comprises of both piles and raft [1–3]. Piled raft foundations proved to be more economical in case of high rise buildings on clay and can provide safe bearing capacity and serviceability requirements [4]. Strategically placed piles help in resisting the excessive deflection of the raft. Usually load carried by the piles in piled-raft foundation is about 70% of the total load and rest is carried by the raft. In practice, the pile is rigidly connected to the raft and hence the raft over the piles is essentially a pile cap that holds the group of piles together [5]. However function of the raft in piled raft foundation is not only limited to act as a pile cap, butalso it transfers a considerable amount of loads to the top soil. Piles

S. V. Krishna (✉) · B. R. Jayalekshmi
Department of Civil Engineering, NITK, Surathkal, Mangalore 575025, India
e-mail: vinodkrishnasondi@gmail.com



T. G. Sitharam et al. (eds.), *Seismic Design and Performance*, Lecture Notes in Civil Engineering 120, https://doi.org/10.1007/978-981-33-4005-3_13

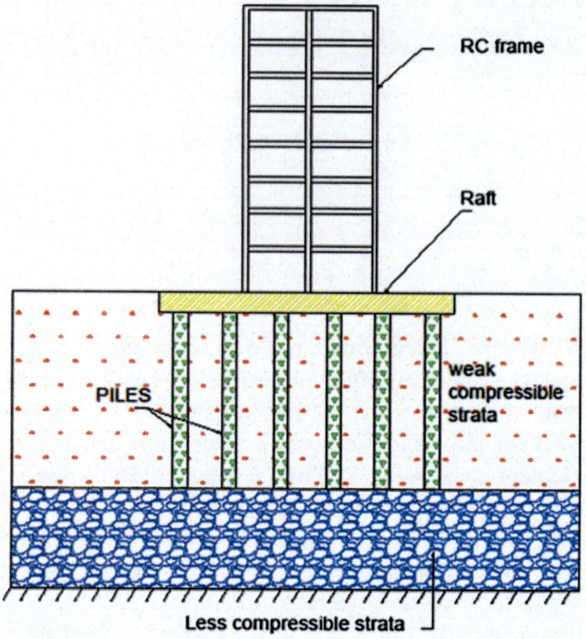


**Fig. 1** Typical sketch of piled raft-superstructure system

can be either end bearing or friction piles. Schematic representation of the piled raft foundation is shown in Fig. 1.

Dynamic soil-structure interaction (DSSI) is an important phenomenon to be considered in seismic design of high rise structures; several investigators have even called it the "missing link" in conventional design. Many research are being carried out on the effect of dynamic soil-structure interaction globally. Generally, dynamic analysis of structures supported by pile is done considering the base of the structure as fixed at the superstructure's base level. Various codes assumed that incorporation of DSSI effect may result in beneficial response of the structure; which is not true most of the times. Figure 1 shows a typical sketch of piled raft-superstructure system.

Some of the previous earthquakes showed that ignorance of the effect of DSSI during the design of structures was the actual reason behind the failure of the structures. In particular, forensic analysis of the Hansin expressway failure during the 1995 Kobe earthquake revealed the detrimental effect of DSSI. Difficulty in modelling and analysis of the building is also a reason for ignoring the effect of DSSI. It is seen from some experimental and numerical analysis results that SSI influences the structural behaviours of unbraced frames in structures subjected to an earthquake excitation. Soil types having shear-wave velocity less than 180 m/s is one of the most important parameters. As per NEHRP guidelines such soils are classified as soft soils. Soft soil has a significant impact on the seismic responses of moment-resisting frames because of its large deformation during earthquake. The present study focuses on the response variation of an RC frame on a piled raft foundation in soft soil by varying the length of the piles.

## 2 Methodology

### 2.1 Model Geometry

Building has 3 bays in either direction, length of each bay is 4 m and storey height is 3 m. The dimension of the raft is 14 m x 14 m × 0.75 m and diameter of the pile is 1 m. The size of soil considered is 110 m × 110 m with non-transmitting boundaries placed at lateral boundaries. Piled raft foundation consists of 16 piles of 1 m diameter having centre to centre distance between the piles as 4 m. These piles were classified as interior piles and exterior piles based on their position in plan view. The exterior piles are those which present at the edges of the raft. There were 12 exterior piles in the considered piled raft system. Four other piles which were present inside the exterior piles are referred as interior piles. Figure 2 shows plan view of piled raft foundation indicating the positions of interior and exterior piles.

Five models were considered for the study based on the lengths of interior and exterior piles. Lengths of the piles in different models are shown in Table 1.

Relative raft stiffness ($k_{rs}$) for the present model was calculated using expression 1. Relative stiffness depends on the dimension of raft and for the current model it was estimated as 0.0446.

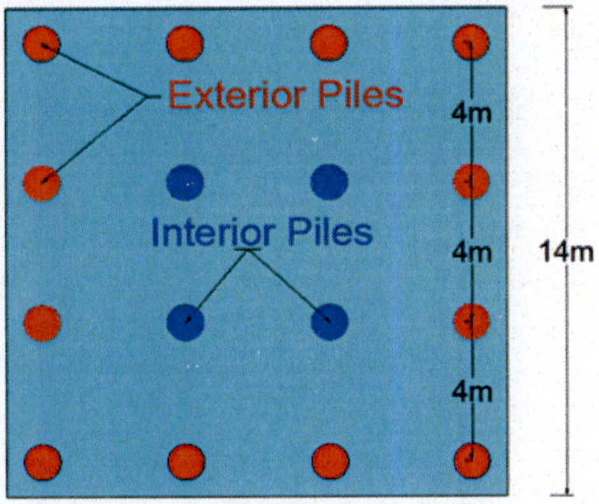


**Fig. 2** Plan view of piled raft foundation

**Table 1** Pile dimensions

| Model no | Pile configuration | Length of exterior pile l (m) | Length of interior piles (L) (m) | Ratio (l/L) |
|---|---|---|---|---|
| 1 | M1 | 30 | 10 | 3 |
| 2 | M2 | 20 | 10 | 2 |
| 3 | M3 | 10 | 10 | 1 |
| 4 | M4 | 10 | 20 | 0.5 |
| 5 | M5 | 10 | 30 | 0.33 |

$$k_{rs} = \frac{4E_r B_r t_r (1 - \mu^2)}{3\pi E_s L_r^4 (1 - \mu_r^2)} \tag{1}$$

where $E_r$ and $E_s$ are the Young's modulus of raft and soil, respectively, $L_r$ and $B_r$ are the raft length and breadth, respectively, $t_r$ is the raft thickness, $\mu_r$ and $\mu$ are the Poisson's ratio of raft and soil, respectively.

## 2.2 *Numerical Modelling*

The soil model, which has dimensions of 110 m x 110 m x 30 m was discretised using 8 noded linear brick reduced integration element i.e. SOLID 185. Sixteen numbers of fixed headed floating piles, were also modelled using SOLID 185 element. A conventional linear elastic model, available in the ANSYS finite element program was used to model the wave propagation. Viscous boundaries were assigned on lateral sides to eliminate the problem of reflection of the waves back into the model from boundaries. The base was restrained in all directions. Square raft was modelled using plate element. The span of the frames was taken as 4 m and the storey heights were taken as 3 m and it was modelled with two-node linear beam element i.e. BEAM 188. The cross section of the beams and columns were taken as 300 × 450 mm and 500 × 500 mm respectively. Figure 3 shows model with discretized finite-element mesh.

The material of the strucute, concrete is assumed to be homogenous, isotropic and elastic material. The elastic properties of materials are taken as per IS 456:2000 [6]. The modulus of elasticity (E) of concrete as per IS 456: 2000 is $5000 * \sqrt{fck}$.

Where, *fck* is the characteristic compressive strength of concrete at 28 days of curing.

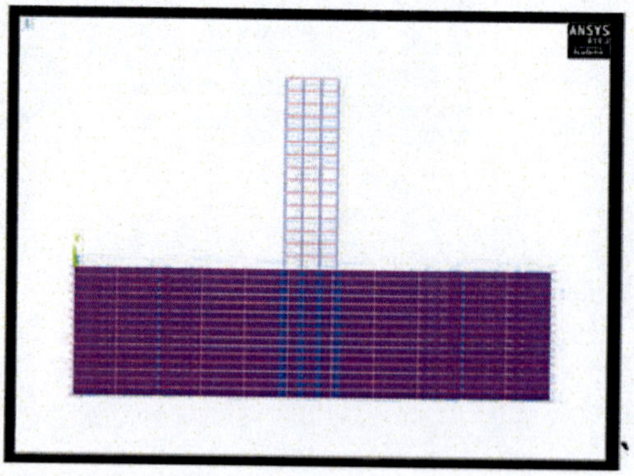


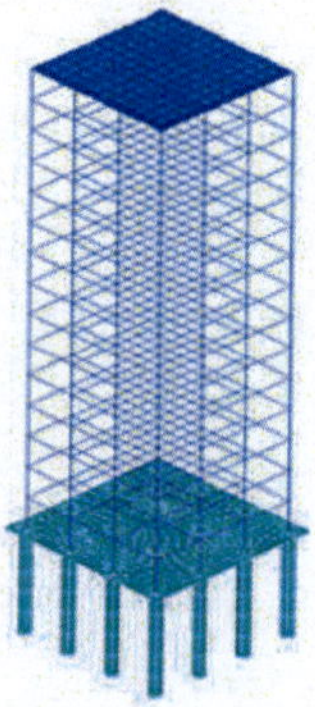

**Fig. 3** Finite element model of super structure—piled raft—soil system

**Table 2** Material properties

| Specification | Dimension | Modulus of elasticity (MPa) | Poisson's ratio | Density (ɣd) (kg/m$^3$) |
|---|---|---|---|---|
| Soil | 110 m*110 m*30 m (B*L*H) | 35 | 0.4 | 1800 |
| Raft | 14 m*14 m*1 m (B*L*H) | 25,000 | 0.15 | 2500 |
| Pile | Shown in Table 1 | 25,000 | 0.15 | 2500 |

The materials considered for modelling and dimensions of various components of model such as soil, pile and raft are shown in Table 2.

### 2.3 *Idealization of Viscous Boundary*

Viscous boundary was provided at the lateral boundaries of the soil block to simulate the effect of continuous soil. COMBIN14 element was used to model the viscous dampers at the boundaries of soil. Damping coefficient values were calculated as follows.

$$\mathrm{Vp} = \sqrt{\frac{K + \frac{4}{3}G}{\rho}} \tag{2}$$

$$\mathrm{Vs} = \sqrt{\frac{G}{\rho}} \tag{3}$$

where G is the Shear modulus, K is the Bulk modulus and ρ is the density of the soil

$$C_n = A_n\rho \cdot V_p \qquad C_{t1} = A_{t1}\rho \cdot Vs \qquad C_{t2} = A_{t2}\rho \cdot Vs \tag{4}$$

where, Vp is the Dilatational velocity, Vs is the Shear wave velocity and $A_n$, $A_t$ = Area controlling viscous dampers. $C_n$ is the damping coefficient in the direction of motion of P-waves and $C_{t1}$ and $C_{t2}$ are the damping coefficients perpendicular to the direction of P-waves. The subscripts n and t are the normal and tangential directions in the boundary.

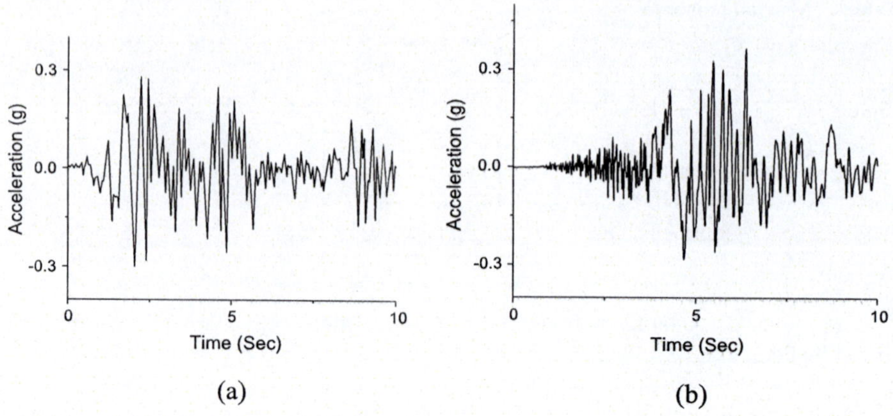


**Fig. 4** Input ground acceleration record of: **a** El centro and **b** Loma prieta earthquakes

### *2.4 Seismic Analysis*

Numerical seismic analysis was carried out on RC frame over piled raft foundation by subjecting soil- piled raft- superstructure system to input ground motions corresponding to El Centro and Loma Prieta earthquake which were having a PGA of 0.294 g and 0.367 g. System is analysed for initial 10 s of earthquake data as PGA occurs well before 10 s in both earthquake events. Acceleration time history of El Centro and Loma Prieta earthquake are shown in Fig. 4a, b. Responses of the RC frame over all the 5 pile configuration is studied for two ground motions. Time history of acceleration, time history of deflections at the roof level of RC frames as well as the natural frequencies of the different configurations of piles were analysed under the earthquake excitations.

## 3 Results and Discussions

Seismic responses of RC frames were studied for roof level of buildings. Results show that, seismic responses in RC frame slightly changes for different pile length combinations.

### *3.1 Natural Frequency*

Modal analysis was carried out for soil-pile-raft-structure system for various pile length combinations. Fundamental natural frequencies for different models is shown in Fig. 5. It is observed from the results that, M1 and M5 models have higher natural

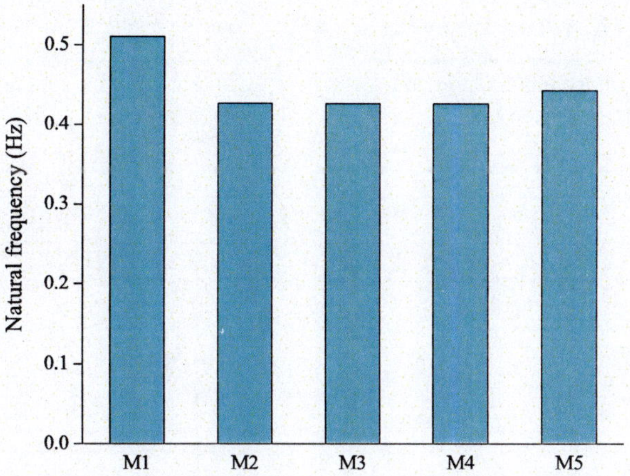


**Fig. 5** Fundamental natural frequency of models

frequency compared to models M2, M3 and M4. It is to be noted that depth of penetration of piles is more in M1 and M5 models. Also there is no much difference in the frequencies of M2, M3 and M4 models.

## *3.2 Acceleration at the Roof of RC Frame*

Acceleration response of the 15 storey RC frame at the roof for various pile configurations under Loma prieta earthquake motion and El Centro earthquake input motions are shown in the Figs. 6 and 7 respectively. It is observed from the results that, for both input ground motions RC frame resting on M1 piled raft configuration found to

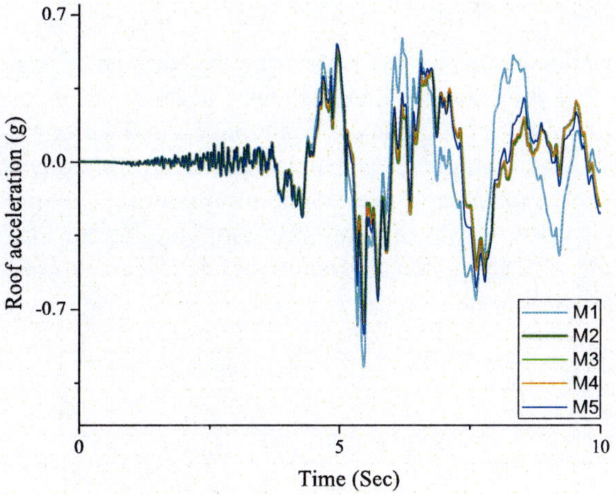


**Fig. 6** Time history of acceleration at roof for Loma Prieta input motion

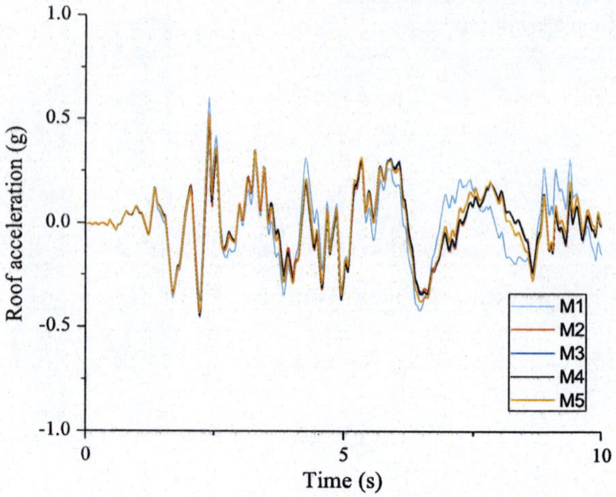


**Fig. 7** Time history of acceleration at roof for El centro input motion

accelerate at slightly higher magnitude at the roof. Acceleration response of the frame over all other pile configuration considered in this study closely matched with one another. Maximum acceleration occurred at roof of RC frame which is resting over M1 pile configuration, and minimum acceleration was observed in the frame which is present over M2 pile configuration for both input motions. Maximum acceleration at roof for M1 model is 27.6% more than the acceleration obtained for M2 model under El Centro input motion.For Loma prieta earthquake input motion maximum acceleration at roof for M1 model is 12.9% more than the acceleration obtained for M2 model.

### *3.3 Deflection at the Roof of RC Frame*

Deflection of the roof top during the earthquake excitation is shown in this section. Roof deflection for Loma prieta input acceleration is shown in Fig. 8. Roof deflection for El Centro earthquake input motion is shown in Fig. 9.

Maximum displacement is observed for M5 pile configuration when input acceleration reaches the peak for both earthquake input motions. Minimum roof displacement was observed in M1 pile configuration for El Centro input motion. For Loma prieta input motion, minimum roof displacement was observed in RC frame over M2 pile configuration.

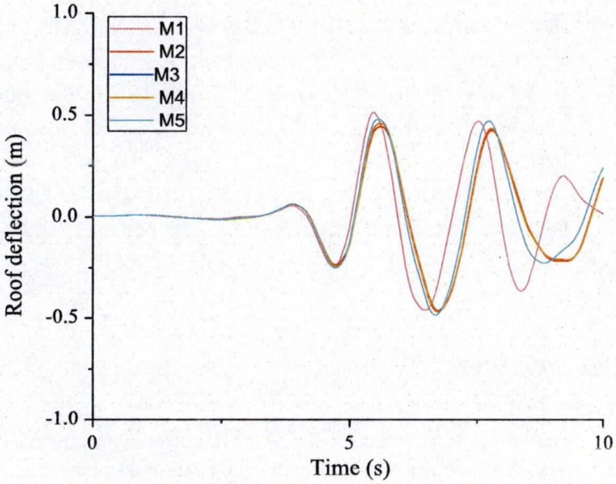


**Fig. 8** Time history of roof displacement for Loma Prieta input motion

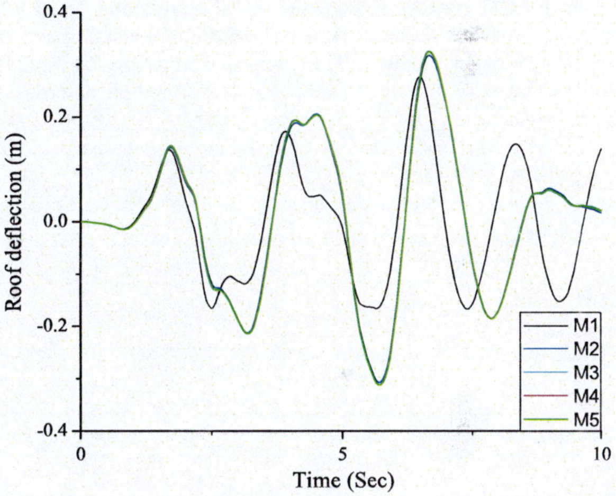


**Fig. 9** Time history of roof displacement for El centro input motion

## 4 Conclusion

Following conclusions are drawn from the three dimensional numerical analyses of the RC frame, which is carried out using ANSYS software to investigate the behaviour of a RC frame over piled raft foundation under time history dynamic loading conditions.

1. For input accelerations below 0.2 g response time history of the RC frame follows same path for all the combinations of pile length. When the input acceleration surpasses 0.3 g, response time history curve follows different paths for rest of the time, even though the input acceleration reduces.

2. Acceleration at roof of RC frame over M1 piled raft foundation configuration is maximum and is 27.6% more than the acceleration obtained for M2 model under El Centro input motion. For Loma Prieta earthquake input motion maximum acceleration at roof for M1 model is 12.9% more than the acceleration obtained for M2 model.
3. Roof deflection is higher for RC frame over M5 foundation compared to other types of piled raft considered in the present study.

## References

1. Katzenbach R, Moormann C (2001) Recommendations for the design and construction of Piled rafts. Proceedings of the fifteenth international conference on soil mechanics and geotechnical engineering vol 2. Istanbul, Turkey
2. Katzenbach R, Arslan U, Moormann C (2004) Piled raft foundation projects in Germany. Thomas Telford, Design Applications of Raft Foundations, pp 323–392
3. Katzenbach R, Schmitt A, Turek J (2005) Assessing settlement of high-rise structures by 3D simulations. Comput-Aided Civil Infrastruct Eng 20(3):221–229
4. Elwakil AZ, Azzam WR (2016) Experimental and numerical study of piled raft system. Alexandria Eng J 55(1):547–560
5. Kurian NP (2005) Design of foundation systems: principles and practices (3rd ed) Narosa publishing house
6. IS 456 (2000) Plain and reinforced concrete—code of practice, bureau of Indian standards, New Delhi

# State of the Art on Combined Pile Raft Foundation

**Shailja Gupta, V. A. Sawant, and P. K. Gupta**

**Abstract** In recent years, the demand of high-rise buildings has increased significantly. For high-rise buildings, pile raft foundation is proved as the most economic and efficient foundation. But in seismically active regions for better safety and durability, it is significant to have a knowledge of dynamic characteristics of piled raft. With the use of pile raft foundation, risk of sudden collapse of these buildings is also controlled. In addition to this, total and differential settlement is also reduced in comparison to unpiled raft and conventional group foundation. To investigate the effect of piled raft under dynamic loading, the relevant available literature is critically reviewed. From the review, it is concluded that pile head fixity has a considerable influence on load sharing mechanism between individual components and less influence on settlement.

**Keywords** Pile · Raft · Piled raft · Load sharing · Settlement

## 1 Introduction

For high-rise buildings, combined pile raft foundation (CPRF) has gained popularity as one of the most efficient and rational foundations from the last two decades. Also, in comparison to unpiled raft and conventional group foundation, it is recognized as the most economic foundation. In CPRF, the use of piles below the raft has resulted into reduction of total and differential settlements, i.e. piles acts as settlement reducer. Also, in CPRF, the ultimate load carrying capacity of both the foundation components (pile and raft) is utilized hence; it leads to considerable savings without compromising the safety and performance of the foundation system. In comparison

S. Gupta (✉) · V. A. Sawant · P. K. Gupta
Department of Civil Engineering, IIT Roorkee, Roorkee 247667, India
e-mail: sgupta9@ce.iitr.ac.in

V. A. Sawant
e-mail: vishwas.sawant@ce.iitr.ac.in

P. K. Gupta
e-mail: pramod.gupta@ce.iitr.ac.in

 
T. G. Sitharam et al. (eds.), *Seismic Design and Performance*, Lecture Notes in Civil Engineering 120, https://doi.org/10.1007/978-981-33-4005-3_14

to conventional group pile foundation, less number of piles are installed below the raft as raft will also share the load. The major advantage of CPRF is that by the use of raft with piles, the risk of sudden catastrophic collapse of the entire building can be controlled. Previously, in highly seismic prone areas, the existence of pile foundations is ignored in pile raft foundation and the foundation system was designed as raft alone foundation. But in these areas for the safety and the better performance of high-rise buildings, dynamic characteristics of pile raft foundation plays a significant role. Hence, in this paper, an attempt is made to study the behaviour of pile raft foundation under dynamic loading. Relevant literature on dynamic response of CPRF is classified in different groups based on the methodology.

## 2 Literature Review

In highly seismic prone areas, the behaviour of CPRF under dynamic loading is important for safety and performance of high-rise buildings. Hence, many researchers have investigated the dynamic characteristics of CPRF. Number of available studies are broadly divided into experimental, numerical and analytical investigations. If required, further sub-division is applied to have proper insight into literature.

### *2.1 Experimental Studies*

For the dynamic characteristics of CPRF, some researchers have performed shake table tests and centrifuge tests. Accordingly, experimental studies are divided into 1-g model study, centrifuge study and case study.

**1-g model study.**

Shake table tests are widely used to determine the dynamic characteristics of a particular domain in many fields. Shake table test is 1-g test. Matsumoto et al. [1] conducted shake table tests at 1-g field on model CPRFs. Frequency of input acceleration and height of centre of gravity of superstructure were two model parameters considered in the study. Toyoura sand was used. Corresponding to the height of centre of gravity of superstructure as 49.3 mm, 123.4 mm and 187.5 mm, the vertical load shared by piles is 71.9%, 71.2% and 78.8%, respectively. At input frequency greater than 30 Hz, the response of raft was significant. As the height of centre of gravity increased, horizontal load was increased and pile resistance was decreased slightly. Corresponding to different height of centre of gravity of superstructure, the proportion of horizontal load carried by piles lies in the range of 30 – 80%. With increase in height of the centre of gravity, bending moment is increased. However, the resonant frequency and horizontal load proportion of piles decreased.

Baziar et al. [2] conducted shake table tests on two different superstructures with different frequencies which was rested on CPRF in a sandy soil. The results of shake table tests showed that performance of pile was largely influenced by frequency of superstructures. In comparison to 4 Hz frequency pile, bending moments located under a 7 Hz superstructure frequency were larger. It can be inferred that shake table tests cannot simulate the correct prototype stress conditions.

**Centrifuge model study.**

Centrifuge tests are better in simulating the stress conditions. To observe the dynamic characteristics of CPRF, few researchers have performed centrifuge tests. Therefore, in this section, relevant papers are discussed briefly.

Horikoshi et al. [3] performed shake table tests on CPRF models by using a centrifuge. In the test, two CPRF models with different pile head connection conditions as rigid, hinged connection and free—standing pile group model were used. Toyoura sand was used in the study. Centrifugal acceleration of 50 g was applied. Also, the total vertical load was applied as 2298 N. Corresponding to this load and applied centrifugal acceleration, piles carried 40% and 45% of the total load for the rigid and hinged connection respectively. For rigid and hinged connection model, the peak values of acceleration were obtained as 135 $m/s^2$ and 160 $m/s^2$, respectively. Thereafter, acceleration decreased with time and become stable in each model. As compared to rigid connection model, final settlement of hinged connection model was 20% larger. In case of horizontal load, piles carried higher load unlike the case of vertical load.

Kang et al. [4] performed centrifuge tests to examine the seismic response of CPRF. In this study, two types of piles (flexible and stiff) were used. Normally, consolidated soft clay was used. Centrifugal acceleration of 50 g was used in all the experiments. Three scaled earthquakes comprising of small (0.022 g), medium (0.052 g) and large earthquake (0.13 g) were fired in every two cycles. In case of stiff piles, the resonance period is lower, whereas in flexible piles, the resonance period is not too different from the structure with the soil around it. Maximum bending moments developed near the pile head. As the stiff piles have higher flexural rigidity, bending moments developed in stiff piles are also higher than that of flexible piles.

Zhang et al. [5] investigated the behaviour of CPRF in soft kaolin clay using seismic centrifuge model tests. Various small-scale CPRF models were designed for pile groups ranging from $2 \times 1$ to $4 \times 3$. The far-field earthquake condition was represented by the excited long duration ground motion of about 200 s. The bending moment of pile was significantly affected by the size of pile group. The shadowing effect was dominant for $4 \times 3$ pile group resulting in different maximum bending moments between outer and inner piles.

Baziar et al. [2] performed centrifuge tests on two different superstructures with different frequencies which was rested on CPRF in a sandy soil. Centrifugal acceleration of 50 g was applied. The test results showed that amplification and horizontal displacement of the superstructures were dependent on the system frequency at the lower level base acceleration (0.14 g). In comparison to the tests at 1 Hz frequency, the internal forces developed in piles at 2 Hz frequency were more than 1.5 times.

Sahraein et al. [6] carried out centrifuge tests to examine the response of oil tank rested on CPRF in liquefiable saturated and non-liquefiable dry sand. In this study, two types of foundation were modelled as slab foundation and CPRF. Centrifugal acceleration of 50 g was applied. In case of dry sand, piles beared most of the tank load for CPRF. Settlement of CPRF was much smaller than slab foundation. As compared to slab foundations, CPRF reduced tank rotation effectively. The liquefaction significantly reduced the stiffness of saturated sand due to which the input motion was attenuated in the short period range, whereas the natural period of ground was increased in the long period range. Due to the reduction of effective stress caused by liquefaction, pile resistance of CPRF decreased significantly and bearing load of raft increased, thereby increasing the effective pore water pressure rapidly.

**Case study.**

Few researchers done case study to determine the seismic behaviour of CPRF.

Yamashita et al. [7] investigated the seismic characteristics of CPRF supporting a 12-storey building in Tokyo. To counter the effects of liquefaction in layered soil, a CPRF with grid-form cement deep mixing walls was used. Between the raft and the bottom floor of the building, a base isolation system was placed. Postconstruction settlement of CPRF was found to be 14.3 mm. Just before the Tohoku earthquake, the settlement increased slightly to 17.3 mm and the load sharing ratio by the piles was 0.669. Near the end of the event (Tohoku earthquake), the load sharing ratio of piles decreased to 0.66 and increased to 0.667 after four days of the event.

Yamashita et al. [8] investigated the seismic performance of a CPRF supporting a 7-storey building on soft ground in Tokyo. As a countermeasure against liquefaction, grid-form cement deep mixing walls were employed. Before the event, the measured settlements lied in the range of 15–29 mm. After the event, the measures settlements were 14–33 mm. Just before the earthquake, the load sharing ratio of piles was 0.72. After the earthquake, it increased to 0.76 and 0.81 after three years of event. Lateral stiffness of raft was larger in comparison to piles, and hence, most of the horizontal load was carried by the raft.

## 2.2 *Numerical Studies*

Some researchers have performed numerical studies for examining the dynamic characteristics of CPRF. Further, numerical studies are classified into case study, finite element modelling and finite difference analysis. Accordingly, the following relevant papers are discussed briefly in this section.

**Finite element modelling.**

Finite element is an accurate method to examine the behaviour of CPRF under any load combination. Hence, few researchers have adopted the same to determine the dynamic characteristics of pile raft.

Eslami et al. [9] examined the seismic response of CPRFs and pile groups. For finite element modelling, ABAQUS software was used. To model the radiational damping, Rayleigh's method was used. $\alpha$ and $\beta$ coefficients of Rayleigh damping were chosen as 5%. A nonlinear elasto-plastic model with a Mohr–Coulomb yield criterion was adopted. In the software, three different models were analysed using the mesh dimensions of 0.25 m × 0.25 m × 0.25 m for regular cubic elements having eight integration points. The first one was a simple raft foundation. The second and third models were free standing pile group and CPRF, respectively. Toyoura sand was used in this study. For seismic loading on CPRF, El Centro time history was chosen. Maximum acceleration in CPRF was obtained as 9.44 m/s$^2$. Also, the settlement of CPRF and pile group was obtained as 4.5 cm and 10.3 cm, respectively. As compared to pile group acceleration, response of CPRF was 26% less. Maximum bending moment was reduced by 15% for CPRF.

Kumar et al. [10] investigated the effect of pile head fixity on CPRF by using PLAXIS3D. Response of CPRF is determined under the available earthquake loadings, like 2001 Bhuj, 1989 Loma Prieta and 1995 Kobe. The two layered soil was modelled using hardening soil model for top 1 m of Toyoura sand and linear elastic model for remaining 1 m of brick base. Raft was modelled by plate element and piles were modelled by ten-noded tetrahedral embedded pile element. In case of vertical load, pile head fixity has little influence on settlement. Raft shared 30–54% of load in hinged and rigid connection model, respectively. Initially, piles carried majority of the vertical load. In case of horizontal load, raft carried 23–31% of load in rigid and hinged connection model, respectively, which further increased on increasing the load. Earthquake induced load was replaced by equivalent static load, and in the study, it was referred as pseudo-static load. As pseudo-static load increases, the load sharing by raft decreases. In rigid connection model, raft shared 49, 42 and 13% of load, whereas in hinged connection, it shared 82, 76 and 14% of load corresponding to 2001 Bhuj, 1989 Loma Prieta and 1995 Kobe earthquake, respectively.

Ladhane and Sawant [11] analysed the dynamic response of pile groups using 3D finite element modelling. Soil and piles have been discretized into 3D isoparametric continuum elements. Kelvin elements were used at the transmitting boundary. Parametric study was carried out to examine the effect of pile spacing, number of piles, arrangement of pile and soil modulus on the response of pile group. To model the soil, eight-noded continuum element was used. 20-noded elements for modelling of piles and pile cap. For maintaining continuity between these two elements, 12-noded and 9-noded elements were formulated. Transmitting boundary was considered at 10$D$, and far-field boundary at 14$D$. Associated flow rule was considered. von Mises yield criteria was used to model the soil. After successful validation, parametric study was carried out. Three configurations of piles were considered, namely series, parallel and square arrangement. As piles spacing increased, peak amplitude decreases. In comparison to linear response, the fundamental frequencies were reduced in nonlinear response. With decrease in $L/D$ ratio, frequencies corresponding to peaks decrease. As the soil modulus decreases, fundamental frequencies corresponding to peaks decrease and maximum amplitude increases. With decrease in diameter, maximum amplitude increases and higher response was obtained. As compared to pile in series, arrangement piles in parallel arrangement offered more resistance.

Kumar et al. [12] examined the behaviour of CPRF under pseudo-static and dynamic load with the use of PLAXIS3D. The soil was modelled by the use of ten-noded tetrahedral elements. To model the Toyoura sand, Mohr–Coulomb constitutive model was used in the study. Square raft was modelled as a plate element. Piles were modelled with the use of embedded pile element. The loading of 6862.5 kN was applied. Piles provided total vertical resistance of 2760 kN. CPRF coefficient was obtained as 0.47. Maximum vertical settlement of about 2.1 cm was obtained. Now the pseudo-static load of 1979 El Centro, 1989 Loma Prieta, 2001 Bhuj and 2011 Sikkim earthquake was applied simultaneously on the model after successful validation. Piles shared 62–76% of pseudo-static load under these loading conditions. Unlike vertical loading conditions, piles shared maximum percentage of load. Corresponding to Bhuj 2001, Sikkim 2011, Loma Prieta 1989 and El Centro 1979, pseudo-static loading maximum horizontal displacement of piles was 0.7, 2, 3.8 and 4.5% of pile diameter, respectively. Damping was modelled with Rayleigh damping option. The value of damping was chosen as 5%. Fundamental frequency and time period of CPRF were taken as 1.27 Hz and 0.78 s, respectively. At the top of CPRF, maximum acceleration was obtained as 2.8 m/s$^2$ when sinusoidal acceleration of magnitude 1 m/s$^2$ and frequency of 1 Hz were applied at the base of soil model. Corresponding to Bhuj 2001, Sikkim 2011, Loma Prieta 1989 and El Centro 1979 earthquake input motions, the maximum acceleration obtained at the raft was 0.24, 0.78, 0.79 and 1.04 g, respectively. For 1989 Loma Prieta earthquake, motion resonance condition occurred as the predominant period of the earthquake was close to the fundamental frequency of CPRF. Corresponding to different earthquake motions, the maximum horizontal displacement was obtained as 5.5, 2.8, 13.3 and 3.2% of pile diameter.

Sawant and Ladhane [13] performed static nonlinear analysis of pile group subjected to lateral load with six different configurations by developing finite element based program. The response of pile group using von Mises, Drucker–Prager and Mohr–Coulomb yield criteria was analysed. Soil and pile media have been discretized into 3D isoparametric continuum elements. Interface elements were introduced at the pile–soil interface to simulate stress transfer between soil and pile under lateral load. Associative flow rule was assumed. The effect of pile spacing and pile group configuration on the nonlinear response of pile groups was parametrically examined upon successful validation. As compared to the parallel configurations, piles in series configurations had higher displacements. Higher displacements were predicted for 2 $\times$ 2 pile group against 3 $\times$ 3 pile group. With increase of piles in pile group, effect of nonlinearity was increased.

Kumar and Choudhury [14] examined the load sharing mechanism of CPRF for different connection conditions under seismic loads by using PLAXIS3D. Two different connection conditions, i.e. fully rigid and fully hinged connection, were used in numerical simulation. Toyoura sand was used in the study. Soil model was divided into two layers. Top 1 m of Toyoura sand was modelled by hardening soil model and remaining 1 m of base brick was modelled by linear elastic model. Soil model was developed by ten-noded tetrahedral elements. Square raft was modelled by using plate element. In case of rigid and hinged connection model, maximum vertical settlement was obtained as 0.043–0.037% of raft width, respectively. Piles shared 46–70% of vertical load in case of rigid and hinged connection, respectively. After successful validation, horizontal loads were applied on CPRF. Corresponding to rigid and hinged connection, piles shared 77–69% of horizontal load, respectively. Thereafter, pseudo-static load of Bhuj 2001, Loma Prieta 1989 and Kobe 1995 was applied simultaneously at the level of raft. As compared to rigid connection, raft mobilizes its ultimate resistance at a faster rate in hinged connection model. After the fully mobilization of raft resistance, piles carried additional lateral loads. Major part of bending moment was developed near the pile head.

Zhang et al. [5] performed finite element analysis to validate the results of centrifuge tests by using ABAQUS software. Piles were modelled as beam elements embedded within solid elements. Perfect bonding among elements was ensured at the soil–pile interface. Hyperbolic hysteresis soil constitutive model was used in the study. For each pile group configuration, numerical simulations satisfactorily replicated the maximum bending moment profiles.

**Case study.**

Few researchers have performed case study on the pile raft foundation supporting superstructure like oil tank. Hence, the relevant paper is discussed briefly in this section.

Kumar and Choudhury [15] performed dynamic soil–structure interaction analysis for the foundation of an oil tank supported on weak soil at Kafza site in Iraq by using Flac3D. Local site comprises of layered soil. Soil was modelled by the use of eight-noded brick elements. Based on Mohr–Coulomb failure criterion, a conventional constitutive soil model was chosen. Lateral dimensions of the soil were kept as four times the radius of the pile cap from its periphery. Soil model was created using radial brick option. Piles were modelled as structural elements with 6 degrees of freedom per node. Pile cap was modelled with shell elements. After successful validation, the same model was adopted for further analyses. With increase in load and change in soil, stratification bending moment decreased significantly. Differential settlement of about 7 cm was obtained. At the pile head, the axial load varies with maximum 697 kN and minimum 478 kN. Duration and peak ground acceleration of developed earthquake motion was 54 s and 0.15 g, respectively. At the end of 54 s long input earthquake motion, pore water pressure ratio of 0.85 was obtained at 5 m depth of soil. At depths of 10 and 15 m, the pore water pressure ratios were obtained as 0.65 and 0.55, respectively. After the dynamic analysis, 6.4 mm differential settlement was obtained.

**Finite difference analysis.**

Some researchers have analysed the seismic behaviour of piled raft by finite difference analysis. Therefore, the relevant paper is discussed in brief.

Kumar and Choudhury [15] performed dynamic soil–structure interaction analysis for the foundation of an oil tank rested on weak soil at Kafza site in Iraq by using Flac3d. Local site comprises of layered soil. Soil was modelled by the use of eight-noded brick elements. Based on Mohr–Coulomb failure criterion, a conventional constitutive soil model was chosen. Lateral dimensions of the soil were kept as four times the radius of the pile cap from its periphery. Soil model was created using radial brick option. Piles were modelled as structural elements with 6 degrees of freedom per node. Pile cap was modelled with shell elements. After successful validation, the same model was adopted for further analyses. With increase in load and change in soil stratification, bending moment decreased significantly. Differential settlement of about 7 cm was obtained. At the pile head, the axial load varies with maximum 697 kN and minimum 478 kN. Duration and peak ground acceleration of developed earthquake motion was 54 s and 0.15 g, respectively. At the end of 54 s long input earthquake motion, pore water pressure ratio of 0.85 was obtained at 5 m depth of soil. At depths of 10 and 15 m, the pore water pressure ratios were obtained as 0.65 and 0.55, respectively. After the dynamic analysis, 6.4 mm differential settlement was obtained.

### *2.3 Analytical Studies*

Some researchers have done analytical studies to examine the behaviour of piled raft under dynamic loading.

Liu et al. [16] developed a simplified solution to determine the dynamic response of pile group partially embedded in a layered saturated soil which was subjected to horizontal harmonic loading. To derive the dynamic pile–soil–pile interaction factor, Winkler's model was used. As the length of pile increases, maximum impedance decreases significantly. On decreasing pile spacing, pile groups become more flexible. On increasing the modulus of upper soil, impedance of pile groups increases.

Roy et al. [17] developed an exact analytical solution to determine the natural frequency of CPRF resting on Winkler foundation which was subjected to harmonic excitation. After successful validation, parametric study was carried out. Piles were assumed to behave as Euler's beam element supporting on linear Winkler foundation. As the length of pile increases, natural radial frequency of the system decreases. Further, with increase in pile diameter, natural radial frequency of the system decreases till optimum diameter, and thereafter, it increases. Now the natural frequency was determined including the superstructure, and hence, CPRF system was replaced by an equivalent mass spring system. After successful validation of developed methodology, parametric study was carried out by considering three-storied building supported on $4 \times 4$ piled raft foundation. With the inclusion of superstructure stiffness flexibility of the system increases and hence natural frequency of the system decreases. With increase in pile diameter from 0.5 to 1.5 m, percentage difference between natural radial frequency reduced from 12 to 28%. On increasing the number of stories of superstructure, natural radial frequency of the system decreases.

## 3 Critical Appraisal

After reviewing the relevant papers, it is concluded that several works have been performed on determining the load sharing ratio and settlement of pile raft. Several research papers are available to study the effect of pile head fixity, pile stiffness and its geometrical properties on the response of pile raft. Based on relevant case studies, other applications of CPRF supporting various superstructures need to be explored apart from conventional uses of CPRF. Experimental and numerical review has suggested that the effect of variation of both geometrical and geotechnical parameters on the behaviour of CPRF can be studied. Effect of superstructure on the dynamic characteristics of CPRF can be investigated. Corresponding to different centrifugal accelerations, stress conditions should be simulated and compared. Comprehensive analysis (experimental, numerical and analytical) of CPRF behaviour under identical conditions may be done. Lastly, cost-effective analysis to determine the economic viability of CPRF can also be performed.

## 4 Conclusions

The development of CPRF has involved considerable experimental, numerical and analytical studies on its dynamic characteristics which influences the safety and performance of high-rise buildings especially in highly seismic prone areas. Centrifuge model testing has superior results over 1-g model testing while simulating the prototype stress conditions. Fully rigid and fully hinged connections of pile raft can be modelled both experimentally and numerically; however, semi-rigid and semi-hinged connections can only be modelled by experimental means. Effect of geotechnical and geometrical parameters of CPRF for different scaled earthquake conditions in liquefiable, non-liquefiable and stratified soil still needs to be explored. Collaborative parametric response on various dynamic parameters like frequency, impedance and acceleration can be done using efficient finite element techniques. Hardening soil model, von Mises, Drucker–Prager and Mohr–Coulomb yield criteria have shown considerable efficiency in modelling the soil. Conventional Winkler model approach is generally used for analytical study of CPRF subjected to dynamic loading. The future researches may be focused on evolving a simplified and efficient approach for analysing the dynamic response of CPRF.

## References

1. Matsumoto T, Fukumura K, Horikoshi K, Oki A (2004) Shaking table tests on model piled rafts in sand considering influence of superstructure. Int J Phys Model Geotech 3:21–38
2. Baziar MH, Rafiee F, Azizkandi AS, Lee CJ (2018) Effect of super-structure frequency on the seismic behavior of pile-raft foundation using physical modelling. Soil Dyn Earthq Eng

104:196–209
3. Horikoshi K, Matsumoto T, Hashizume Y, Watanabe T (2003) Performance of piled raft foundations subjected to dynamic loading. Int J Phys Model Geotech 2:51–62
4. Kang MA, Banerjee S, Lee FH, Xie HP (2012) Dynamic soil-pile-raft interaction in normally consolidated soft clay during earthquakes. J Earthq Tsunami 6(3):1250031:1–12
5. Zhang L, Goh SH, Liu H (2017) Seismic response of pile-raft-clay system subjected to a long duration earthquake: centrifuge tests and finite element analysis. Soil Dyn Earthq Eng 92:488–502
6. Sahraeian SMS, Takemura J, Seki S (2018) An investigation about seismic behavior of piled raft foundation for oil storage tanks using centrifuge modelling. Soil Dyn Earthq Eng 104:210–227
7. Yamashita K, Hamada J, Onimaru S, Higashino M (2012) Seismic behavior of piled raft with ground improvement supporting a base isolated building on soft ground in Tokyo. Soils Found 52(5):1000–1015
8. Yamashita K, Hamada J, Tanikawa T (2016) Static and seismic performance of a friction piled raft combined with grid-form deep mixing walls in soft ground. Soils Found 56(3):559–573
9. Eslami MM, Aminikhah A, Ahmadi MM (2011) A comparative study on pile group and piled raft foundations (PRF) behavior under seismic loading. Comput Methods Civil Eng 2(2):185–199
10. Kumar A, Choudhury D, Katzenbach R (2015) Behaviour of combined pile-raft foundation (CPRF) under static and pseudo-static conditions using PLAXIS3D. In: 6th international proceedings on earthquake geotechnical engineering. New Zealand
11. Ladhane KB, Sawant VA (2015) Effect of pile group configurations on nonlinear dynamic response. Int J Geomech–ASCE 16(1):04015013:1–14
12. Kumar A, Choudhury D, Katzenbach R (2016) Effect of earthquake on combined pile–raft foundation. Int J Geomech–ASCE 16(5):04016013:1–16
13. Sawant VA, Ladhane KB (2016) Nonlinear FEA of pile group subjected to lateral load. Open J Civil Eng 6:19–30
14. Kumar A, Choudhury D (2017) Load sharing mechanism of combined pile-raft foundation (CPRF) under seismic loads. Geotech Eng J SEAGS AGSSEA 48(3):95–101
15. Kumar A, Choudhury D (2016) DSSI analysis of pile foundations for an oil tank in Iraq. In: proceedings of the institute of civil engineers. ICE Publishing. pp 129–138
16. Liu Y, Wang X, Zhang M (2014) Lateral vibration of pile groups partially embedded in layered saturated soils. Int J Geomech–ASCE 15(4):04041063:1–11
17. Roy J, Kumar A, Choudhury D (2018) Natural frequencies of piled raft foundation including superstructure effect. Soil Dyn Earthq Eng 112:69–75
18. Zhang L (2009) Nonlinear analysis of laterally loaded rigid piles in cohesionless soil. Comput Geotech 36:718–724

# Analytical Study on Dynamic Coupling Interaction Effects Between Adjacent Building Structures

**S. V. Chandramouli and N. Munirudrappa**

**Abstract** The objective of the present study is to investigate the dynamic coupling interaction effects between adjacent building structures and the underlying soil when subjected to artificial seismic excitation. In this paper, the dynamic structure–soil–structure interaction (SSSI) between adjacent buildings idealized as a 2D-discrete model and underlying or surrounding soil is coupled by a rotational interaction spring and the non-stationary Kanai–Tajimi accelerogram is employed as artificial seismic excitation input. Based on the numerical investigation, the results showed that constructing an adjacent building taller to an existing building generally increases the response, thereby amplifies the seismic risk although in the taller adjacent building comparative seismic risk reduction in noted based on its own response.

**Keywords** Structure–soil–structure interaction · Horizontal ground motion · 2D-discrete model system · Non-stationary Kanai–Tajimi accelerogram

## 1 Introduction

The world is urbanizing at a greater rate than in history. The urban cities are complex, bringing a host of transportation, infrastructure, health and sustainability issues before and after a major disaster such as an earthquake. During an earthquake, structures interact with the surrounding soil beneath their foundations; this problem is generally known as soil–foundation–structure interaction (SFSI) or Structure–soil interaction, or as it is more commonly known, soil–structure interaction (SSI). Accounting for soil flexibility in the analysis of structure does not generally affect the response significantly. In general, for stiff, heavy buildings (and other structures)

S. V. Chandramouli (✉)
Department of Civil Engineering, PES University, Bangalore 560085, India
e-mail: chandramouli@pes.edu

N. Munirudrappa
Department of Civil Engineering, DSCE, Bangalore 560078, India
e-mail: munirudrappa@gmail.com

 
T. G. Sitharam et al. (eds.), *Seismic Design and Performance*, Lecture Notes in Civil Engineering 120, https://doi.org/10.1007/978-981-33-4005-3_15

soil flexibility can change the response due to SSI. SSI is an important research topic in earthquake engineering that has attracted extensive attention for several decades.

An extension of SSI is an interaction of adjacent structures. The present-day cities' existence of a high raised and high density of buildings inevitably results in the possibility of seismic interaction of adjacent buildings through the underlying soil. During an earthquake, the presence of one structure may affect the dynamic response of another, resulting in structure–soil–structure interaction (SSSI). The SSSI is important for the performance assessment of buildings in dense urban areas and is receiving some considerable attention in recent years. Although, the limited studies on SSSI are well-documented field and/or experimental data available but still no related provisions are made in any standard code of practice.

A recent review of the SSSI problem can be found in Menglin et al. [1]. Due to the difficulties involved in modelling the multiple interactions, previous studies of the dynamic SSSI phenomena have been explored analytically and also numerically based either on finite element (FE) or boundary element (BE) methods or on coupled FE/BE procedure and provided qualitative evidence of the dynamic interaction effects between adjacent structures [2–9]. Based on these studies on SSSI, the dynamic interaction between adjacent buildings with consideration of the underlying or surrounding soil influences each other through the soil during earthquakes and exhibits dynamic behaviours different from those of isolated building.

In this paper, a numerical investigation is carried out on adjacent buildings idealized as a simplified two-dimensional discrete model and underlying or surrounding soil coupled by a rotational interaction spring and the non-stationary Kanai–Tajimi accelerogram is employed as artificial seismic excitation input to study the dynamic of SSSI phenomena. The dynamic response of two adjacent buildings in the coupled SSSI system due to artificial seismic excitation input is compared and the parameters of the two adjacent buildings that affect the dynamic coupling interaction between adjacent building structures in the coupled SSSI system are also investigated.

## 2 Equation of Motion of the Coupled SSSI System

In this paper, the coupled SSSI system consists of two adjacent buildings modelled as a simple 2D-discrete model system with six degrees of freedom (DOF) and underlying or surrounding soil is coupled by a rotational interaction spring. The assumption made to derive the equation of motion of the coupled SSSI system in the present study is as follows, (i) the inter-building spacing, i.e. space between the two adjacent buildings is assumed to be large enough to avoid pounding, (ii) the total height of the two adjacent buildings can vary and be different but is idealized as 2-DOF lumped mass system, (iii) both the buildings have a similar square plan area of $b^2$, and the average building density, $\rho_b$ is kept constant, (iv) the soil profile under the two adjacent buildings remains same, and the radius of gyration of the soil-cylinder (directly under the rigid foundation) is calculated according to the Newmark's empirical expression $r_b \approx 0.33$. Therefore, $r_1 = r_2 = 0.33b$. [10] (v) the wave passage effects and spatially heterogeneous ground motion are assumed to be negligible; therefore, the foundation of the two adjacent buildings is applied with same ground motion.

The idealized 2D-discrete model of the coupled SSSI system is shown in Fig. 1. The 2D-discrete model system of the two adjacent buildings ($j \in [1, 2]$) consists total of six degrees of freedom (DOF). Each building superstructures consists of two translational DOFs relative to the ground, and the soil/foundation system of each building has one rotational DOF at the foundation level $\theta_j$. $m_{b1}$and $m_{b2}$ are total masses; $k_{b1}$and $k_{b2}$ are lateral stiffnesses; $m_{b1}.r_1^2$ and $m_{b2}.r_2^2$ are foundation/soil mass polar second moments of area; $k_{s1}$and $k_{s2}$ are rotational stiffnesses of foundation–soil beneath each building; $r_1$ and $r_2$ are radii of gyration of soil semi-cylinders; $h_1$ and $h_2$ are the total height; $b_1$ and $b_2$ are the width of the buildings foundation of building 1 and 2, respectively, $z$ is the non-dimensional inter-building distance, $\kappa$ is the stiffness of the inter-building soil rotational spring and $\ddot{x}_g$ is the horizontal acceleration ground motion.

The Euler–Lagrange equation of motion in its general form describing the dynamics of the coupled SSSI system can be derived using standard procedure and is formulated in Eq. (1).

$$\hat{\mathbf{M}}\ddot{\mathbf{x}} + \hat{\mathbf{C}}\dot{\mathbf{x}} + \hat{\mathbf{K}}\mathbf{x} = \hat{\mathbf{P}}\ddot{x}_g \tag{1}$$

where the system matrices are defined as follows

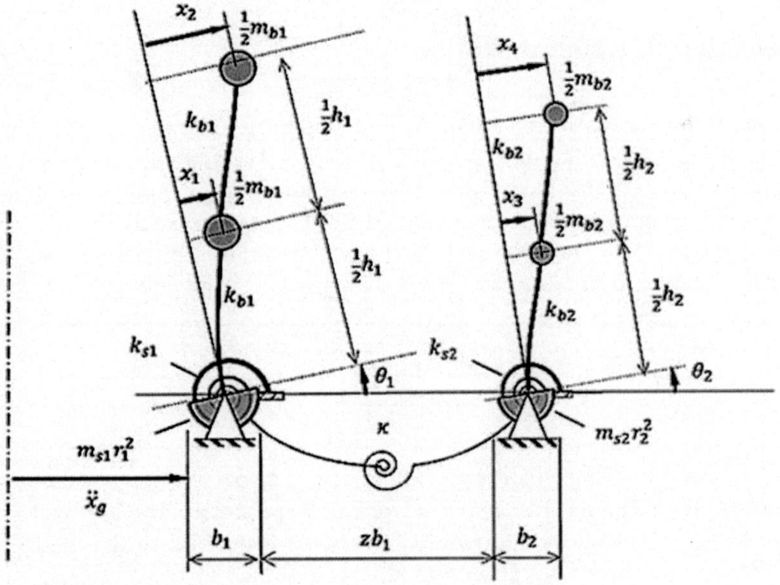


**Fig. 1** Idealized 2D-discrete model of the coupled SSSI system

$$\hat{\mathbf{M}} = \begin{bmatrix} \frac{1}{2}m_{b1} & 0 & -\frac{1}{4}m_{b1}h_1 & 0 & 0 & -0 \\ 0 & \frac{1}{2}m_{b1} & -\frac{1}{2}m_{b1}h_1 & 0 & 0 & 0 \\ -\frac{1}{4}m_{b1}h_1 & -\frac{1}{2}m_{b1}h_1 & \frac{5}{8}m_{b1}h_1^2 + m_{s1}r_1^2 & 0 & 0 & 0 \\ 0 & 0 & 0 & \frac{1}{2}m_{b2} & 0 & -\frac{1}{4}m_{b2}h_2 \\ 0 & 0 & 0 & 0 & \frac{1}{2}m_{b2} & -\frac{1}{2}m_{b2}h_2 \\ 0 & 0 & 0 & -\frac{1}{4}m_{b2}h_2 & -\frac{1}{2}m_{b2}h_2 & \frac{5}{8}m_{b2}h_2^2 + m_{s2}r_2^2 \end{bmatrix}$$

$$\hat{\mathbf{K}} = \begin{bmatrix} 2k_{b1} & -k_{b1} & 0 & 0 & 0 & 0 \\ -k_{b1} & k_{b1} & 0 & 0 & 0 & 0 \\ 0 & 0 & k_{s1}+\kappa & 0 & 0 & -\kappa \\ 0 & 0 & 0 & 2k_{b2} & -k_{b2} & 0 \\ 0 & 0 & 0 & -k_{b2} & k_{b2} & 0 \\ 0 & 0 & -\kappa & 0 & 0 & k_{s2}+\kappa \end{bmatrix}, \; \hat{\mathbf{p}} = \begin{bmatrix} -\frac{1}{2}m_{b1} \\ -\frac{1}{2}m_{b1} \\ \frac{3}{4}m_{b1}h_1 \\ -\frac{1}{2}m_{b2} \\ -\frac{1}{2}m_{b2} \\ \frac{3}{4}m_{b2}h_2 \end{bmatrix}, \; \mathbf{x} = \begin{bmatrix} x_1 \\ x_2 \\ \theta_1 \\ x_3 \\ x_4 \\ \theta_2 \end{bmatrix}$$

The Caughey orthogonal damping matrix $\widehat{\mathbf{C}}$ of the system defined in Eq. (1) assumes that six eigenvalues of the linear unforced system, each natural mode n $\in$ [1, 6] is damped at 2% of critical damping, i.e. $\xi_n = 0.02$. $\Phi_n$ is the eigenvector for mode $n$, $\omega_n$ are the natural frequencies of the system. Thus, the damping matrix $\widehat{\mathbf{C}}$ can be calculated as [11]:

$$\hat{\mathbf{C}} = \hat{\mathbf{M}}\left(\sum_{\text{n}=1}^{6} \frac{2\xi_n\omega_n}{\Phi_n^T M \Phi_n}\Phi_n\Phi_n^T\right)\hat{\mathbf{M}} \tag{2}$$

## 3 Selection of Ground Motion

In this paper, the non-stationary Kanai–Tajimi accelerogram is employed to simulate the artificial seismic excitation [12–14]. This artificial seismic accelerogram generated is employed as the horizontal acceleration ground motion input to the simplified idealized 2D-discrete model of the coupled SSSI system to study the dynamic of SSSI phenomena. Guo, and Kareem, [14] developed non-stationary Kanai–Tajimi spectrum as shown in Eq. (3).

$$F(s) = \text{s}_o \frac{\omega_g^4 + (2\xi_g\omega_g \text{s})^2}{(\omega_g^2 - s^2)^2 + (2\xi_g\omega_g \text{s})^2}, \quad \text{s}_o = \sigma_g^2 \frac{2\xi_g}{\pi\omega_g(4\xi_g^2 + 1)} \tag{3}$$

where $\xi_g$, $\omega_g$ are the ground damping and frequency, respectively. $\sigma_g$, is the standard deviation of the excitation (assuming a two sided spectrum). The artificial ground acceleration is produced by applying a Gaussian white noise to the filter. In this study, the duration of strong ground motion is assumed to be 40 sec, $\sigma_g = 0.3$, $\xi_g = 0.3$ and $\omega_g = 10\pi rad/s$ are used in the numerical simulations of non-stationary Kanai–Tajimi accelerogram (Fig. 2).

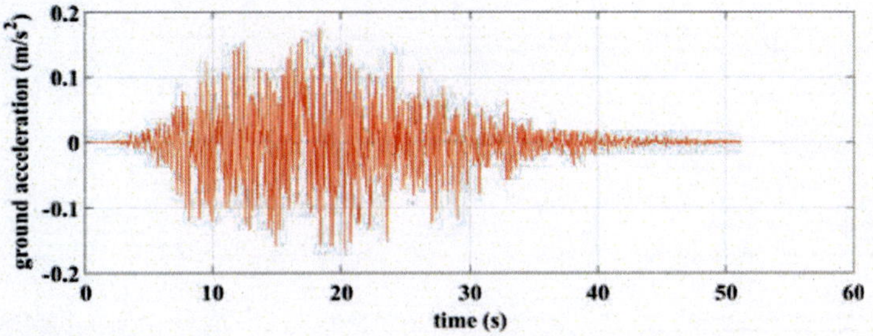


**Fig. 2** Artificial seismic accelerogram generated based on non-stationary Kanai–Tajimi filter

## 4 Numerical Simulation

The second-order ODE of the coupled SSSI system in Eq. 1 is solved as a state space model in time domain. The numerical simulations are conducted using MATLAB. The following parameters are adapted for the coupled SSSI system in the present study. The total height $h_1 = 6.40$ m and $h_2 = 6.00$ m ; $b_1 = 20$ m and $b_2 = 20$ m, is the width of the buildings foundation of building 1 and 2, respectively, $m_s = 0.35b^3\rho_s$ is the dynamic mass of soil beneath buildings and $m_{bj} = \rho_b h_j b^2$, the mass of the buildings, where $\rho_s = 1600kg/m^3$ and $\rho_b = 800kg/m^3$ are the average densities of soil and building, respectively. In the present study, the two buildings are placed in very close proximity to each other, i.e. z, the ratio of the distance between buildings to building width is 0.01, $i.e.$, $z = 0.10$. The values of foundation rotational spring $k_{s1} = k_{s2} = k_s q_2$, and the interaction spring stiffness $\kappa$ are modelled as an inverse cube function of non-dimensional inter-building separation distance $z$, [15]. The rotational stiffness spring coefficient $k_s$ is obtained by using the empirical formula in the absence of building interaction [16].

$$\kappa = q_k q_2 k_s, \, q_k = -\frac{0.25}{(1+z)^3}, \, q_2 = 1 + \frac{0.5}{(1+z)^3}, \, k_s = \frac{1}{2}\frac{G_s b^3}{1-\mu}$$

where $G_s$ is the elastic shear modulus of the soil, the shear wave velocity of the soil $V_s = \sqrt{G_s/\omega_s}$ in 250 m/s and the Poisson's ratio of the soil as $\mu = 0.30$.

## 5 Results and Discussions

The results are obtained by numerical simulations performed in MATLAB. Figure 3 provides the acceleration, velocity and displacement response of the idealized 2D-discrete model of the coupled SSSI system. Figures 4 and 5 provide the acceleration responses of the building model 1 and 2 of the coupled SSSI system both in time and frequency domain, respectively. In general, the maximum displacement and

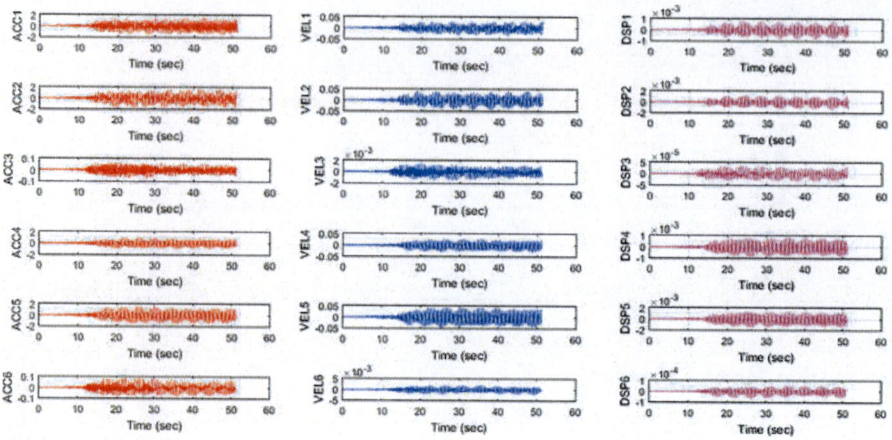


**Fig. 3** Acceleration, velocity and displacement response of the Idealized 2D-discrete model of the coupled SSSI system

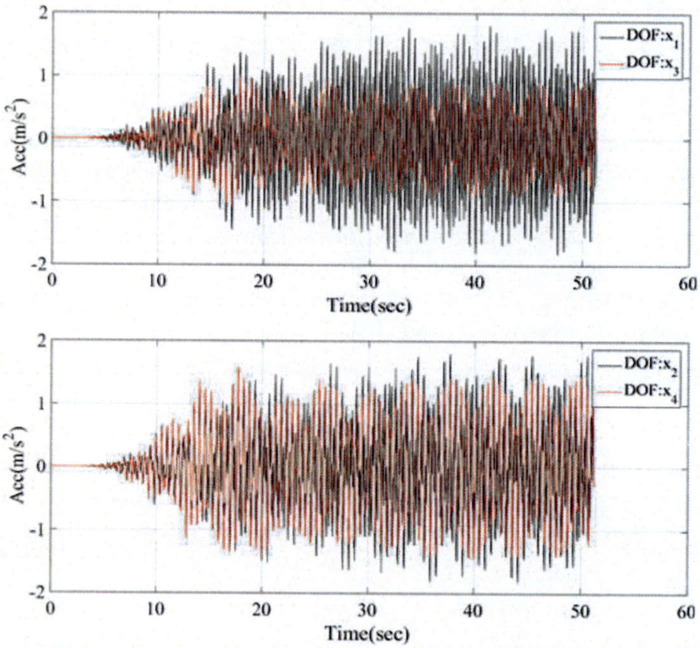


**Fig. 4** Acceleration response at the mass levels of the building model 1 and 2 of the coupled SSSI system in time domain

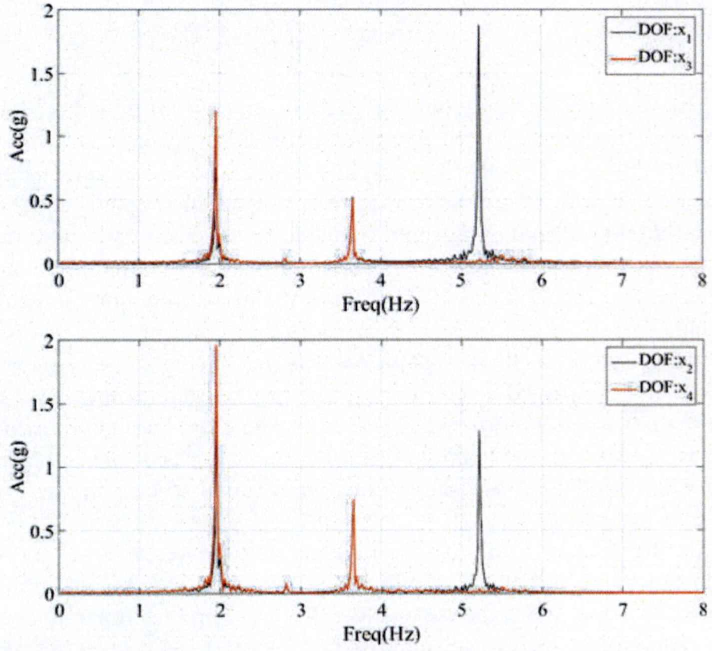


**Fig. 5** Acceleration response at the mass levels of the building model 1 and 2 of the coupled SSSI system in frequency domain

acceleration of buildings increases for the SSSI case and the interaction are found to be amplified when the buildings are very closely spaced.

Table 1 provides the summary of peak acceleration, peak velocity and peak displacement of the building models of the coupled SSSI system. Based on the results, the responses of the building model 2 increased to that of the responses of the building model 1, indicating the amplification of response due to interacting dynamic coupling interaction effects between taller adjacent building structure and the underlying soil when subjected to artificial seismic excitation.

**Table 1** Peak acceleration, peak velocity and peak displacement of the building models of the coupled SSSI system

| Peak Acceleration in g | | | | Peak Velocity in cm/sec | | | | Peak Displacement in cm | | | |
|---|---|---|---|---|---|---|---|---|---|---|---|
| Building 1 | | Building 2 | | Building 1 | | Building 2 | | Building 1 | | Building 2 | |
| $x_1$ | 0.1808 | $x_3$ | 0.0901 | $x_1$ | 9.218 | $x_3$ | 7.4705 | $x_1$ | 51.0011 | $x_3$ | 51.1481 |
| $x_2$ | 0.1719 | $x_4$ | 0.1318 | $x_2$ | 10.2341 | $x_4$ | 9.4778 | $x_2$ | 52.057 | $x_4$ | 51.7481 |
| $\theta_1$ | 0.0066 | $\theta_2$ | 0.0078 | $\theta_1$ | 0.3154 | $\theta_2$ | 0.4396 | $\theta_1$ | 0.0356 | $\theta_2$ | 0.0386 |

# 6 Conclusions

In this paper, numerical investigations are carried out on a simplified model to study the dynamic of structure–soil–structure interaction (SSSI) phenomena. As expected, the simplified model is effective to study the dynamic coupling interaction effects between adjacent building structures and underlying or surrounding soil.

Based on the initial trails, the eigen frequencies for the coupled and uncoupled system are very similar, i.e. there is a maximum of 11% variation in the natural frequencies between the SSI and SSSI systems, therefore an important feature of the SSSI systems.

The building space has been a major contributing factor to the responses, when the two buildings are placed in very close proximity to each other (i.e. 0.1 d) which is large enough to avoid pounding but is found to be close enough to maximize the SSSI effects. In general, the maximum displacement and acceleration of buildings increase for the SSSI case and the interaction is found to be amplified when the buildings are very closely spaced.

Based on the present study, results indicate that the response of the structure adjacent to the taller building is significantly increased and thereby amplifies the seismic risk; however, the response of the taller building is comparatively reduced indicating reduction of its own seismic risk.

The present study is based on the analyses undertaken for limited parameters and needs further investigation.

## References

1. Lou M, Wang H, Chen X, Zhai Y (2011) Structure–soil–structure interaction: literature review. Soil Dyn Earthq Eng 31(12):1724–1731
2. Hans S, Boutin C, Ibraim E, Roussillon P (2005) In situ experiments and seismic analysis of existing buildings. Part I: experimental investigations. Earthq Eng Struct Dyn 34(12):1513–1529
3. Kitada Y, Hirotani T, Iguchi M (1999) Models test on dynamic structure–structure interaction of nuclear power plant buildings. Nucl Eng Des 192(2–3):205–216
4. Lee TH, Wesley DA (1973) Soil-structure interaction of nuclear reactor structures considering through-soil coupling between adjacent structures. Nucl Eng Des 24(3):374–387
5. Lysmer J, Seed HB, Udaka T, Hwang RN, Tsai CF (1975) Efficient finite element analysis of seismic soil structure interaction. Report No. EERC 75–34. Earthquake engineering research center, University of California, Berkeley, CA
6. Murakami H, Luco JE (1977) Seismic response of a periodic array of structures. J Eng Mechan Division 103(5):965–977
7. Roesset JM, Gonzalez JJ (1978) Dynamic interaction between adjacent structures. In: Dynamic response and wave propagation in Soils. Rotterdam: AA balkema. pp 127–166
8. Wong HL, Trifunac MD (1975) Two-dimensional, antiplane, building-soil-building interaction for two or more buildings and for incident planet SH waves. Bull Seismol Soc Am 65(6):1863–1885
9. Yano T, Kitada Y, Iguchi M, Hirotani T, Yoshida K (2000) Model test on dynamic cross interaction of adjacent buildings in nuclear power plants. In: 12th world conference on earthquake engineering. New Zealand

10. Newmark NM, Rosenb lueth E (1971) Fundamentals of earthquake engineering, Prentice Hall. Inc. Englewood Cliffs, NJ, USA
11. Chopra AK (2000) Dynamics of structures: theory and applications to earthquake engineering, 2nd edn. Prentice Hall, NJ
12. Lin YK, Yong Y (1987) Evolutionary kanai-tajimi earthquake models. J Eng Mechan 113(8):1119–1137
13. Rofooei FR, Mobarake A, Ahmadi G (2001) Generation of artificial earthquake records with a nonstationary Kanai-Tajimi model. Eng Struct 23(7):827–837
14. Guo Y, Kareem A (2016) System identification through non stationary data using time–frequency blind source separation. J Sound Vib 371:110–131
15. Vicencio F, Alexander NA (2018) Higher mode seismic structure-soil-structure interaction between adjacent building during earthquakes. Eng Struct 174:322–337
16. Gorbunov-Posadov MI, Serebrajanyi V (1961) Design of structure upon elastic foundations. In: 5th international conference on soil mechanics and foundation engineering, Paris. pp 643–648

# A Comparative Study on Predictive Capability of Different SSI Models

**Monu Lal Burnwal and Prishati Raychowdhury**

**Abstract** Accurate and reliable modeling of the soil-foundation interface poses the utmost importance in the seismic soil-structure interaction (SSI) analysis of a structure. The issue is even more critical when the structure rests on a soft clay or loose sand deposits, where foundation deformations may be large and permanent, leading to material and geometric nonlinearity at the soil-foundation interface. The present study focuses on evaluating the uncertainty in the modeling approach of the subsoil and soil-foundation interface and its effect on the overall structural response. Three different modeling approaches, namely discrete Winkler-based spring model, continuum finite element model:—rigid base and compliant base, are considered in the analysis. The analysis is carried out in *OpenSees* (Open System for Earthquake Engineering Simulation). Nonlinear material models such as *q-z*, *p-y* and *t-z* springs are used for the Winkler-based discrete model, whereas *PressureDependMultiYield* material models are used for the continuum model. The different models are applied to a typical steel moment-resisting frame building resting on loose Ganga sand bed, and the seismic response of the building is estimated using various models mentioned above. Finally, the numerical results are compared with a series of shake table experiments conducted at IIT Kanpur on a reduced scale model of the building and efficiency of the different SSI models is compared based on their respective predictive capabilities.

**Keywords** Soil-structure interaction · Nonlinear dynamic analysis · Discrete spring model · Continuum finite element model · Macro-element model

M. L. Burnwal (✉) · P. Raychowdhury
Department of Civil Engineering, IIT Kanpur, Kanpur 208016, India
e-mail: monulal@iitk.ac.in

P. Raychowdhury
e-mail: prishati@iitk.ac.in



T. G. Sitharam et al. (eds.), *Seismic Design and Performance*, Lecture Notes in Civil Engineering 120, https://doi.org/10.1007/978-981-33-4005-3_16

# 1 Introduction

## *1.1 Modeling*

Modeling soil with all bits of details is a subjective job as with such kinds of complexities and nonlinearity; no perfect models fit in all types of conditions. Over the years, many researchers have reported the dynamic soil-structure interaction effect with various analytical and numerical (discrete and continuum models) approaches (References [1–8], etc.) and through dynamic prototype and model experiments [9–13]. Practical responses include consistent and accurate geometry modeling, soil-foundation interfaces, appropriate constitutive material models that simulate proper soil behavior and nonlinearity, field boundary conditions, representative loading, and types of simulations. Uncertainty in all these abovementioned considerations influences the soil responses from small to significant variations.

Uncertainty on the predictability of soil-structure responses may be due to the following:

- Selection of model its geometry, element types, mesh type.
- The choice of material models requires skills and literature search to judge for the analysis type and loading conditions one is considering and whether it is feasible to apply (like clay and sand material model).
- Use of field boundary conditions that can be simulated through—fixed, discrete w, reflective, viscous base boundary conditions.
- Different kinds of approaches applied to solve a sort of problem like—macro-element model, discrete BNWF springs, or continuum approach for different base boundaries.
- Use of the various sort of interfaces while modeling like soil-structure interfaces, soil-soil interfaces, structure-footing interfaces, etc.
- Choice of loading conditions and its pattern: (a) static (whether constant or linear) or (b) transient.

Many researchers in this decade have studied the modeling of SSI systems with soft and loose soils. The correct, reliable modeling requires the use of accurate constitutive material models that describe the soil and its properties in an authentic way. Few studies have compared the SSI modeling of the continuum and discrete responses. A comparative study has been conducted to know about the performance and predictability of soil and SMRF structures. Responses using three different SSI approaches—beam on nonlinear Winkler foundation (BNWF) model (discrete springs), soil continuum model with rigid base (reflective boundaries), and soil continuum model with compliant base (viscous dashpot boundaries)—have been studied.

All the three models are prepared in finite element based open-source platform OpenSees (see Fig. 2). It provides the flexibility of having an insight into how the model behaves at every step of analysis and about the failure with different types of inbuilt classes and material being built by various developers throughout the world.

## 2 Model and Materials

### *2.1 Structure*

The structure is a steel moment-resisting frame building with 3 m ISMB600 column and 6 m ISMB 500 beam modeled as nonlinear beam column element. The square footing of 1 × 1m is designed for each column and is modeled with elastic beam column elements. The prototype building is in Zone III and is designed as per [14] and [15]. The imposed mass at each floor level is kept to be 20,246 kg, except for the top floor mass, which is selected as 0.85 times that of other typical floor masses. The structure is made to vibrate with a weak beam strong column concept. Three-story structure model in the shake table experiment is made of steel sections and is having the reduced sections, area and moment of inertia are considered for the comparison (developed by Vivek et al. [13]). The scale for the prototype to model ratio is ten and is following the similitude as presented in Iai [16] for the shake table experiments.

### *2.2 Soil*

Loose Ganga sand with a relative density of 40% has been considered for the study Table 1 shows the properties of Ganga sand.

Discrete Winkler approach for modeling the soil-foundation interface [17] can capture both material and geometric nonlinearity. Material nonlinearity should represent soil yielding and degradation, while geometric nonlinearity is for uplifting,

**Table 1** Properties of Ganga sand used in the test and numerical simulations

| Soil properties | Ganga sand | |
|---|---|---|
| Specific gravity | G | 2.67 |
| % Soil | % Sand | 98.19 |
| | % Fines | 1.81 |
| | $D_{10}$ | 0.15 mm |
| Effective particle size | $D_{30}$ | 0.195 mm |
| | $D_{60}$ | 0.25 mm |
| Coefficient of uniformity | Cu | 1.667 |
| Coefficient of curvature | Cc | 1.014 |
| Maximum void ratio | $e_{max}$ | 0.99 |
| Minimum void ratio | $e_{min}$ | 0.70 |
| Unit weight | $\Upsilon$ | 14.07 kN/m$^3$ |
| Friction angle | $\phi$ | 32.5° |
| Shear modulus | $G_{soil}$ | 8.18 MPa |
| Poisson's ratio | $\nu$ | 0.2 |

gapping, and sliding off the foundation. BNWF model considers the soil-foundation interface modeling using zero-length elements to represent the footing and a series of nonlinear springs to describe the soil. The zero-length elements are used to create soil springs and are consigned to different uniaxial material objects in the lateral and vertical directions. The spring nodes are generated with two dimensions and two translational degrees of freedom.

Springs oriented in the lateral direction represent p-x and t-z springs, and the q-z springs are vertically placed along the length of the footings. P-x springs elements represent the passive earth pressure resistance of soil-footing interface, t-z springs represent frictional resistance along the length of the footing, and q-z spring represents the bearing capacity of footing. The PySimple1 and TzSimple1 material objects are incorporated in the x-direction (direction of seismic loading), while the QzSimple1 is incorporated in the z-direction (vertical direction). The details of these spring elements, including their backbone equations, validation, and applications, can be found in Boulanger et al. [17] and Boulanger [18]. Every material model comprises a visco-elastic component that describes far-field behavior and a plastic, drag, and closure component that represents near-field displacement as shown in Fig. 1. Soil-foundation gapping or separation can be represented by a gap component, which is ideally present in QzSimple2 and PySimple1 material models. BNWF involves discrete springs to represent soils, but it has certain disadvantages over continuum as the spring is discrete and unidirectional. Limitations like stress–strain response, amplification due to layered soil, and other continuous responses of soil profile exist with BNWF modeling. Continuum modeling in FEM removes all such limitations, but sensitivity and constitutive models must be chosen very carefully, especially for loose and soft soils.

OpenSees library includes nDMaterial as pressure-dependent and independent models to simulate the sand and clay in its true fashion by taking 10–20 parameters, which can be changed during the course of analysis. Continuous quad elements are used for modeling loose sand, using "PressureDependMultiYield" (PDMY02) ND-Material, proposed by Yang et al. [19], that characterizes the stress–strain relationship at Gauss points. "PDMY02" is an elastic–plastic material that simulates the properties and responses of pressure-sensitive sand material under normal loading conditions. It includes dilatancy and non-flow liquefaction effects during monotonous or cyclic loading.

Most SSI studies in the past have focused on either the discrete approach or the continuum approach. The discrete method is a fast analysis technique of replacing soils with its equivalent springs and dashpots to take the load. Different phenomena

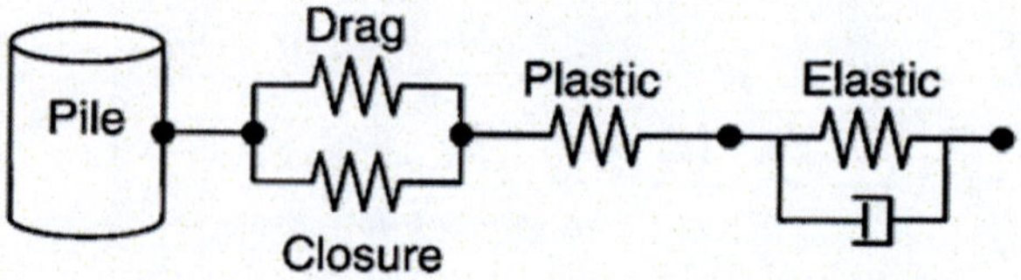


**Fig. 1** P-y material model suggested by Boulanger et al. [17]

of soils are represented by different types of uniaxial springs arranged in such a way that it represents the true soil conditions.

## 3 Methodology

Staged analysis procedures have been performed for static and transient analysis. The steps for the analysis are as follows:

Firstly, the soil mesh is generated with all the parameters that match with the loose sand at field conditions, like soil profile height, the weight of soil by soil density, element size, soil domain length. Constitutive models like the one which is used for analysis require parameters like unit weight shear modulus, Poisson's ratio, friction angle, phase transformation angle, peak shear strain, dilation constants, and liquefaction constant.

Secondly, the boundary condition for lateral and base boundaries is specified: standard fixities have been provided for defining the boundaries, i.e., the base has been fixed in both x- and y-direction, and side is fixed for only in the x-direction to provide vertical displacement during the elastic geostatic stage. This is done to generate the same initial state of stress as the soil was subjected to when no structure was present. Eigenvalues analysis is performed to know the time period associated with the soil.

Thirdly, the soil is set to first run in the elastic stage, and then the material is updated to plastic to know whether it satisfies elastoplastic equilibrium. Once the equilibrium is attained, the structure with its footing is added to the soil surface. Gravity loads are applied to the structure in an increasing linear pattern. The model is analyzed to satisfy the static equilibrium.

For the seismic analysis step, the horizontal boundaries fixities are removed and are replaced by the corresponding reactions that we get from the static equilibrium stage. Boundary conditions which prevail during the different stage are:

For continuum reflective base (CRB) boundary: Side boundaries are fixed in both the directions and base boundaries are allowed to move in ground shaking direction, i.e., in the x-direction, the fixities are removed. Reflective base boundaries provide the best results for input motion as displacement time history as compared to velocity and acceleration time history.

For continuum viscous base (CVB): Fixities at side boundaries are removed in the x-direction and are replaced corresponding by viscous dashpots that represent the soil free field. The vertical fixities are provided in the y-direction to simulate the laminar box conditions. Base boundary fixities are removed in the horizontal direction and are replaced corresponding by viscous dashpots that represent the elastic rock. As dashpots are involved for simulating the absorbent boundary condition, the force is calculated using the equation

$$f(t) = C_d \text{Velocity}(t)$$

where $C_d$ is the dashpot coefficient, f(t) is the force–time history, and velocity(t) is the velocity–time history. So, we require a velocity–time history profile for the viscous compliant base analysis.

For the BNWF discrete springs model, the input motion is provided in time-series acceleration form by applying the excitation uniformly throughout the model (Fig. 2).

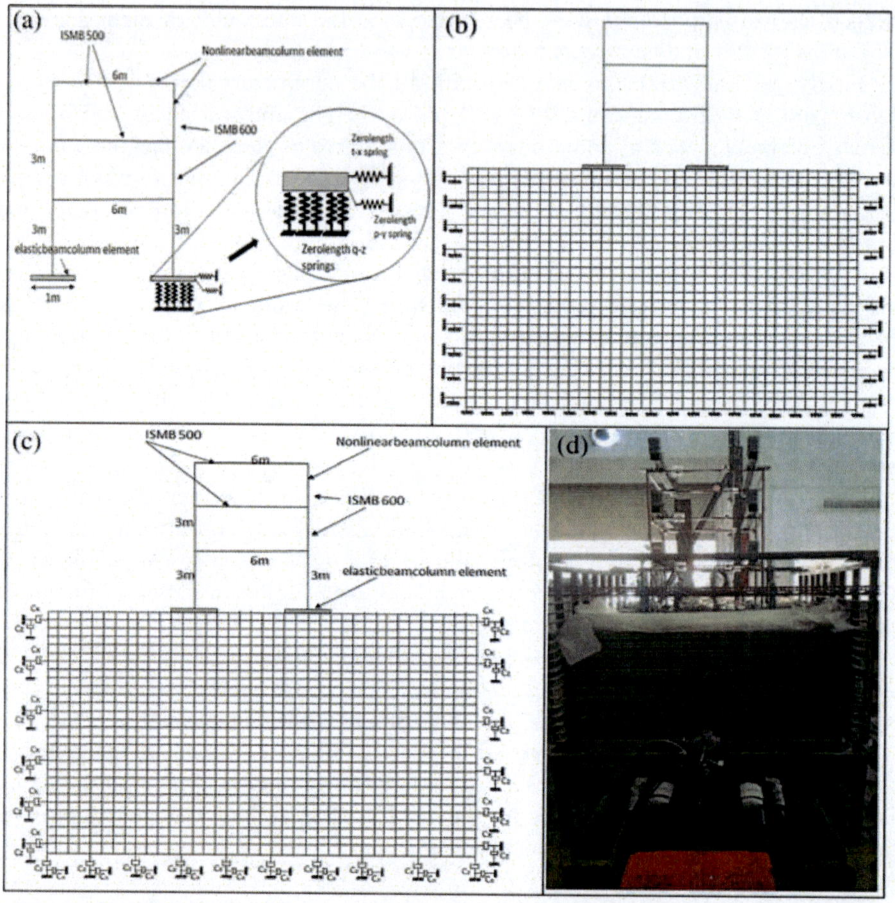


**Fig. 2** **a** SMRF model details with discrete BNWF springs, **b** continuum model with reflective(rigid) boundaries, **c** continuum model with viscous(compliant) barriers, **d** three-story shake table tests using the laminar box (Vivek et al. [20])

# 4 Results and Discussion

## *4.1 Input Motion*

Figure 3 shows the response spectrum of input and scaled harmonic motion. 3-story SMRF structure is modeled with a scale ratio(λ) of 10 (i.e., prototype to model scaling is 10) and is compared with the three models. The models are analyzed for input motion of a sinusoidal wave having a displacement of 1 mm and frequency of 5 Hz, which was an input to the shake table experiment. The experimental model was the replica of a prototype structure. The frequency has been downscaled from 5 Hz in model test to 1.75 Hz in prototype by applying a factor of $\sqrt{\lambda}$ using dynamics similitude. Figure 3 shows the input motion being taken for the shake table experiments. The input motion is scaled by scaling the frequency to match the fundamental period of the full-scale structure. The frequency scaled response spectrum is also plotted in Fig. 4 with red color. Amplitude scaling of 10 is applied to frequency scaled acceleration time history so that the response spectrum amplitude for the prototype is the same as the model. It is done to get the same amplitude of force as it was in the experiments.

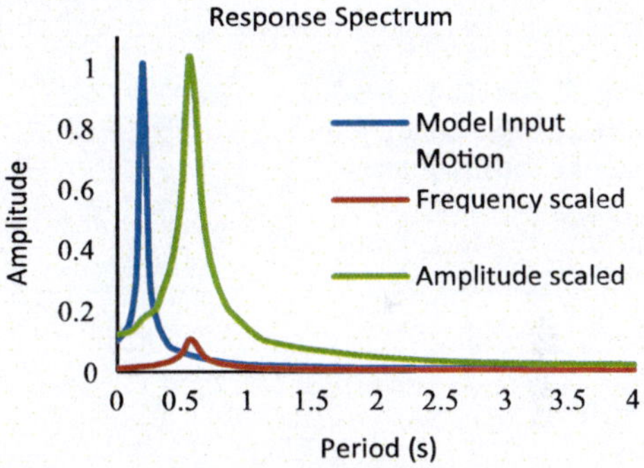


**Fig. 3** Response spectrum of input and scaled sinusoidal motion

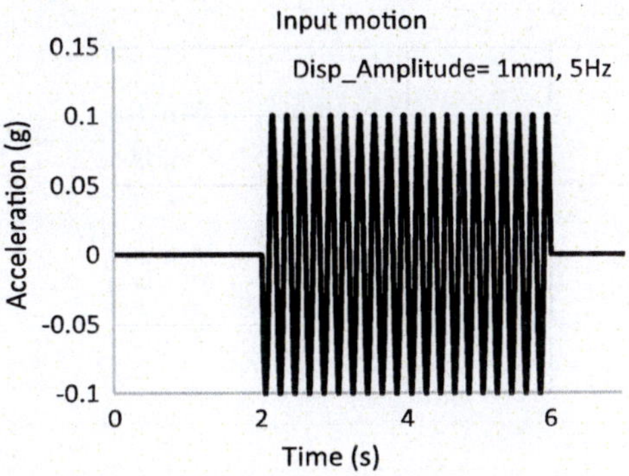


**Fig. 4** Input motion applied for shake table test with disp = 5 mm and freq = 5 Hz

## 4.2 Eigenvalue Analysis

Eigenvalue analysis has been performed for all the models with different base conditions. The effect of SSI on the fundamental period has been observed. The fixed base time period of full-scale building is 0.46 s, which is replicated in all the models. The fundamental frequency for the discrete Winkler spring base (DWB), continuum reflective base (CRB), continuum viscous base (CCB) models, and shake table (ST) experiments is found to be 1.014, 1.087, 1.087, and 0.950, respectively. It has been perceived that there has been a shift of fundamental period from 1.04 s for discrete spring base to 1.087 s for the continuum base (Fig. 5). Thus, an increase of about 7.2% is noted in the fundamental period from the DWB model to the soil continuum model with a period ratio of 1.07. Also, there is an increase in the first mode period from 6.7 to 12% for DWB and CRB or CVB from the ST experiment. The periods for different models and compared with the model ST tests results and % error are listed in Table 2.

Table 3 shows the period and frequency for the first, second, and third modes of the different models. Continuum base period is larger than the discrete base due to the rigidity in the discrete springs base model. Since the stiffness of the DWB model is much higher than the SSI model, the circular frequency ($\omega_n$) will increase, which will enhance the period.

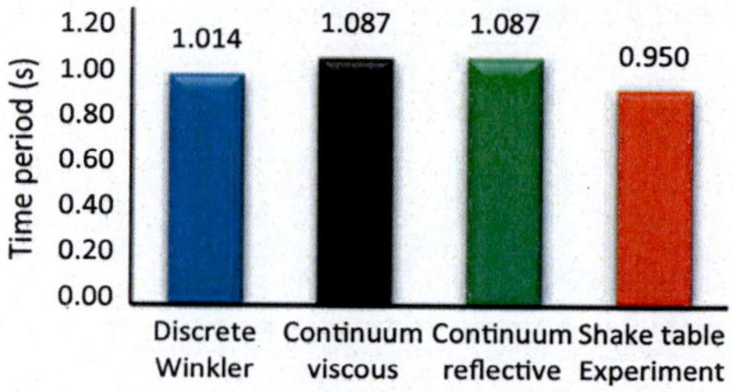


**Fig. 5** Bar charts showing change of frequency with models and experiments

**Table 2** Fundamental time period of models and percent error from ST test

| Models | Time period (secs) | ST test period (s) | Percent error (ST Test) |
|---|---|---|---|
| Discrete Winkler | 1.014 | 0.950 | 6.747 |
| Continuum viscous | 1.087 | 0.950 | 12.634 |
| Continuum reflective | 1.087 | 0.950 | 12.634 |

**Table 3** Time period and frequency for various modes

| Model | Mode | Time period (secs) | Frequency (Hz) |
|---|---|---|---|
| Discrete Winkler | 1st | 1.014 | 0.986 |
| | 2nd | 0.193 | 5.194 |
| | 3rd | 0.103 | 9.681 |
| Continuum viscous | 1st | 1.087 | 0.920 |
| | 2nd | 0.303 | 3.296 |
| | 3rd | 0.230 | 4.356 |
| Continuum reflective | 1st | 1.087 | 0.920 |
| | 2nd | 0.303 | 3.296 |
| | 3rd | 0.230 | 4.356 |
| ST Experiment | 1st | 0.950 | 1.053 |
| | 2nd | 0.240 | 4.167 |

## 4.3 Displacement Time History

Displacement time history at the top story for all the models is shown in Fig. 6. Displacement (mm) is plotted for a time interval of 3.5 secs. DWB model is 180° and 230° out of phase with CRB and CVB. Continuum viscous and reflective models are 45° and 90° out of phase from the input motion. Response for amplitude and phase of CVB and CRB is dependent on the damping ratio of the system.

In a viscous soil-structure system, resonance occurs at

$$\omega = \omega_n \sqrt{(1 - 2\xi^2)}$$
$$\Rightarrow \frac{\omega}{\omega_n} = \sqrt{(1 - 2\xi^2)}$$

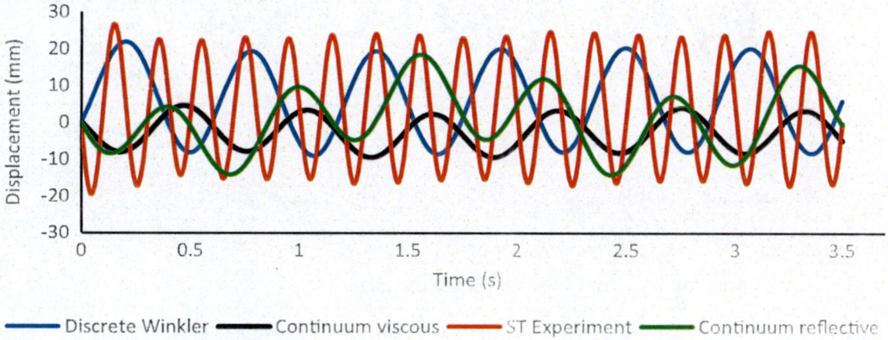


**Fig. 6** Displacement time history at top story for all the models

so, if damping ratio $\xi$ is very small than 1; $\omega$ is close to $\omega_n$ which makes the system to resonate and the phase difference is $-90°$. Before resonance, it is close to zero, and after that, the phase becomes close to $-180°$. So, as $\xi$ is increased in the system, phase lag and amplitude are reduced at resonance.

### 4.4 Column Moment Time History

Moment time history at the base of the structure for all the models and the ST test is shown in Fig. 7. The column moment profile shows a decreasing trend to the top of the structure with the maximum value of the moments at the base. The base moment increases with an increase in the intensity of excitation as well as the foundation rigidity for all the foundation cases.

The variation of the overturning moment is due to the flexure of columns, which depends on the mass and stiffness of the structure. It can be observed that the peak moment from the simulation's underestimates the experimental peak moment values for all the cases. The base moment decreases from DWB to CVB models as the flexibility of the models is increased from discrete spring to reflective to viscous dashpot. Soil continuum having viscous dashpots at boundaries absorbs the waves emanating from the dynamic analysis. Absorbent soil produces less vibration of soil, which makes the base moment to be least. CRB is showing a significant moment as compared to other models. It may be due to wave energy, which gets accumulated during the analysis, which increases the force applied to the model.

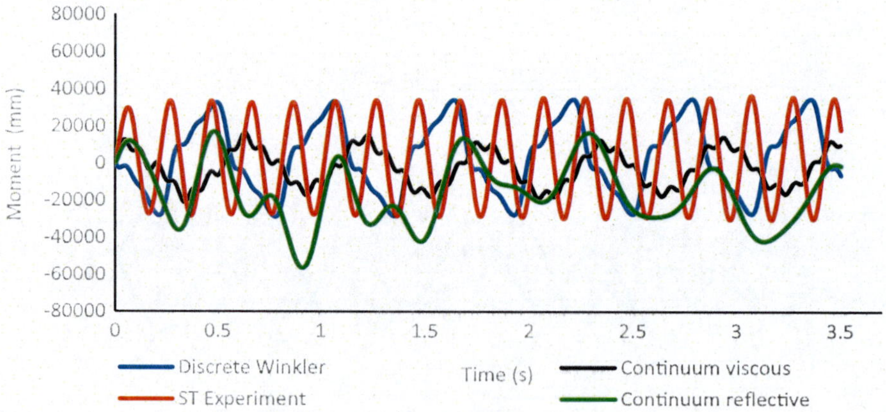


Fig. 7 Moment time history at top story for all the models

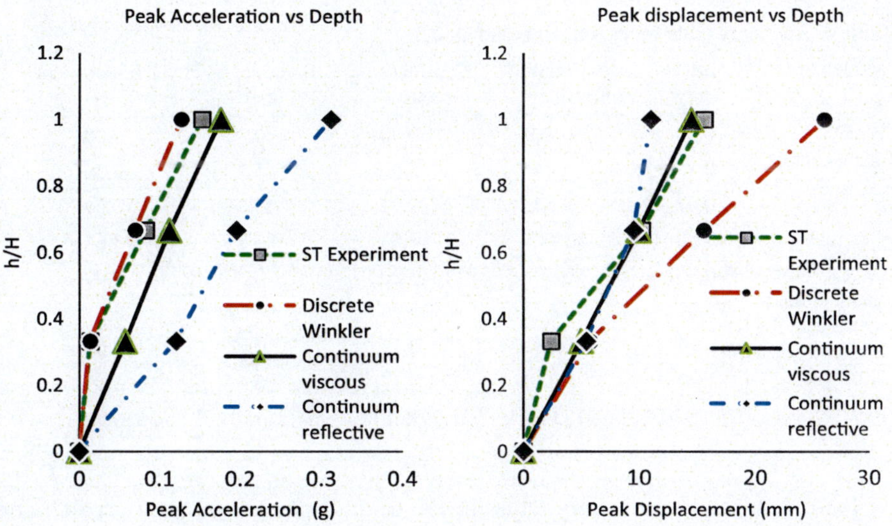


**Fig. 8** **a** Peak ground acceleration with floor height, ratio **b** peak displacement with floor height ratio

## 4.5 *Peak Displacement*

Figure 8 shows a comparison of peak story displacement of the structures for DWB, CRB, and CVB model with the amplitude scaled model shake table results. For a flexible foundation case, the total story displacement is contributed from the sliding, rigid body rotation, and structural movement due to flexure. For the prototype structure with untied nonlinear base shake table experiment for the top story, peak displacement is 15.64 mm. DWB, CRB, and CVB models show a relative movement 26.05 mm 11.0 mm and 14.53 mm for the top story from the footing base. It is observed that story displacement is higher for flexible base cases. The peak displacement gets enhanced as we move from base to top story. It is observed that the CVB model was able to capture the peak displacement more accurately, as in the experimental model.

## 4.6 *Peak Acceleration*

Figure 8 shows a comparison of peak acceleration of the structures for DWB, CRB, and CVB models with the ST model. DWB, CRB, and CVB models show a relative acceleration of 0.13 g, 0.17 g, and 0.31 g for the top story. It is observed that the CVB model was able to capture the peak acceleration more accurately, as in the experimental model.

**Table 4** Summary table for experiments and models

| Models | h/H | Continuum reflective | Continuum viscous | Discrete Winkler | ST Experiment |
|---|---|---|---|---|---|
| Peak displacement | 0.00 | 0.00 | 0.00 | 0.00 | 0.00 |
| | 0.33 | 5.44 | 5.00 | 5.77 | 2.38 |
| | 0.67 | 9.54 | 9.95 | 15.61 | 10.24 |
| | 1.00 | 11.00 | 14.53 | 26.05 | 15.64 |
| Peak acceleration | 0.00 | 0.00 | 0.00 | 0.00 | 0.00 |
| | 0.33 | 0.12 | 0.06 | 0.01 | 0.01 |
| | 0.67 | 0.20 | 0.11 | 0.07 | 0.08 |
| | 1.00 | 0.31 | 0.17 | 0.13 | 0.15 |
| Peak moment | | 56,313.06 | 21,097.88 | 34,850.07 | 36,830.00 |

Table 4 shows the summary table for the peak displacement, peak acceleration with depth by height ratio, and base moment for the three-story structure for the models and experiments. Peak displacement and acceleration get amplified as we move from base to the top of the structure.

## 5 Conclusion

The study focuses on modeling and analysis procedures for continuum and a discrete spring model on the soil-foundation-structure system using different base boundary conditions. Frequency, and amplitude scaling of 10, of the input harmonic motion, is applied as an input to the numerical models. The modeling approach provides comparable response quantities. Eigenvalue and seismic analysis have been carried out, and the following observations have been listed from the investigations:

- Eigenvalue analysis shows an increase of about 7.2% in the fundamental period when foundation flexibility is introduced using soil continuum (DWB to CRB or CVB).
- Displacement time history results suggest as damping ratio ($\xi$) is increased in the system; phase lag and amplitude are reduced at resonance.
- As the flexibility of the model is increased (DWB to CVB), the base moment is decreased. However, CRB has a large moment due to trapping of waves in the soil.
- The effect of SSI modeling and boundary conditions is significant, which increases the displacement demand of the model.
- With the CVB model, peak displacement and peak acceleration are predicted accurately for the ST experiments.

## References


1. Bielak J (1977) Dynamic behaviour of structures with pile-supported foundations. 6th World Conf Earthq Eng 3:1576–1581
2. Stewart JP, Seed RB, Fenves GL (1998) Empirical evaluation of inertial soil-structure interaction effects by Environment Engineering
3. Allotey N, El Naggar MH (2008) An investigation into the Winkler modeling of the cyclic response of rigid footings. Soil Dyn Earthq Eng 28:44–57. https://doi.org/10.1016/j.soildyn.2007.04.003
4. Pitilakis D, Dietz M, Wood DM, Clouteau D, Modaressi A (2008) Numerical simulation of dynamic soil-structure interaction in shaking table testing. Soil Dyn Earthq Eng 28:453–467. https://doi.org/10.1016/j.soildyn.2007.07.011
5. Raychowdhury P (2011) Seismic response of low-rise steel moment-resisting frame (SMRF) buildings incorporating nonlinear soil-structure interaction (SSI). Eng Struct 33:958–967. https://doi.org/10.1016/j.engstruct.2010.12.017
6. Raychowdhury P, Singh P (2012) Effect of nonlinear soil-structure interaction on seismic response of low-rise SMRF buildings. Earthq Eng Eng Vib 11:541–551. https://doi.org/10.1007/s11803-012-0140-2
7. Zamani N, El Shamy U (2012) Analysis of the seismic response of soil-foundation-structure systems using a microscale framework. Soil Dyn Earthq Eng 43:398–412. https://doi.org/10.1016/j.soildyn.2012.07.010
8. Torabi H, Rayhani MT (2014) Computers and geotechnics three dimensional finite element modeling of seismic soil—structure interaction in soft soil. Comput Geotech 60:9–19. https://doi.org/10.1016/j.compgeo.2014.03.014
9. Yoshimi Y, Tokimatsu K (1977) Settlement of buildings on saturated sand during earthquakes. Soils Found 17:23–38. https://doi.org/10.3208/sandf1972.17.23
10. Liu L, Dobry R (1997) Seismic response of shallow foundation on liquefiable sand. J Geotech Geoenvironmental Eng 123:557–567. https://doi.org/10.1061/(ASCE)1090-0241(1997)123:6(557)
11. Algie TB, Pender MJ, Orense RP (2010) Large scale field tests of rocking foundations on an Auckland residual soil. pp 57–65
12. Tsukamoto Y, Ishihara K, Sawada S, Fujiwara S (2012) Settlement of rigid circular foundations during seismic shaking in shaking table tests. Int J Geomech 12:462–470. https://doi.org/10.1061/(ASCE)GM.1943-5622.0000153
13. Vivek B, Raychowdhury P (2017) Influence of SSI on period and damping of buildings supported by shallow foundations on cohesionless soil. Int J Geomech 17:04017030. https://doi.org/10.1061/(ASCE)GM.1943-5622.0000890
14. Is 800–2007: Is 800–2007 (2007)
15. IS:1893 (2002) Criteria for earthquake resistant design of structures. Indian stand. 1–44. https://doi.org/10.1111/j.1540-4560.1987.tb02330.x
16. Iai S (1989) Similitude for shaking table tests on soil-structure-fluid model in 1g gravitational field. Soils Found 29:105–118. https://doi.org/10.3208/sandf1972.29.105
17. Boulanger RW, Curras CJ, Kutter BL, Wilson DW, Abghari A (1999) Seismic soil-pile-structure interaction experiments and analyses. J Geotech Geoenvironmental Eng 125:750–759. https://doi.org/10.1061/(ASCE)1090-0241(1999)125:9(750)
18. Boulanger RW (2000) The pysimple1, qzsimple1 and tzsimple1 Material Documentation. Documentation for the OpenSees Platform. http://opensees.berkeley.edu
19. Yang Z, Elgamal A, Parra E (2003) Computational model for cyclic mobility and associated shear deformation. J Geotech Geoenvironmental Eng 129:1119–1127. https://doi.org/10.1061/(ASCE)1090-0241(2003)129:12(1119)
20. Vivek B (2016) Seismic Soil-structure interaction of low to medium-rise buildings: experiments and simulations. IIT Kanpur, India.

# Seismic Performance of Building Frames Considering Soil–Structure Interaction

**M. G. Kalyanshetti and S. A. Halkude**

**Abstract** It is a fact that response of the soil influences the motion of the structure and the motion of the structure in turn influences the response of the soil which is known as soil–structure interaction (SSI). The present study aims to evaluate the seismic performance of building frames considering SSI effect on scale-down steel building frame model prepared corresponding to RC prototype building frames of various height which are designated with reference to H/B ratio (H and B being height and width of building, respectively) as low-rise building for H/B less than 3.5, mid-rise building for H/B in between 3.5 and 5.25, and high-rise building for H/B higher than 5.25. The building frames considered are square in plan with single bay in both directions resting on GW soil. The experimental setup for fixed and flexible base condition is developed in the laboratory. The scaled-down models are subjected to El Centro time history using uniaxial servo hydraulic shake table. The effect of SSI on various dynamic parameters, i.e. acceleration, velocity, and displacement are studied. The study reveals that SSI effect is not significant in the low-rise building frame. However, with the increasing H/B ratio, SSI effect goes on becoming significant and predominant with nonlinear response. The study further reveals that mid-rise building frames is severely affected than high-rise building frame. In high-rise building frame, lower 30% portion is not affected by SSI. The middle 30% portion is producing almost same results for both the base condition. However, upper 40% portion of the building is significantly affected by SSI. Therefore, it is inferred that, not all, but certain building frames get affected by SSI depending upon their configuration such as H/D ratio and plan geometry apart from ground input motion (El Centro) supported on soil (GW) soil. Hence, it is recommended to incorporate SSI in the analysis of mid-rise and high-rise building frames to identify its vulnerability against safety and stability.

M. G. Kalyanshetti (✉)
Civil Engineering Department, Walchand Institute of Technology, Solapur, India
e-mail: mgkalyanshetti@gmail.com

S. A. Halkude
P. A. H. Solapur University, Solapur, India

 
T. G. Sitharam et al. (eds.), *Seismic Design and Performance*, Lecture Notes in Civil Engineering 120, https://doi.org/10.1007/978-981-33-4005-3_17

**Keywords** Soil–structure interaction (SSI) · Fixed base · Flexible base · Shake table · Scaled-down model · Time history

## 1 Introduction

Soil–structure interaction (SSI) includes a set of mechanisms accounting for the flexibility of the foundation support beneath a given structure resulting in altering the ground motion around the vicinity of the foundation compared to the free field. It determines the actual loading experienced by the soil–structure system resulting from the free-field seismic ground motions. The 1985 Mexico City earthquake and 1995 Kobe earthquake clearly illustrate the importance of local soil properties on the performance of structure. Many computational and analytical approaches for solving soil–structure interaction problems were identified [1]. Traditionally, it has been considered that SSI can be conveniently neglected for conservative design and also neglecting SSI tremendously reduces the complication in the analysis of the structures tempting designers to neglect the effect of SSI in the analysis [2]. Unfortunately, the assumption does not always hold true. During theoretical analysis, certain set of assumptions regarding idealization of the material, boundary conditions, etc. are required to be made to simplify the analysis to simulate the field conditions. However, these are not always true representation of what is happening on field. Therefore, in order to investigate the realistic behaviour, experimental study needs to be carried out which gives a valuable insight with respect to limitations of various assumptions made while carrying out theoretical analysis [3]. Halkude et al. investigated the SSI effect on various type of soil by winklers model and elastic continuum model. It is inferred that the SSI effect is greatly influenced by soft soil [4]. S. Hamid Reza et al. have studied an enhanced numerical soil–structure model which treats the behaviour of soil and structure with equal rigour. The proposed numerical soil–structure model has been verified and validated by performing experimental shaking table tests [5]. Akanshu Sharma et al. tested a 1:5 scaled reinforced concrete (RC) framed structure on shake table under dynamic loads to study effects of masonry infill panels and TLD in reducing the seismic response of structure [6]. Jayalekshmi B. R. evaluated the SSI effects on the seismic response of structures founded on Shedi soil of Dakshina Kannada by conducting experimental study on 1:10 scaled single-bay three-dimensional multistorey building models made of aluminium. The experimental results were presented and the modifications in dynamic characteristics due to the incorporation of soil flexibility were presented [7].

Present experimental study aims to compare the seismic performance of building frames under the conventional fixed base conditions and flexible base conditions, on scale-down experimental steel model prepared using similitude laws. The buildings are assumed to rest on GW soil and are subjected El Centro time history motion.

Following are the objectives of proposed study.

1. To validate the experimental study and thereby study the impact of SSI on the seismic performance of building frame of various heights by studying impact on various dynamic parameters such as acceleration, velocity, displacement.
2. To formulate general guidelines regarding incorporation of SSI in the analysis by identifying SSI sensitivity of building frames with reference to their geometrical configuration.

## 2 Prototype RC Building Frame Considered for the Analysis

For the present study, three RC building frames of G+3, G+5, and G+7 are considered square in plan having single bay in both directions with 'H' being the total height of building and 'B' being the width of the building. These buildings are designated as low-rise building to G+3 with H/B equal to 3.5, mid-rise building to G+5 with H/B equal to 5.25, and high-rise building to G+7 with H/B equal to 7. The geometry of building frames is decided based on the feasibility of experimental study on scale-down steel model and the guidelines of I.S. codes [8, 9]. The details of dimensional characteristics are illustrated in Table 1.

**Table 1** Properties of RC building frames

| Sr. No. | Contents | Description | | |
|---|---|---|---|---|
| 2 | No. of stories | G+3 | G+5 | G+7 |
| 3 | Storey height | 3.5 m | 3.5 m | 3.5 m |
| 4 | Grade of concrete | M 25 | M 25 | M 25 |
| 5 | Grade of steel | Fe415 | Fe415 | Fe415 |
| 6 | Bay width | 4 m | 4 m | 4 m |
| 7 | Slab thickness | 0.15 m | 0.15 m | 0.15 m |
| 8 | Size of column | 0.45 m × 0.3 m | 0.45 m × 0.3 m<br>*Top three stories*<br>0.5 m × 0.3 m<br>*Bottom three stories* | 0.45 m × 0.3 m<br>*Top four stories*<br>0.55 m × 0.3 m<br>*Bottom four stories* |
| 9 | Size of beam (DxB) | 0.4 m × 0.3 m | 0.4 m × 0.3 m | 0.4 m × 0.3 m |
| 10 | Floor finish | 0.6 $kN/m^2$ | 0.6 $kN/m^2$ | 0.6 $kN/m^2$ |
| 11 | Live load | 4 $kN/m^2$ | 4 $kN/m^2$ | 4 $kN/m^2$ |
| 12 | Seismic zone | III | III | III |

# 3 Scaled-Down Model

The important part for experimental study is to develop an experimental model which will able to represent identical behaviour as that of prototype structure. However, it is found that every aspect of the structure could not be scaled due to limitation of pay load capacity of Shake Table. But it should not be considered as a matter of big concern, since the objective of the test is not to qualify the prototype structure by testing but to study the behaviour of structures, in general, subjected to dynamic loading. The major task in the scaling down process is to achieve 'dynamic similarity' where model and prototype experience homologous forces. For this purpose, approach suggested by Meymand [10] is adopted. According to this approach, three principal test conditions establish many of the scaling parameters. The first condition is that testing is conducted in 1-g environment, which defines model and prototype accelerations to be equal. Secondly, a model with similar density to the prototype is desired, fixing another component of the scaling relations. Thirdly, the test medium is primarily composed of GW soil, whose undrained stress–strain response is independent of confining pressure, thereby simplifying the constitutive scaling requirements. In addition to the three principal test conditions, Meymand [10] pointed out that the natural frequency of the prototype should be scaled by an appropriate scaling relation. By defining scaling conditions for density and acceleration, the mass, length, and time scale factors can all be expressed in terms of the geometric scaling factor (S), and a complete set of dimensionally correct scaling relations (ratio of prototype to model) can be derived for all variables being studied. Based on above discussion, following two conditions are established for preparing scaled-down model:

(1) Natural frequency of the prototype shall be scaled by an appropriate scaling relation to that of model.
(2) Density of the prototype and model shall be similar.

## 3.1 Scale Factor

Adopting appropriate geometric scale factor is one of the important steps in scale modelling on shake table. Although small scale models could save cost, the precision of the results could be substantially reduced. Considering the specifications of shake table, the c/c distance between two columns is set as 0.32 m leading to a linear scale factor, of $4.0/0.32 = 12.5$ (column spacing in prototype structure is 4 m). Therefore, employing geometric scaling factor of 1:12.5 as explained above, height, length, and width of the structural model is obtained as 1.120 m, 0.32 m, and 0.32 m, respectively. The scaling relations for the various parameter adopted in this study are shown in Table 2 [11].

**Table 2** Scaling relations in terms of geometric scaling factor [11]

| Parameters- | Mass density | Acceleration | Frequency | Time | Length | EI | Strain |
|---|---|---|---|---|---|---|---|
| Scale factor- | 1 | 1 | $S^{-1/2}$ | $S^{1/2}$ | S | $S^5$ | 1 |

## 3.2 Scaling Down of Building Frames

The scaled-down models of all the three building frames are prepared based on similitude laws. However, the scale-down procedure of G+3 building frame is discussed here as a representative case.

According to the first condition, the relation between natural frequency of model and prototype as per Table 2 is

$$\begin{aligned} f_m/f_p &= S^{-1/2} \\ &= 3.54 \end{aligned} \tag{1}$$

Natural frequency of the G+3 prototype structure as calculated by application software (modal analysis) is, $f_p = 1.4792$ Hz. Therefore, required frequency of the model ($f_m$) in accordance with Eq. (1) is 5.24 Hz.

Density of the prototype structure ($\rho_m$) is 264.01 kg/m$^3$. Therefore, as per the second condition, the mass of the structural model ($M_m$) is estimated as:

$$\begin{aligned} M_m &= \rho_m \times V_m \\ &= 264.01 \times (1.12 \times 0.32 \times 0.32) \\ &= 30.27\ \text{Kg} \end{aligned} \tag{2}$$

The dimensions of column and slab of scaled-down steel model are determined so that the weight of model nearly equals to 30.27 kg, and also, it satisfies stiffness and flexural rigidity as required by simulated laws. Considering all above, the dimensions of scaled-down steel model are worked out. The details are presented in Fig. 1a. Similar calculations were done for G+5 and G+7 steel models and the details of all the three models are presented in Table 3.

## 3.3 Validation

The accuracy of the scaled-down models is ascertained by comparing the roof acceleration of G+3 scaled-down model with the roof acceleration of same model developed using ETAB application software subjected to El Centro ground excitation. The model developed in ETAB is shown in Fig. 1b. The comparison of roof acceleration corresponding to experimental study and analytical study (using application software) is presented in Fig. 2. It is observed from Fig. 2 that the roof acceleration

**Fig. 1** G+3 model

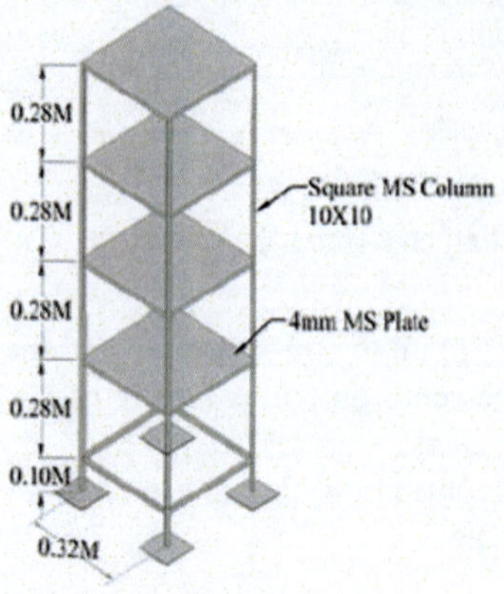


(a) Experimental model

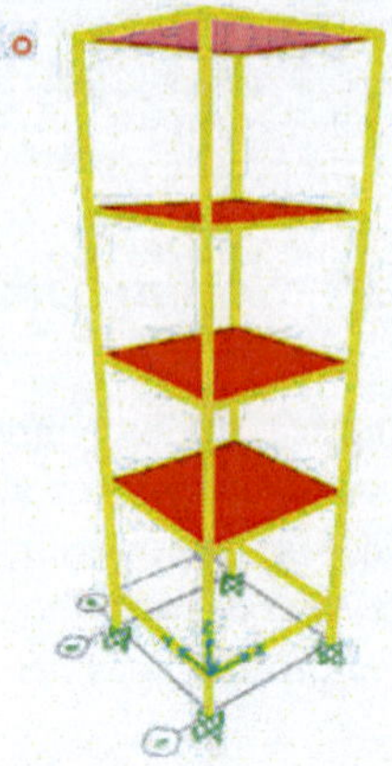

(b) ETAB model

**Table 3** Geometric and material properties of steel scaled-down model

| Sr. No | Contents | Description | | |
|---|---|---|---|---|
| 1 | Building frame | G+3 | G+5 | G+7 |
| 2 | Storey Height | 280 mm | 280 mm | 280 mm |
| 3 | Grade of Steel | Fe250 | Fe250 | Fe250 |
| 4 | Bay width | 320 mm | 320 mm | 320 mm |
| 5 | Slab thickness | 4 mm | 3 mm | 3 mm |
| 6 | Size of Column | 10 mm × 10 mm | 12 mm × 12 mm | 12 mm × 12 mm |
| 7 | Size of Plinth Beam | 10 mm × 10 mm | 12 mm × 12 mm | 12 mm × 12 mm |

**Fig. 2** Comparison of roof acceleration

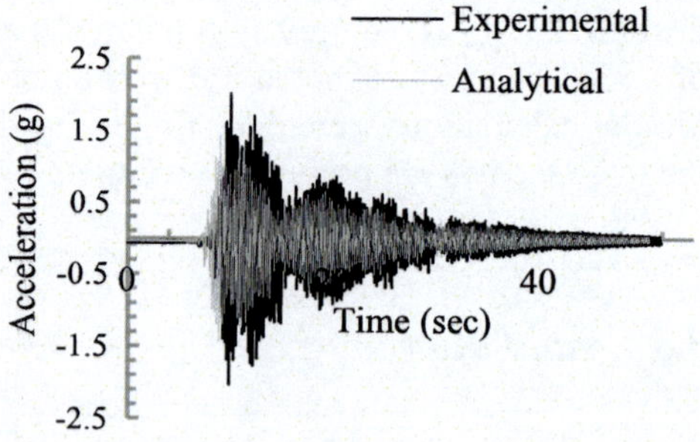


produced in experimental study is almost 10 to 15% higher than that of analytical study. This variation is obvious due to limitation in reproducing the replica model. In the analytical study, analysis is carried out by applying uniaxial excitation and acceleration is obtained in the same direction by restraining the movement of structure in other direction. However, in the laboratory even though the excitation is uniaxial but the model is not restrained to respond in only one direction; hence, the acceleration produced in the laboratory is higher than that produced by analytical study. Therefore, variation in the acceleration is considered to be reasonably within accepted limit, and thus, the scale-down model is validated. The same ideology is adopted to validate G+5 and G+7 model.

## 4 Experimental Setup on Shake Table

The shake table at the Civil Engineering Department, Walchand Institute of Technology, is uniaxially driven. The table size is 2 m × 2 m with maximum payload capacity of 30 kN with an operating frequency range of 0.01–50 Hz. Objective of the present study is to evaluate the change in the various responses of structure such as acceleration, velocity, displacement, etc. for fixed base and flexible base condition. Accordingly, experimental setups are developed in the laboratory to produce fixed base condition and flexible base condition. These are described in following sections.

### *4.1 Fixed Base Condition*

The fixed base response of the scaled-down model is investigated by securing the foundation of the model directly to the platform of the shaking table. Four accelerometers are used to acquire the acceleration data. Accelerometer no. 1 is placed at bottom of shake table (actuator) to verify the preciseness of simulations of the ground acceleration. Accelerometer no. 2, 3, and 4 are used to record accelerations of various floor levels. The fixed base setup is shown in Fig. 3.

### *4.2 Flexible Base Condition*

Flexible base conditions refer to the case, wherein the foundation along with the sub soil is considered. To simulate this condition in the laboratory, a laterally confined soil mass beneath the footing is required to be used. Therefore, a container made of steel plates is used to produce this lateral confinement. The experimental scaled-down model is square in plan; therefore, in order to have equal quantity/mass of soil all around the model on all sides, the square container is used. The plan dimensions of this container are kept to the maximum possible extent considering the size limitation of shake table (2 × 2 $m^2$) and adequate space required for the fixtures for mounting the container on shake table. The depth of the container is decided based on the requirement to account for embedment depth [12]. Payload capacity (30 kN) of shake table is also considered while developing the setup. Thus, in view of all these, the steel container of size 1.5 m × 1.5 m × 0.7 m is used. The soil is filled in and

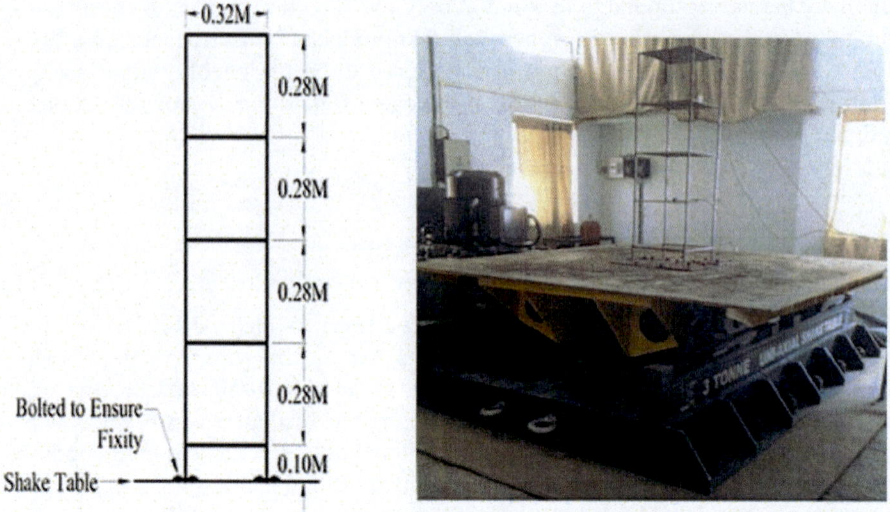


**Fig. 3** Experimental setup of fixed base condition

compacted in layers of 10 cm thickness with 95% of MDD. Then the experimental model is kept on the soil mass. The complete setup of the flexible base condition is shown in Fig. 4.

The locally available soil is used for the study. The various properties of soil used are determined in the laboratory in accordance with respective IS codes [13–16]. These properties are presented in Table 4.

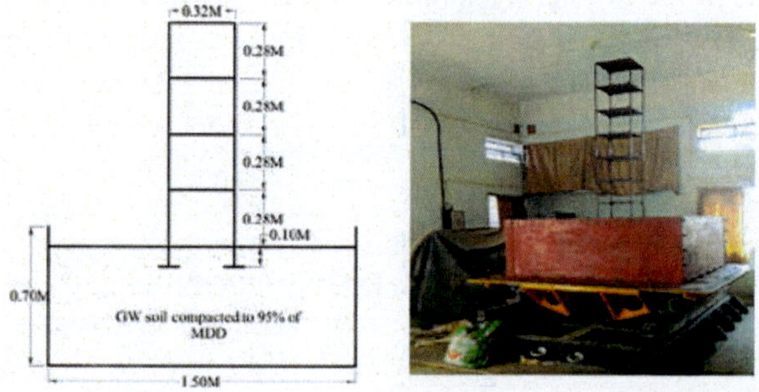


(a) Size details (b) Laboratory set up

**Fig. 4** Experimental setup of flexible base

**Table 4** Properties of foundation soil

| Sr. No. | Name of test/property | Result |
|---|---|---|
| 1 | Specific gravity | 2.63 |
| 2 | Compaction properties | |
| | Maximum dry density ($KN/m^3$) | 19.2 |
| | Optimum moisture content (%) | 14 |
| 3 | Strength parameters | |
| | Soil internal frictional angle (Ø) | 28° |
| | Cohesion 'C'($KN/m^2$) | 12 |
| 4 | Particle size distribution analysis | |
| | % of gravel, % of sand and % of fines | 71, 25, and 4 |
| | $C_u$, $C_c$ | 12.86, 1.69 |
| 5 | Classification of soil | GW(Well-graded gravel) |
| 6 | Modulus of elasticity 'E' ($kN/m^2$) | 50,000 |
| 7 | Poisson's ratio 'μ' | 0.3 |

### *4.3 Input Time History*

For the seismic analysis of structure, one of the most important tasks is to select the proper real ground motion. Among the many existing ground motions, the severest ground motion should be selected that can drive the structure to its critical response and thereby result in the highest damage potential. El Centro 1940 (E-W) record obtained during the Imperial Valley, California earthquake of 18 May 1940 is one of such severest ground motion record that is used extensively in research studies. Therefore, in the present study also, the El Centro time history used. The acceleration spectrum is presented in Fig. 5.

## 5 Results and Discussion

The scaled-down models are used in the laboratory to evaluate their responses for fixed and flexible base condition to investigate SSI effect. The laboratory setup as described in Sect. 4 is developed. All the models are subjected to El Centro time history. The acceleration is recorded by accelerometers, velocity is calculated by integrating acceleration, and further displacement is calculated by integrating velocity. Thus, these parameters at each floor level of all the building frames with fixed and

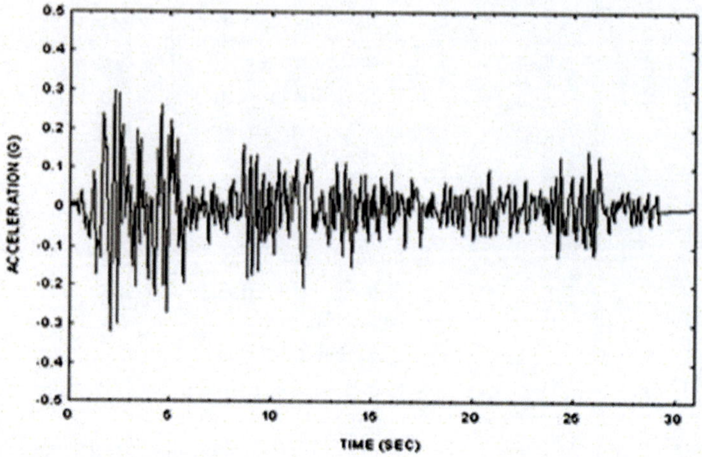


**Fig. 5** El Centro time history [17]

flexible base conditions are calculated and plots are developed to represent the variation of these parameters with respect to time. Typical plots of G+3 building only are shown below for fixed base condition and flexible base condition in Figs. 6 and 7, respectively. The plots are of responses at roof level.

It is observed from Fig. 6 that it is observed that fixed base condition is producing maximum acceleration of 2.10 g, maximum velocity of 0.4 m/s, and maximum displacement of 9.2 mm, whereas from Fig. 7, the same responses for flexible base condition are 1.23 g, 0.25 m/s and 5 mm, respectively. This shows that flexible base is producing almost 55–65% lower responses in comparison with fixed base condition. However, this response is of roof level only. Such plots of all floor levels are obtained to identify the effect of flexible base all over the building frame. Such kind of plots for G+5 and G+7 building frames are also obtained at all floor levels. The maximum acceleration, velocity, and displacement corresponding to fixed and flexible base of all building frame are compared to identify SSI effect. These plots are presented in next section.

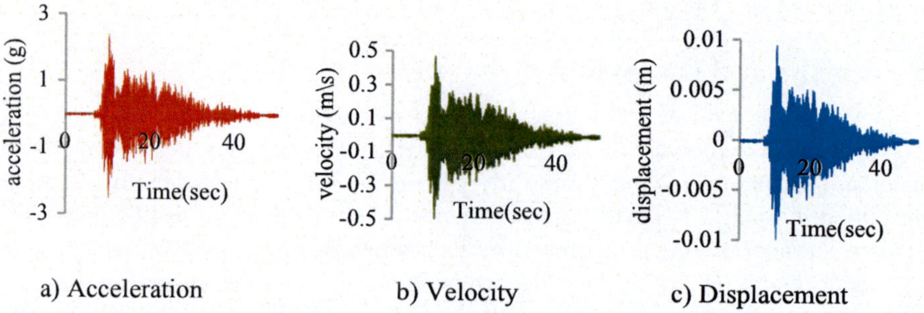


**Fig. 6** Fixed base response of G+3 building frame

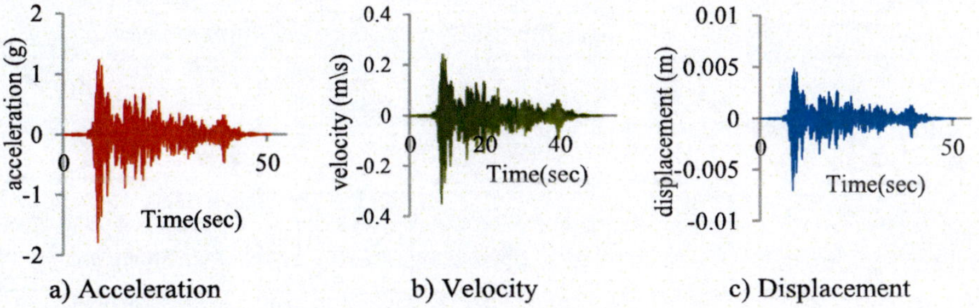


**Fig. 7** Flexible base response of G+3 building frame

## 5.1 Comparison of Fixed and Flexible Base Response

The comparison of dynamic response corresponding to fixed base and flexible base of G+3 (low-rise), G+5 (mid-rise), and G+7 (high-rise) building frame is presented in Figs. 8, 9 and 10, respectively.

(a) **G+3 (low-rise) building frame**:
It is observed from Fig. 8a, b that the acceleration and velocity corresponding to fixed base condition goes on increasing nonlinearly from first floor to fourth

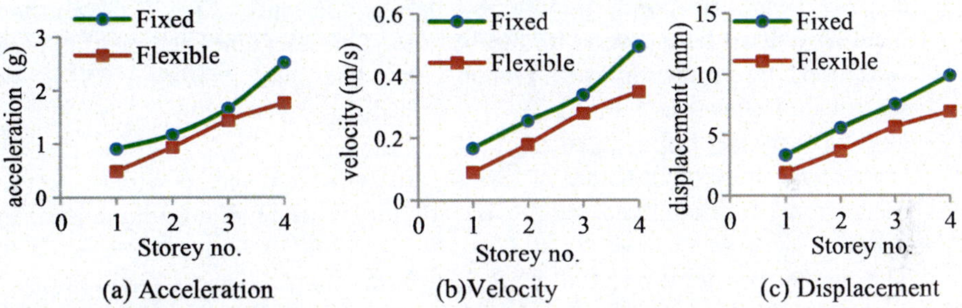


**Fig. 8** Comparison of dynamic parameters for G+3 (low-rise) building frame

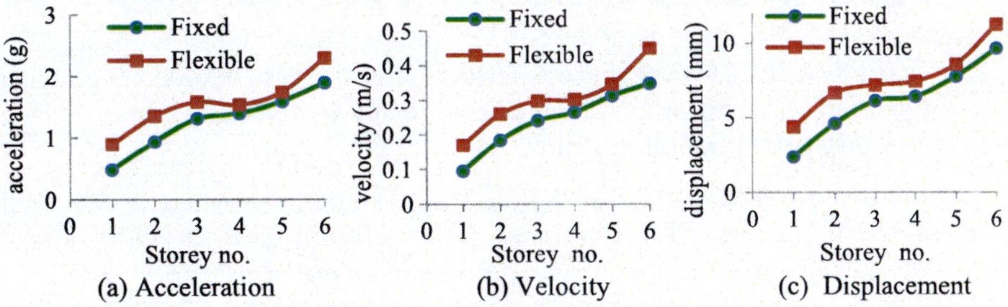


**Fig. 9** Comparison of dynamic parameters for G+5 (mid-rise) building frame

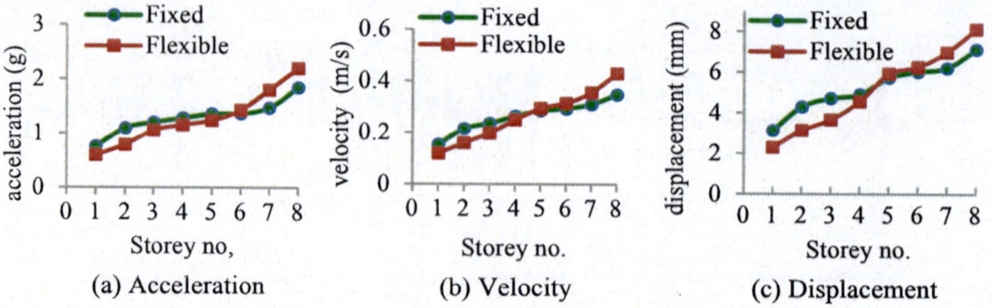


**Fig. 10** Comparison of dynamic parameters for G+7 (high-rise) building frame

floor. The rate of increment is observed to be in the range of 20% to 30% up to third floor level, which further increases to in the range of 45% to 50% from third to fourth storey. From Fig. 8c, it is observed that the displacement is increasing almost linearly at all floor levels in the range of 50% to 55% in comparison with previous floor.

The comparison between fixed and flexible base conditions presented in Fig. 8a–c reveals that flexible base condition is producing lower responses than fixed base. However, this difference is not the same in all floors. The acceleration and velocity are almost 10–15% lower than fixed base up to third floor, and suddenly, this increase becomes steeper up to 30% from third to fourth floor. However, displacement corresponding to flexible base is constantly 30–35% lower than fixed base at all floor levels. It is to be noted here that since the flexible base conditions are producing lower response than fixed base condition, SSI is not significant in low-rise building.

(b) **G + 5 (mid-rise) building frame**:

Figure 9a–c shows that nature of variation in dynamic response with respect to no. of storey for both fixed and flexible conditions are of alike and nonlinear in nature.

The comparison between fixed and flexible base conditions presented in Fig. 9a–c reveals that the response is reversed as compared to G+3 building. Flexible base condition is producing higher response than fixed base. Flexible base is producing almost 30–40% higher response in the bottom (upto 3rd floor) and top floors (floor 6–8), whereas in the middle floors (floor 3–6), the response is 20–25% higher than the fixed base condition. Thus, middle floors are least affected by SSI. However, since at all floor, flexible base is producing higher response than fixed base, SSI is significant in mid-rise building frame.

(c) **G + 7 (high-rise) building frame**:

It is observed from Fig. 10 that the reversal of trend is observed @ 5th floor. Flexible base is producing lower response than fixed base up to @ 5th floor, and beyond this, flexible base is producing higher response than fixed base.

The comparison between fixed and flexible base conditions presented in Fig. 10a–c reveals that unlike G+5 building frame, SSI effect is not significant at all floors.

In the lower 30% of total height, flexible base condition is producing 20–30% lower response than fixed base condition. In the mid-height from 30 to 60% of total height, the fixed and flexible base conditions are producing almost similar response with a marginal variation. In the remaining upper 40% of total height, flexible base condition is producing 30–40% higher response than fixed base condition. This indicates that SSI is significant in top 40% height of building frame.

## 6 Conclusion and Recommendation

In the present study, SSI effect is investigated experimentally using G+3, G+5, and G+7 scaled-down steel building frames which are designated as low-rise building (H/B equal to 3.5), mid-rise building (H/B equal to 5.25), and high-rise building (H/B equal to 7) respectively. It is well understood that SSI is influenced by structural configuration, supporting soil and the ground excitation. However, in the preset study supporting soil (GW soil) and the ground excitation (El Centro) being constant, the study is focused on assessment of performance of building frame only. Based on the obtained data, the following conclusions are drawn.

1. SSI effect is not significant for low-rise building frames. It is due to the fact that these buildings are not influenced by seismic forces. Hence, the conventional fixed base analysis is recommended for such buildings.
2. The study reveals that mid-rise and high-rise buildings are influenced by SSI. In mid-rise building, SSI effect is prominently observed as compared to high-rise buildings. In high-rise building, SSI effect is not significant in almost lower 60% of height, whereas only upper 40% of height is affected by SSI. This reveals that the structural configuration plays an important role. Due to this fact, in the present study, it is observed that the mid-rise buildings are greatly affected than high-rise building frame.
3. In case of any variation in the parameters, the performance of building frame is likely to vary.
4. The above discussions reveal that conventional analysis excluding SSI may not be adequate to ensure the structural safety. Therefore, it is recommended to incorporate SSI in the analysis of at least mid-rise and high-rise building frames.

## References

1. Gazetas G (1991) Formulas and charts for impedances of surface and embedded foundations. J Geotech Eng 117(9):1363–1381
2. Dutta SC, Bhattacharya K, Roy R (2004) Response of low-rise buildings under seismic ground excitation incorporating soil–structure interaction. Soil Dyn Earthq Eng 24:893–914
3. Hirave V, Kalyanshetti M (2018) Seismic response of steel braced building frame considering soil structure interaction (SSI): an experimental study. J Inst Eng India, Ser A 99(1):113–122

4. Halkude SA, Kalyanshetti MG (2014) Soil structure interaction effect on seismic response of R.C. frames with isolated footing. Int J Eng Res Technol 03(1):2767–2775
5. Tabatabaiefar SHR, Fatahi B, Samali B (2014) Numerical and experimental investigations on seismic response of building frames under influence of soil-structure interaction. Advan Struct Eng 17(1):109–130
6. Sharma A, Reddy GR, Vaze KK (2012) Shake table tests on a non-seismically detailed RC frame structure. Struct Eng Mech 41(1):1–24
7. Jayalekshmi BR, Lohith K, Shivashankar R, Venkataramana K (2008) Experimental investigation on dynamic characteristics of structures founded on dispersive soil. The 12th international conference of international association for computer methods and advances in geomechanics (IACMAG) Goa, India, pp 2717–2724
8. IS 1893 (part 1) (2002) Indian standard criteria for earthquake resistant design of structures, Bureau of Indian Standards, New Delhi
9. IS 456 (2000) Plain and reinforced concrete—Code of practice, Bureau of Indian Standards, New Delhi
10. Meymand PJ (1998) Shaking table scale model tests of nonlinear soil-pile-superstructure interaction in soft clay. Ph.D. thesis in civil engineering. University of California, Berkeley
11. Sabnis GM, Harris HG, White RN, Mirza MS (1983) Structural modeling and experimental techniques, prentice hall Inc., Engelwood Cliff, New Jersey
12. Bowles JE (1996) Foundation analysis and design. Tata McGraw-Hill, New Delhi
13. IS-2720 (Part 3) (1980) "Methods of test for soil—determination of specific gravity" Bureau of Indian Standards, New Delhi
14. IS-2720 (Part 7) (1980) Methods of test for soils—determination of water content-dry density relation using light compaction. Bureau of Indian Standards, New Delhi
15. IS-2720 (Part 13) (1986) Methods of test for soils—direct shear test. Bureau of Indian Standards, New Delhi
16. IS-2720 (Part 5) (1985) Methods of test for soils—determination of liquid and plastic limit. Bureau of Indian Standards, New Delhi
17. Chopra AK (2003) Dynamics of structures, theory and application to earthquake engineering. Prentice Hall, New Delhi

# Seismic Analysis of Reinforced Concrete Structure Resting on Liquefiable Soil

**M. Sankara Narayanan and Visuvasam Joseph Antony**

**Abstract** Soil-foundation-structure interaction has a great impact on reinforced concrete structures founded on liquefiable soil during seismic event. The main objective of this study is to investigate the effects of liquefaction on soil-foundation-structure systems. Non-linear time history analysis is performed on framed structures resting on soil-foundation system with and without considering the effects of liquefaction using finite element program. The results are compared with response of the structure with rigid base. It is observed that the increase in pore water pressure of sandy soil resulted in decrease of effective vertical stress and shear strain, eventually leading to liquefaction. The input acceleration is found to be amplified at the top of the structure causing the structure to settle and tilt to a larger extent.

**Keywords** Soil-structure interaction · PLAXIS3D · Excess pore pressure · Liquefaction

## 1 Introduction

India has encountered major earthquakes (Bhuj-2001) in the past decades leading to heavy damages and complete collapse of the structures. In high seismic prone areas with loose cohesionless soil like sand, the response of the superstructure to dynamic load is highly influenced by the substructure's response. Such phenomenon is called as Soil-Structure Interaction (SSI). Observations from 1989 San Francisco earthquake and 1995 Kobe earthquake indicated that the seismic response of soil-foundation structure has a crucial impact on the super structure's response. Hence, design of structures with SSI effects is significant to provide a suitable mitigation strategy and hence prevent the collapse of structures under earthquake forces.

The importance of SSI effects during a seismic event was accentuated by many researchers in the recent decades (Mylonakis et al. [1]; Stewart et al. [2]; Luco [3]; Khalil et al. [4]; Ayedemir and Ekiz [5] and Chen [6]). An experimental work on

M. S. Narayanan (✉) · V. J. Antony
School of Civil Engineering, Vellore Institute of Technology, Vellore, India
e-mail: msankar2003@gmail.com

 
T. G. Sitharam et al. (eds.), *Seismic Design and Performance*, Lecture Notes in Civil Engineering 120, https://doi.org/10.1007/978-981-33-4005-3_18

the SSI effect of a two storey shear building was carried out by Chang and Kim [7], through which an effective method for a pseudo-dynamic test was proposed that can simulate SSI in a shake table. Haiyang et al. [8] proposed an experimental method using shake table to determine the SSI effect on base isolated structures subjected to dynamic load. A direct approach method of numerical study was put through by Nguyen et al. [9] that detailed the importance of soil-structure-foundation interaction of a moment resting framed building with shallow foundation system. It was observed that the reduction in size of foundation lengthened the natural period the structure thereby significantly reducing the base shear. Kim et al. [10] performed non-linear seismic analysis on raft and pile foundations to determine the influence of soil-structure interaction.

Densification of soil occurs in a loose cohesionless saturated soil layer in undrained condition when subjected to seismic forces. Due to this, there will be an increase in excess pore pressure (EPP) and a reduction in effective stress because of the inability to dissipate EPP quickly. This phenomenon is called as liquefaction where the soil behaves as fluid. Such soil cluster leads to lateral spreading of soil and ground settlement. Moreover, buildings supported on such liquefiable soil will be subjected to high settlement, lateral displacement and tilting due to SSI effects. Significant damages to buildings founded on loose saturated sandy soil due to liquefaction were reported in past earthquake events such as 2001 Bhuj earthquake, 1995 Kobe earthquake and 1964 Niigata earthquake made [11].

Adalier et al. [12] conducted a centrifuge test on earth embankment founded on liquefiable soil to earthquake loads. Results show that the liquefaction caused lateral spreading and formation of large strains in the embankment leading to settlement. Bhatnagar et al. [13] validated the centrifuge results of Adalier et al. using finite element analysis and the results showed a good relevance with the experimental results. Dash et al. [14], made a case study on the failure of Kandla port tower due to 2001 Bhuj earthquake. Findings show that liquefaction could occur to sandy soil cluster even if it is present below a non-liquefiable layer. They reported that the tower tilted toward the seaside about 300 mm and the ground settled about 300 mm resulting in the settlement of raft foundation. Abd-Elhamed et al. [15] studied the behavior of RC structures resting on liquefiable soil by numerical analysis for 1999 Kocaeli, 1994 Northridge and 1978 Tabas ground motions. They also investigated the significance of using stone column technique for improving the liquefiable soil.

In this study, a numerical analysis is performed to evaluate the behavior of RC framed structure resting in a soil profile with a liquefiable soil layer underneath a non-liquefiable soil cluster (Bhuj—Kandla port soil profile). A five storey structure is modelled using Plaxis 3D and subjected to ground motion (2001 Bhuj). The lateral storey displacement, foundation settlement, ground settlement, acceleration amplification and the liquefaction potential of the flexible base structure systems are monitored and compared with rigid base superstructure response.

**Table 1** Seismic Parameters as per IS 1893 Part 1 [17]

| Parameter | Value |
|---|---|
| Importance factor | 1.5 |
| Zone factor | 0.36 |
| Response reduction factor | 5 |
| Time period | 0.4546 s |

## 2 Soil Profile

The soil profile in the vicinity of Kandla Port, Bhuj consists of unconsolidated clusters of clay and sand. Top layer is made of soft clay with liquid limit between 62 and 68% and plastic limit in the range of 26–28%. The lower layer consists of a 12 m thick fine and coarse sand and 12 m thick layer of hard clay with plastic limit in the rand of 39–64%. Groundwater level is present 1.5 m below the top soil surface as referred from Dash et al. [14]. The SPT ($N_{60}$) value for sand deposit is less than 15 and the unit weight is 17 kN/m$^3$.

### *2.1 Superstructure*

A five storey RC moment resisting framed structure of 3 bays in both *x* and *y* directions with a bay length of 4 m each and storey height of 3 m is modelled. The column and beam dimensions are 450 mm × 450 mm and 300 mm × 450 mm respectively adopted for the structure with a slab thickness of 150 mm. As per IS 456 [16], concrete grade of M30 and steel grade of Fe500 is provided for the structural members of the building. A total dead load of 13.5 kN/m is given to each beam and a live load of 4 kN/m$^2$ is provided. Response spectrum and equivalent static analysis are performed as per IS 1893 Part 1 [17] on rigid base superstructure and designed as per IS 456 [16] using SAP 2000. Table 1 shows the seismic parameters adopted for analysis and design.

### *2.2 Foundation System*

For a base shear of 1800 kN and total vertical load of 67,399 kN, a 14 m × 14 m raft foundation of 1.5 m thickness is designed using conventional design procedures as per IS 2950 part 1 [18] and IS 456 [16]. M30 grade concrete and Fe500 grade steel are provided for the raft foundation system with elastic modulus E of 27,000 MPa.

# 3 Numerical Modelling

The numerical modeling is performed using a three-dimensional finite element program Plaxis 3D [19]. It is specially designed to perform a wide range of geotechnical applications like nonlinear dynamic analysis, flow and stability analysis. It provides advanced material models to simulate various behaviors such as pre-consolidation, soil creep, liquefaction, anisotropy and cyclic behavior.

## 3.1 UBC3D-PLM Soil Model

Liquefaction phenomenon in sandy soils is simulated in PLAXIS3D using UBC3D-PLM which is an effective elasto-plastic model and a 3D formulation of UBCSAND model Tsegaye 2010 [20]. This model uses a densification rule along with two yield surfaces to facilitate a perfect transition into liquefied state and precise development of excess pore pressure. Stress dependent stiffness, densification, post-liquefaction stiffness, Mohr-Coulomb failure criterion and plastic straining are the main parameters of this soil model. Parameters adopted for UBC3D—PLM in Plaxis 3D is provided in Table 2.

**Elasto Plastic Behaviour**

The bulk and shear modulus of the saturated soil given by following equations are used to define the elastic behavior of the UBC3D-PLM model (Tsegaye [20]).

$$K = k_B^* p_{\text{ref}} \left( \frac{p'}{p_{\text{ref}}} \right)^{\text{me}} \tag{1}$$

$$G = k_G^* p_{\text{ref}} \left( \frac{p'}{p_{\text{ref}}} \right)^{\text{ne}} \tag{2}$$

where, $k_G^*$ and $k_B^*$ are shear and bulk modulus factors respectively, $p'$ and $p_{\text{ref}}$ are the effective stress and reference pressure respectively, me and ne are the stress dependent stiffness powers.

The plastic behavior is simulated using the Hardening rule which is based on the transition of the shear strength. Tsegaye [20] reformulated the hardening rule which is given by the following Eq. (3).

$$d \sin \Phi_{\text{mob}} = 1.5 k_G^{*p} p_{\text{ref}} \left( \frac{p'}{p_{\text{ref}}} \right)^{np} \frac{p_{\text{ref}}}{p'} \left( 1 - \frac{\sin \Phi_{\text{mob}}}{\sin \Phi_p} R_f \right)^2 d\lambda \tag{3}$$

where, $\Phi_{\text{mob}}$ is the mobilized friction angle, $np$ is the plastic shear modulus exponent, $k_G^{*p}$ is the plastic shear modulus factor, $R_f$ is the failure ratio (0.5–1), $\Phi_p$ is the peak friction angle and $d\lambda$ is the plastic strain increment multiplier.

**Table 2** UBC3D-PLM parameters

| Parameter | Bhuj sand |
|---|---|
| Initial void ratio, $e_{\text{ini}}$ | 0.54 |
| Rate of stress dependency of plastic shear modulus, *np* | 0.4 |
| Cohesion, $c$ (kPa) | 0 |
| Unsaturated unit weight $\gamma_{\text{unsat}}$ (kN/m$^3$) | 17 |
| Saturated unit weight $\gamma_{\text{sat}}$ (kN/m$^3$) | 20.38 |
| Elastic shear modulus factor, $k_B^{*e}$ (kPa) | 654.46 |
| Elastic bulk modulus factor, $k_G^{*e}$ (kPa) | 934.95 |
| Plastic shear modulus factor, $k_G^{*p}$ (kPa) | 380.48 |
| Rate of stress dependency of elastic bulk modulus, *me* | 0.5 |
| Rate of stress dependency of elastic shear modulus, *ne* | 0.5 |
| Rate of stress dependency of plastic shear modulus, *np* | 0.4 |
| Constant volume friction angle, $\Phi_{\text{cv}}$ | 33° |
| Peak friction angle, $\Phi_{\text{p}}$ | 33° |
| Tension cut-off and tensile strength, $\sigma_t$ (kPa) | 0 |
| Corrected SPT value, $(N_1)_{60}$ | 10 |
| Densification factor, $f_{\text{dens}}$ | 1 |
| Post-liquefaction factor, $f_{\text{Epost}}$ | 0.3 |
| Reference pressure, $p_{\text{ref}}$ (kPa) | 100 |
| Failure ration, $R_f$ | 0.778 |

**Flow Rule**

The flow rule proposed by Puebla et al. [21] for UBCSAND model is used in UBC3D-PLM model. Constant volume friction angle $\Phi_{cv}$ defines the plastic volumetric strain similar to the Hardening soil model.

**Densification Law—Pore Water Pressure Calculation**

The densification law is incorporated for precise calculation of excess pore pressure. This property is defined by a modified version of the input parameter, Plastic shear modulus factor $k_G^{*p}$. This factor defines the counting of cycles in secondary loading leading to an increase in EPP until liquefaction has started. This modified factor is given by the following equation Tsegaye 2010 [20].

$$k^{*p}_{G,\text{ Secondary}} = k_G^{*p}\left(4 + \frac{n_{\text{rev}}}{2}\right) \text{hard } f_{\text{dens}} \tag{4}$$

where $n_{rev}$ is the number of shear stress reversals from loading to unloading or vice versa, hard is the correction factor for densification rule and $f_{dens}$ is the adjustment factor for densification rule.

Through this law which is implicitly provided in Plaxis 3D, the pore water pressure is determined at each time step of the analysis. The total stress increment for a saturated soil specimen during loading is computed using the given equation:

$$dp = K_u d\varepsilon_v \tag{5}$$

where $dp$ is the stress increment, $d\varepsilon_v$ is the volumetric soil strain and $K_u$ is the soil bulk modulus. Similarly, the pore water pressure increment is computed through the following equation:

$$dp_w = \frac{K_w}{\eta} d\varepsilon_v d\varepsilon_v \tag{6}$$

where, $dp_w$ is the excess pore pressure, $\eta$ is the porosity of soil and $K_w$ is the bulk modulus of water.

**Post-Liquefaction Rule**

Volumetric locking, a post-liquefaction behavior prohibits the modelling of stiffness degradation of both loose and dense sandy soils. This restraint is subdued by decreasing the plastic shear modulus and thereby decreasing the soil stiffness after the unloading phase. This stiffness degradation is calculated using the below equation (Tsegaye [20]).

$$k^{*p}_{G\,\text{post-liquefaction}} = k^{*p}_{G} E_{\text{dil}} \tag{7}$$

$$E_{\text{dil}} = \max\left(e^{-110\varepsilon\,\text{dil}};\ f_{\text{Epost}}\right) \tag{8}$$

where, $k^{*p}_{G\,\text{post - liquefaction}}$ is the plastic shear modulus during liquefaction, $\varepsilon_{\text{dil}}$ is the plastic deviatoric strain in the dilation of soil and $f_{\text{Epost}}$ is the value that limits the minimum value of $E_{\text{dil}}$.

## 3.2 *Hardening Soil Model*

The top and the bottom clay layers are modelled using hardening soil model. Stress dependency of soil stiffness is an important feature of this model. Other characteristics of the model are plastic straining, elastic unloading/reloading and Mohr-Coulomb failure criterions [22]. The parameters used for the top and bottom layer is given in Table 3.

**Table 3** Hardening soil properties

| Parameter | Soft clay | Hard clay |
|---|---|---|
| Unsaturated unit weight $\gamma_{\text{unsat}}$ (kN/m$^3$) | 16 | 18 |
| Saturated unit weight $\gamma_{\text{sat}}$ (kN/m$^3$) | 17 | 19 |
| Initial void ratio, $e_{\text{ini}}$ | 0.5 | 0.5 |
| Secant stiffness in standard drained triaxial test, $E_{50}^{\text{ref}}$ (kPa) | 2000 | 2500 |
| Tangent stiffness for oedometer loading $E_{\text{oed}}^{\text{ref}}$ (kPa) | 2000 | 2500 |
| Unloading Stiffness, $E_{ur}^{\text{ref}}$ (kPa) | 10,000 | 20,000 |
| Cohesion, $c$ (kPa) | 5 | 10 |
| Angle of internal friction, $\Phi$ | 25 | 25 |
| Angle of dilatancy, $\psi$ | 0 | 0 |
| Power, $m$ | 1 | 1 |

### *3.3 Superstructure, Foundation and Boundary Conditions*

Beams and columns in the building are modelled using 3-noded beam element, while the foundation and floor slab is modelled using 6-noded plate element as shown in Fig. 1. They are modelled with elasto-plastic behavior. The input material parameters for the structural elements are given in Table 4. A damping of 5% is provided to the building using Rayleigh coefficients. 12-node interface elements are created in the vertical boundaries of x direction to simulate free-field boundary condition which absorbs the reflected secondary waves. A compliant base boundary condition is provided along $Z_{\text{min}}$ boundary to have minimum reflection at the bottom boundary.

**Fig. 1** 3D model in PLAXIS 3D showing the soil-foundation-structure system

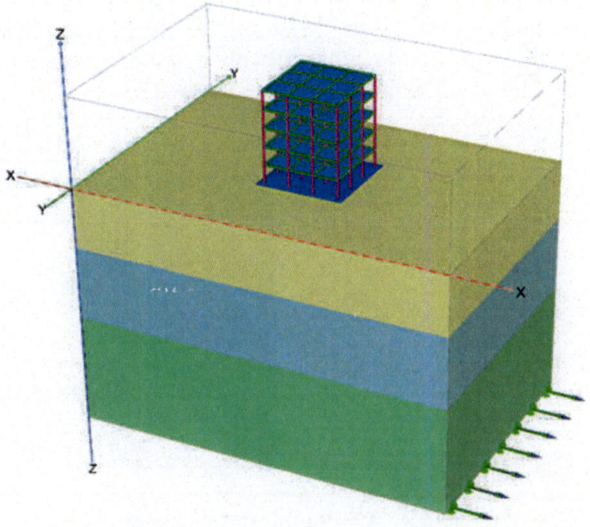

**Table 4** Input properties of structural elements

| Parameter | Raft | Beam, column, slab |
|---|---|---|
| Unit weight (kN/m$^3$) | 25 | 25 |
| Elastic modulus (GPa) | 28 | 28 |
| Poisson's ratio | 0.3 | 0.3 |

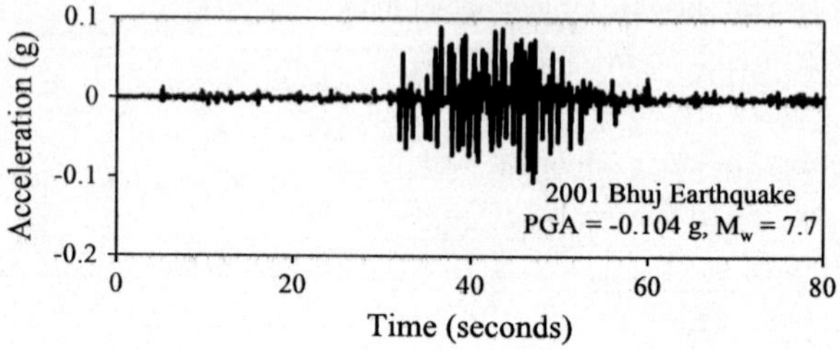


**Fig. 2** Input ground motion—2001 Bhuj

## 3.4 *Input Ground Motion*

The soil-foundation-structure is analyzed for 2001 Bhuj ground motion, which has a magnitude of 7.7 Mw and a peak ground acceleration of −0.104 g. Figure 2 shows the acceleration versus time data.

## 3.5 *Time History Analysis*

Plaxis 3D uses Newmark Time Integration Scheme to perform the non-linear time stepping analysis which is an implicit integration method that produces stable and reliable calculation process with a more accurate result. The displacement and velocity at any point $t + \Delta t$ are denoted by:

$$u^{t+\Delta t} = u^t + \dot{u}^t \Delta t + \left((0.5-\alpha)\ddot{u}^t + \alpha \ddot{u}^{t+\Delta t}\right)\Delta t^2 \quad (9)$$

$$\dot{u}^{t+\Delta t} = \dot{u}^t + \left((1-\beta)\,\ddot{u}^t + \beta\,\ddot{u}^{t+\Delta t}\right)\Delta t \quad (10)$$

where $\alpha = 0.25$ and $\beta = 0.5$ are Newmark coefficients.

The time step of a dynamic analysis depends on the material properties, element size, sampling time of input wave and the number of subsets required for the time history analysis. The critical time step of a single element is given by,

$$\Delta t_{\text{critical}} = I_{\text{min}}/V_s$$

where $I_{\text{min}}$ and $V_s$ are the minimum nodal distance and shear wave velocity of an element. The time step is equal to the average critical time step of all elements. Once

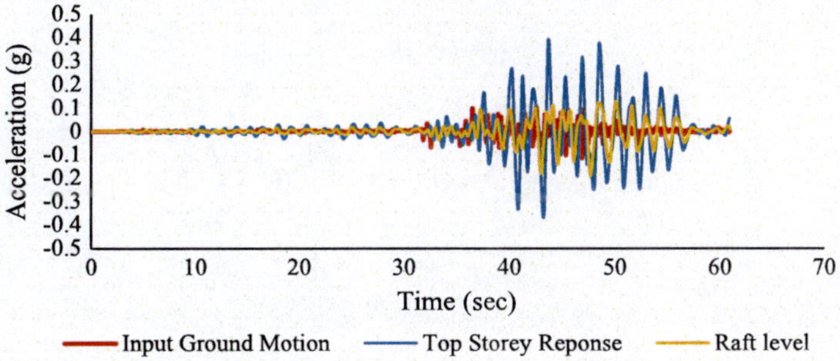


**Fig. 3** Acceleration response for 2001 Bhuj earthquake

the time step is fixed, the dynamic integration coefficients necessary for time history analysis and the computation of displacement, velocity and acceleration at the end of each time step is calculated using Newmark parameters $\alpha$ and $\delta$.

## 4 Results and Discussion

The results are examined by means of top storey displacement, relative storey displacement of each storey, Inter Storey Drift (IDR) ratio in terms of percentage, ground motion response in terms of acceleration of the structure to the input ground motion and excess pore pressure in the UBC3D-PLM layer due to liquefaction.

### *4.1 Acceleration Response*

Figure 3 shows the acceleration response of Liquefaction model for the 2001 Bhuj earthquake. It is observed that a peak of 0.37 g is attained in the 43rd second at the top storey level of the liquified model building, which is 3 times greater than the PGA of input ground motion. The peak acceleration at the soil top is observed to be −0.19 g which is 1.8 times greater than the PGA of input motion. The results clearly show that the acceleration of the ground motion is amplified at various levels.

### *4.2 Storey Displacement*

Figure 4 shows the top storey lateral deflection in x-direction for the input Bhuj earthquake for 3 models such as rock base, without liquefaction and with liquefaction models. The maximum lateral displacement of 390 mm is recorded for the model undergoing liquefaction, while that of the models without liquefaction and rock

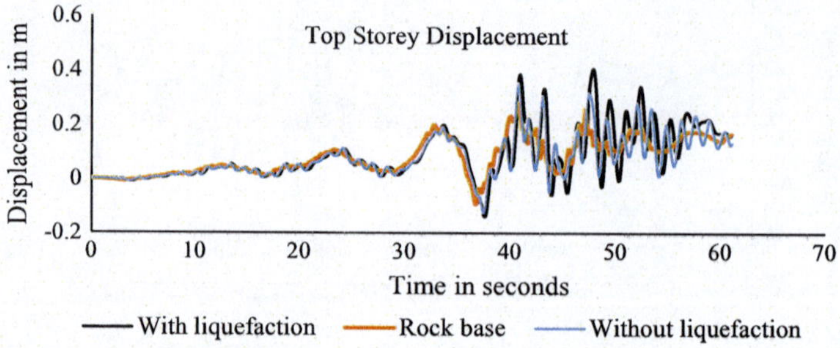


**Fig. 4** Top storey displacement response

base are 320 mm and 260 mm respectively. The lateral displacement is observed to be increased by 50% due to the effect of soil structure interaction. Moreover, it is observed that the effect of SSI is low when the liquefaction phenomenon is absent.

Figure 5a shows the maximum relative displacement at each storey level for all the 3 models. It is noted that the model that doesn't undergo liquefaction shows a higher relative displacement with a maximum of 110 mm in the top storey, which is 426% higher than that of rock base model. It is also observed that the models with and without liquefaction show a linear increase in relative displacement.

Figure 5b shows the comparison of IDR ratio (%) of all the 3 models. The IDR increased from 0.7 to 0.8% when liquefaction occurred denoting a minimal change. Also, the IDR increased by 100% for the model with liquefaction compared to rock base model, which is due to the effect of soil structure interaction (Fig. 6).

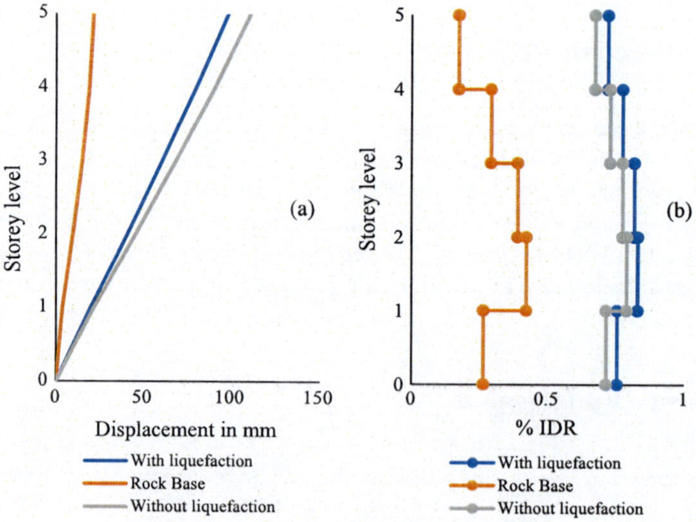


**Fig. 5** **a** Relative storey Displacement **b** Maximum Inter storey drift

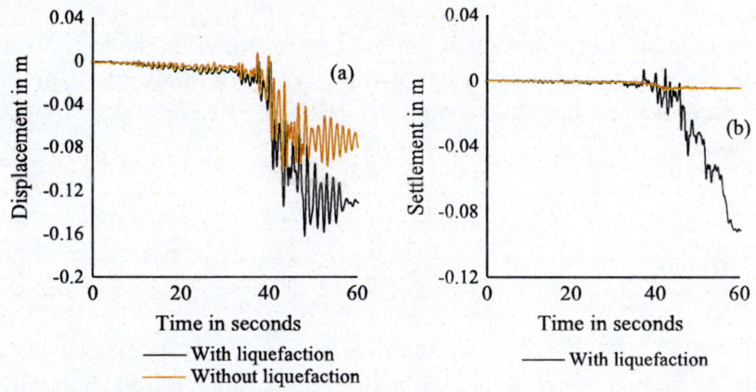


**Fig. 6** **a** Raft settlement **b** Soil settlement

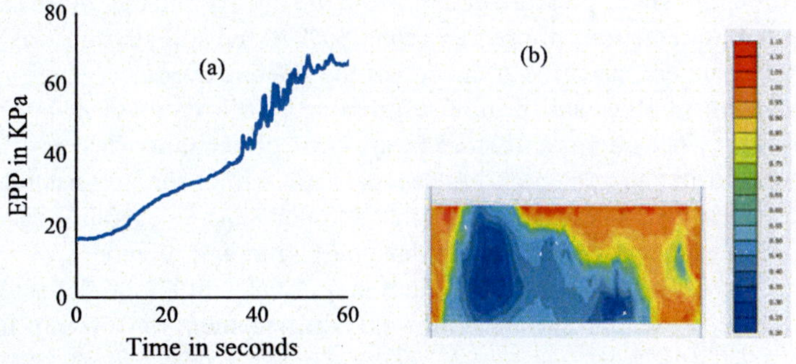


**Fig. 7** **a** EPP at bottom of soil layer 2 **b** Max EPP ratio cross section diagram

## 4.3 Excess Pore Pressure

Figure 7a shows the excess pore pressure recorded at the bottom layer of the liquefied sand directly below the building area. It shows a maximum EPP of 66 kPa, which is responsible for the liquefaction of the soil layer. Figure 7b shows the EPP ratio where most of the soil areas have been liquefied which is indicated by EPP ratio greater than 0.75.

## 4.4 Settlement—Raft and Soil

Figure 6a and b show the raft and soil settlement for the input 2001 Bhuj earthquake. A maximum raft settlement of 162 mm is observed for the liquefied model which is 59% greater than that of the model without liquefaction. The final raft settlement of 124 mm and 66 mm is observed at the end of the seismic event for the liquefied and non-liquefied model. Liquefaction of the soil layer resulted in the top soil settlement of 95 mm at the end of the seismic event, whereas only a 4 mm settlement is observed in the non-liquefied model. Moreover, the raft ended with a 0.8° rocking in the liquefied model.

## 5 Conclusion

The current study suggests a new numerical procedure to determine the response of structure embedded in a liquefiable soil stratum subjected to dynamic loading. Importance of soil structure interaction is clearly depicted from the following conclusions:

- The excess pore water pressure generated in the unconsolidated loose sandy soil, accelerated liquefaction, where more than 50% of the soil stratum was liquefied (EPP ratio greater than 0.75) at the end of the seismic event.
- The inclusion of Bhuj soil profile resulted in a 50% increase in lateral storey displacement when compared to rock base model, on account of SSI consideration.
- Inter-storey drift ratio for rock base model falls within the permissible limit of IS Code 1893 [23], whereas the liquefied and non-liquefied models have a much greater IDR suggesting a revision in design of structural members.
- Liquefaction of the soil stratum resulted in a permanent lateral deflection of the structure greater than 100 mm, with a raft displacement of 150 mm indicating failure of the structure.
- The varying shear wave velocity of the soil layers beneath the structure, resulted in amplification of the seismic wave passing through it up to 2 times and the superstructure further amplified the seismic wave 3 times.
- Soil liquefaction due to seismic load resulted in a permanent settlement of the building by 125 mm and a soil settlement of 100 mm leading to the failure of foundation system.
- Up to 0.8° rocking of the raft foundation due to liquefaction is recorded which lead to the lateral deflection.
- Thus, the effect of SSI and liquefaction is clearly shown with an alternate finite element numerical procedure to perform dynamic analysis.

## References

1. Mylonakis G, Gazetas G (2000) Seismic soil structure interaction: beneficial or detrimental. J Earthquake Eng 4(3):277–301
2. Jonathan SP, Seed RB, Fenves GL (1999) Seismic soil structure interaction in buildings. II: empirical findings. J Geotech Geoenvironmental Eng 125(1):38–48
3. Luco JE, Trifunac M, Wong HL (1988) Isolation of soil structure interaction effects by full-scale forced vibration tests. Earthq Eng Struct Dyn 16(1), 1–21 (1988)

4. Khalil L, Sadek M, Shahrour I (2007) Influence of the soil structure interaction (SSI) on the fundamental frequency of structures. In: 4th Internation conference on earthquake geotechnical engineering. Paper no. 1545
5. Ayedemir ME, Ekiz I (2013) Soil-Structure Interaction effects on seismic behavior of multistrorey structures. Eur J Environ Civil Eng 17(8):635–653
6. Chen S, Jun-Yang S, Yi-Ying W (2019) Structure and infrastructure engineering 1573–2479
7. Chang M, Kwan Kim J (2019) Pseudo-dynamic test for soil-structure interaction analysis using shake tables. KSCE J Civ Eng
8. Haiyang Z, Jisai F, Xu Y, Su C, Xiaohui C (2019) Earthquake responses of a base isolated structure on a multi-layered soft soil foundation by using shaking table tests. Eng Struct 179:79–91
9. Nguyen QV, Fatahi B, Hokmabadi AS (2017) Influence of size and load bearing mechanism of piles on seismic performance of buildings considering soil-pile structure interaction. Int J Geomech 1–22
10. Kim YS, Roesset JM (2004) Effect of nonlinear soil behaviour on inelastic seismic response of structure. Int J Geomech 4(2):104–114
11. Dash SR, Bhattacharya S (2012) Mechanism of failure of three pile-supported structures during three different earthquakes. In: 15th world conference on earthquake engineering
12. Adalier K, Elgamal AW, Martin GR (1998) J Geotech Geoenvironmetnal Eng 124:500–517
13. Bhatnagar S, Kumari S, Sawant VA (2015) Numerical Analysis of Earth Embankment resting on liquefiable soil and remedial measures. Int J Geomech
14. Dash SR, Govindaraju L, Bhattacharya S (2009) A case study of damages of Kandla port and customs office tower supported on a mat-pile foundation in liquefied soils under the 2001 Bhuj earthquake. Soil Dyn Earthq Eng 29:333–346
15. Abd-Elhamed A, Sayed M (2019) Seismic response evaluation of structures on improved liquefiable soil. Eur J Environ Civil Eng
16. IS 456 (2000) Plain and reinforced concrete-code of practice. Bureau of Indian Standards, New Delhi, Inida
17. IS 1893 (Part 1) (2016) Criteria for earthquake resistant design of structures: general provisions and buildings. Bureau of Indian Standards, New Delhi, India
18. IS 2950 (Part 1) (1981) Code of practice for design and construction of raft foundations
19. Rollins KM, Dusty Lane J, Travis Gerber M (2005) Measured and computed lateral response of a pile in sand. J Geotech Geoenvironmental Eng 131(1)
20. Tsegaye AB (2010) Plaxis liquefaction model report No.1, Plaxis Knowledge base
21. Puebla H, Byrne M, Phillips P (1997) Analysis of canlex liquefaction embankments prototype and centrifuge models. Can Geotech J 34:641–657
22. Plaxis Material Models CONNECT Edition V20, Plaxis Knowledge base
23. PLAXIS 3D V20 (Computer Software). PLAXIS BV, Netherlands

# Effect of Soil-Structure Interaction on Seismic Performance of RC Framed Building with Viscous Wall Dampers

**A. Veerapandi and J. Visuvasam**

**Abstract** During earthquake, the structure deform beyond the elastic limit and remain intact only due to their inability to deform in-elastically. Therefore most of the earthquake force is absorbed by the RC Framed structure which leads to the damage of the structure. So, to absorb the earthquake force during the earthquake, energy dissipating devices are used. In this study Viscous wall dampers are used as energy dissipating devices for the RC framed structure. The viscous wall dampers are modelled in the finite element software SAP 2000. Non-linear dynamic analysis with Soil Structure Interaction is done for a 15-storey building with Piled-Raft foundation. Four different configuration (placement) of the viscous wall dampers is used in the RC frame building and four different earthquake data is used to find the performance of the viscous wall dampers. The Performance of the VWD (Viscous Wall Damper) is compared with considering soil-structure interaction (SSI) and with the fixed base. The seismic performance of the RC building with and without viscous wall dampers are also compared. It is found that the displacement, inter-storey drift and acceleration and base shear is increased. Considering SSI for the analysis of the VWD, it has been found that displacement and inter-storey drift increases in comparison with the fixed base. So, SSI should be considered during the analysis of the RC framed structure with viscous wall dampers.

**Keywords** Non-linear time history · Pile-raft · SSI · Viscous wall damper · SAP 2000

A. Veerapandi (✉) · J. Visuvasam
School of Civil Engineering, Vellore Institute of Technology, Vellore 632014, India
e-mail: veerapandi2510@gmail.com

J. Visuvasam
e-mail: visuvasam.j@vit.ac.in



T. G. Sitharam et al. (eds.), *Seismic Design and Performance*, Lecture Notes in Civil Engineering 120, https://doi.org/10.1007/978-981-33-4005-3_19

# 1 Introduction

Major damage of the building is caused majorly due to the natural calamities. The various factors causing the building damage are Earthquake, Fire, Lack of Maintenance, Erosion. Among all these factors Earthquake causes the most devastating damage to the structure. The major earthquake such as (Kobe-1995, El-Centro-1940, Bhuj-2001) caused collapse of many structure. So, to avoid the damage of the structure Energy dissipating structures and devices such as Elastomeric dampers, Mild steel dampers, Viscous dampers, Frictional dampers [1], Base isolation technique [2], Shear walls are used. In [3] Fluid viscous dampers are provided to the retrofitted structure to reduce the overturning of the building and to reduce the displacement of the building. The Optimum placement of dampers is discussed in [4], it is found that the dampers should be placed were the drift is more in the structure. Unnecessary dampers should be removed downward from the top. Shake table test of the RC frames with viscous wall dampers is carried out in [5] and the stress concentration is increased in the column adjacent to the dampers and the frictional dampers is not suitable for low intensity earthquakes.

## *1.1 Soil Structure Interaction*

The Soil-Structure Interaction deals with the collective response of the structure, the foundation and the soil medium surrounding and underlying beneath the foundation.

When the structure vibrates inertia is developed which gives rise to moment, torsion and base shear. These forces generate rotation and displacement at the soil-foundation interface. These rotation and displacement are caused due to the flexibility in soil-foundation system which leads to increase in the period of the structure.

In [6] the structure is modelled as a shear beam supported on a rigid foundation embedded in elastic soft soil. Vertically incident shear waves is given as seismic excitation. It is found that the form of control rule changes as result of rocking of the foundation associated with inertial and kinematic interaction. In [7] two seismically isolated bridges are selected one with heavy superstructure and light substructure and another with light superstructure and large substructure are considered and it is found that the soil structure interaction plays a major in bridges with light superstructure and heavy substructure. Considering SSI increases the time period of the structure and it is found that the SSI does not a considerable effect in stiff soil but SSI effect plays a major in soft soil. In our study we have taken soft soil for soil structure interaction.

## 2 Viscous Wall Damper

A Viscous wall damper consist of a hollow rectangular tank and steel vane as shown in Fig. 1, the steel vane is connected to the top beam and the rectangular tank is connected to the bottom beam of the same storey. The rectangular tank is filled with highly viscous fluid. When the structure vibrates due to the earthquake or dynamic forces, displacement will occur in the structure which in turn creates shear in the vane. Due to which the vane shears t the highly viscous fluid and generates viscous damping. Damping can be increased by increasing area of the vane.

### *2.1 Numerical Modelling of VWD*

The Maxwell Damper model is used to represent the VWDs. The Maxwell model consist of linear spring $K$, in series with the exponential damper represented as $C$. The force in the damper is related to the velocity through the Force-Displacement Relationship

$$F = CV^{\alpha}$$

In the places were dampers are placed the beams are divided into three element such that the length of the centre element is same as the width of the VWD. The centre element are divided into two equal element. Since the elements within the width of the VWD represent the stiffness of the tank and the vane, these beam elements are modelled as very stiff. At mid height of each bay containing VWD a pair of nodes are created as close to each other as in Fig. 2. These pair of nodes should be connected to the center of the upper and lower beam of storey containing the VWD.

**Fig. 1** Viscous wall damper

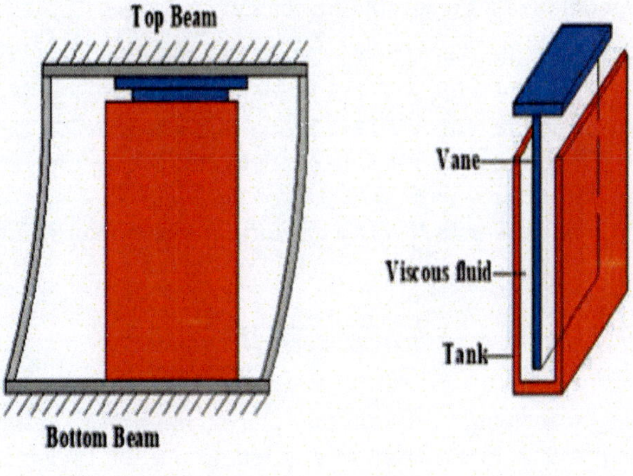

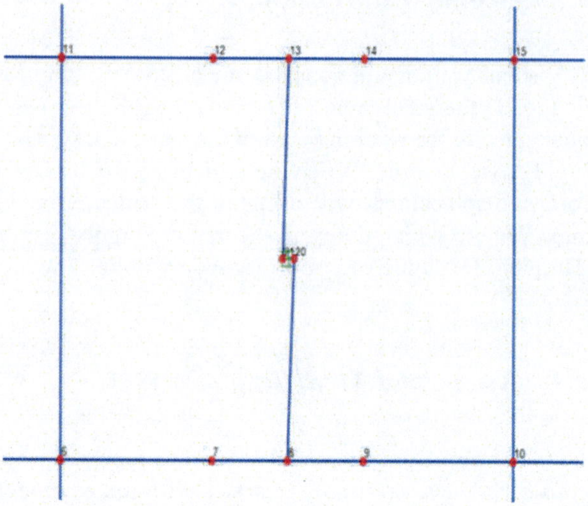

**Fig. 2** Modelling Of VWD in SAP 2000

The Stiffness, Damping co-efficient and Damping exponent are arrived from various test conducted by [8], were $K = 44657$ kN/m, $C = 1953$ [kN-(sec/m)$^{\alpha}$], $\alpha = 0.5$.

## 3 Modelling of the Superstructure

The present study consist of 15-storey RC framed building Fig. 3e, the structure is assumed to be located on zone-V, soft soil. The structure consist of $(4 \times 4)$ bay as shown in the Fig. 3b. Each bay is of 5 m in length. The load in the structure are assigned as per (IS 875). Totally 5 model have been modelled. The red line in the model plan indicates the placement of damper

- Structure with VWD in Model(1) fixed base Fig. 3a
- Structure with VWD in Model(2)fixed base Fig. 3b
- Structure with VWD in Model(3)fixed base Fig. 3c
- Structure with VWD in Model(4)fixed base Fig. 3d
- Structure without VWD fixed base Fig. 3e
- Structure with VWD with Soil Structure Interaction Fig. 3f

### *3.1 Placement of the Wall Dampers*

The placement of the damper place a major role in the damping of the structure. In [9] a $(3 \times 3)$ structure of bay length 6 m and height of the building is 48 m (12-storey). This study [9] suggest the placement of dampers in the center bay on all floors

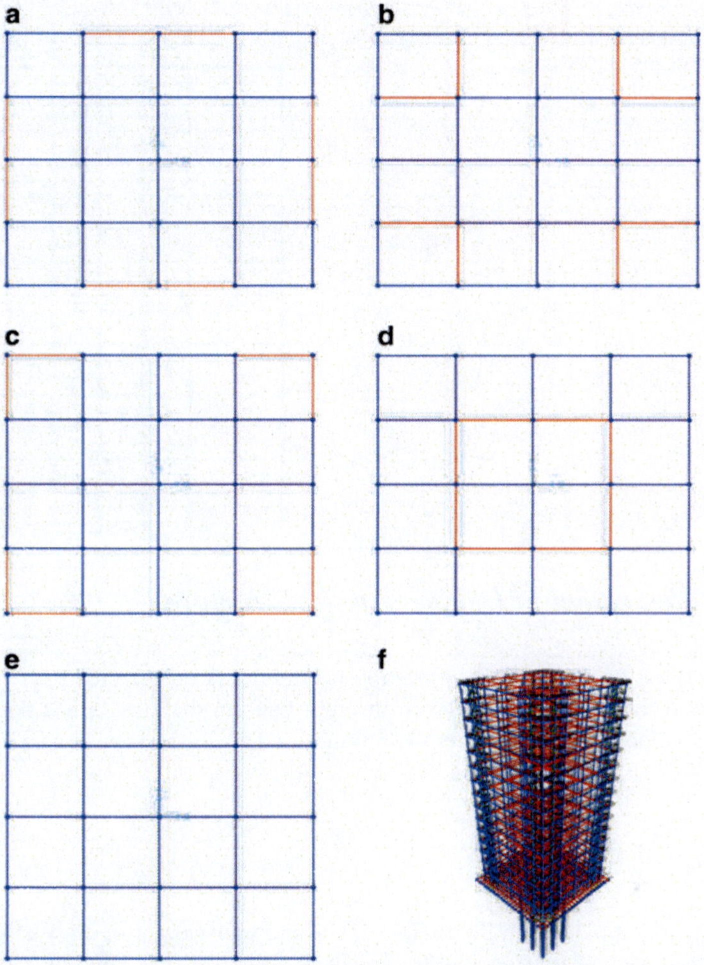


**Fig. 3** **a**. Structure with VWD model (1) (Plan view). **b**. Structure with VWD model (2) (Plan view). **c**. Structure with VWD model (3) (Plan View), **d**. Structure with VWD model (4) (Plan view). **e** Structure without VWD model (5) (Plan view). **f**. Structure with model (3) elevation with PILED-RAFT foundation

has significantly reduced the displacement, inter-story drift, increases acceleration in some cases. So, in this study we have taken dampers on all floors in those bays were the dampers are placed as shown in Fig. 4a In X-axis 4 bays are used for the placement of the dampers and in Y-axis 4 bays are used for the placement of the dampers. So, totally 8 bays are used for the placement of the damper in both the direction.

**Fig. 4** **a**. VWD are placed in all the floors depending upon the model

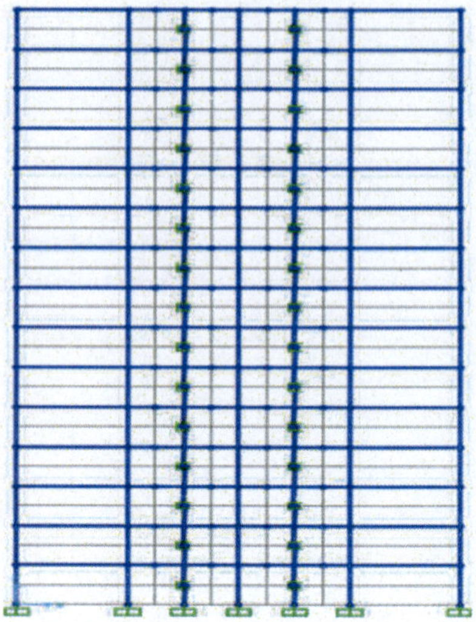

## 3.2 *Configuration of Dampers in the Structure*

In the structure the dampers are placed in all the floors on selected bays. For each configuration the dampers are placed in eight bays in the structure. Like that the dampers are placed in four different configuration (position) as shown in Fig. 3a–d.

# 4 Design of Foundation

The foundation is placed on the soft soil with the unit weight of 17.4 kN/m$^3$ [10], the angle of internal friction is 33°, poison's ratio = 0.25, Young's modulus = 27386 kN/m$^2$.

The total load of the building was 2,02,081 kN, the raft was designed as per Indian standard codes. The area of the raft was (23 × 23 m$^2$) and the maximum factored bending moment in the raft was 5436 Kn-m. The depth of the raft was designed as 1.2 m.

The pile was designed as a friction pile as per IS 2911 and 9 piles was designed for the pile raft foundation and the diameter of the pile was found to be 0.8 m and the length of the pile was 23.8 m and spacing between each pile was taken as six times the diameter of the pile.

**Table 1** Stiffness of the raft foundation

| Stiffness | Values in (Kn/m) |
|---|---|
| $K_{x,\mathrm{emb}}$ | 483.75 |
| $K_{y,\mathrm{emb}}$ | 483.75 |
| $K_{z,\mathrm{emb}}$ | 592.6 |
| $K_{rx,\mathrm{emb}}$ | 39115 |
| $K_{ry,\mathrm{emb}}$ | 40450 |
| $K_{t,\mathrm{emb}}$ | 269064 |

The values in the table are given to the springs in the raft in SAP 2000

### *4.1 Modelling of Piled Raft in SAP 2000 with SSI*

In [11] the study consist of a one story system with a mass supported by four columns. The piles are modeled as beam-column element supported by laterally distributed springs and dampers Fig. 3f. The skin friction values are given to the vertical spring Z as per IS2911, the horizontal spring values are given by the formula

$$K_h(z) = \delta E_S(z)$$

Stiffness of raft for various modes of vibration are given by [12], Table 1 shows the stiffness of springs.

## 5 Non-linear Time History Analysis

In NRH Analysis, instead of lateral forces an earthquake record is applied to the foundation level of the building. The NRH analysis is considered to be more accurate in representation of dynamic loading, since the concentration of damage is considered at the weakest location of the buildings and the redistribution of forces upon the formation of plastic hinges. In this study three ground motions(scaled-matched with the response spectrum curve)are taken they are CHI-CHI (1999), IMPERIAL(1979), KOCAELI(1999).

## 6 Results and Discussion

For the four position of the dampers as shown in Fig. 3a–d, nonlinear time history is done for three earthquakes and the results are compared with structure without VWD and the model which performs better is taken for soil structure interaction (Figs. 5, 6 and 7).

**a**

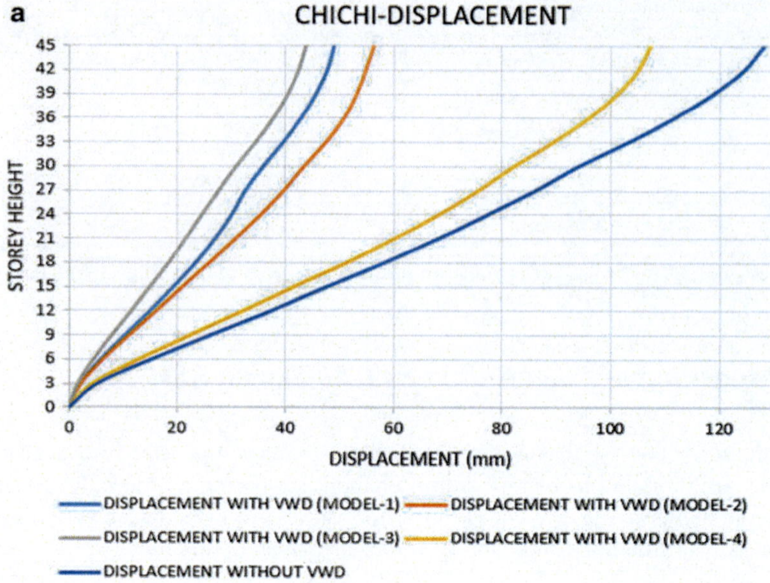


**b**

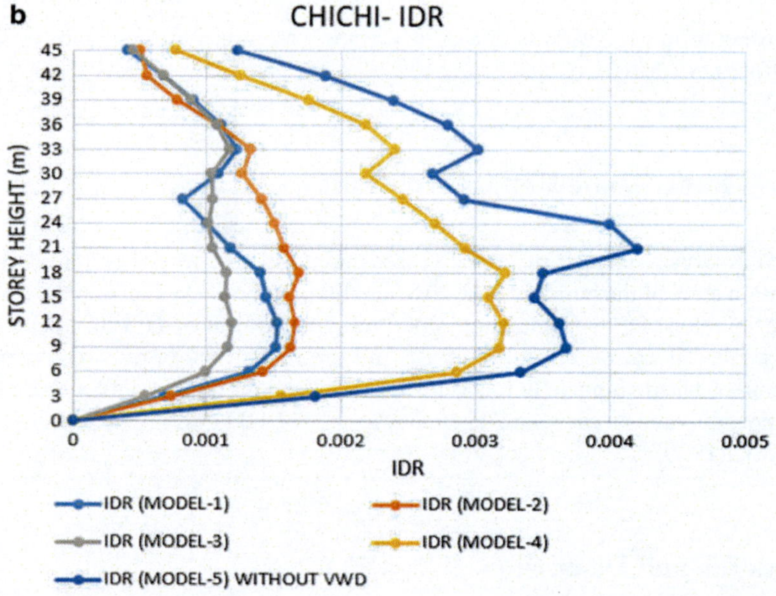


**Fig. 5** **a**. Displacement for CHICHI earthquake for all the four models with VWD and model without VWD. **b**. IDR for CHICHI earthquake for all the four models with VWD and model without VWD

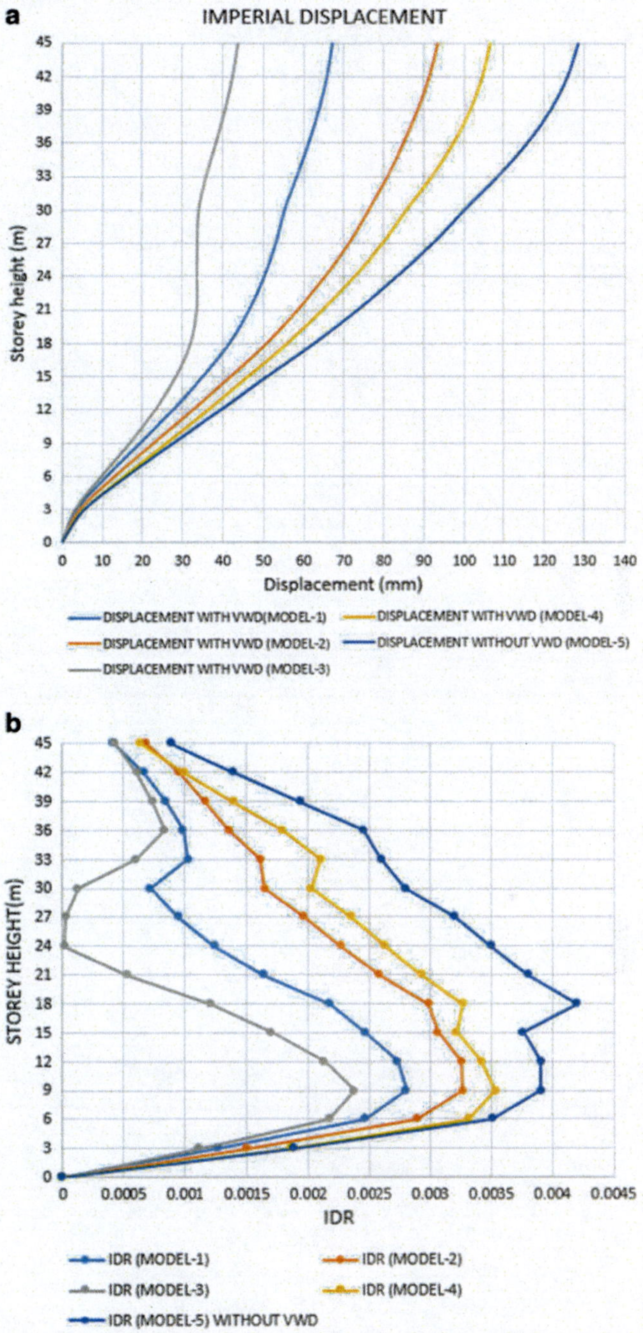


**Fig. 6** **a**. Displacement for IMPERIAL earthquake for all the four models with VWD and model without VWD. **b**. IDR for IMPERIAL earthquake for all the four models with VWD and model without VWD

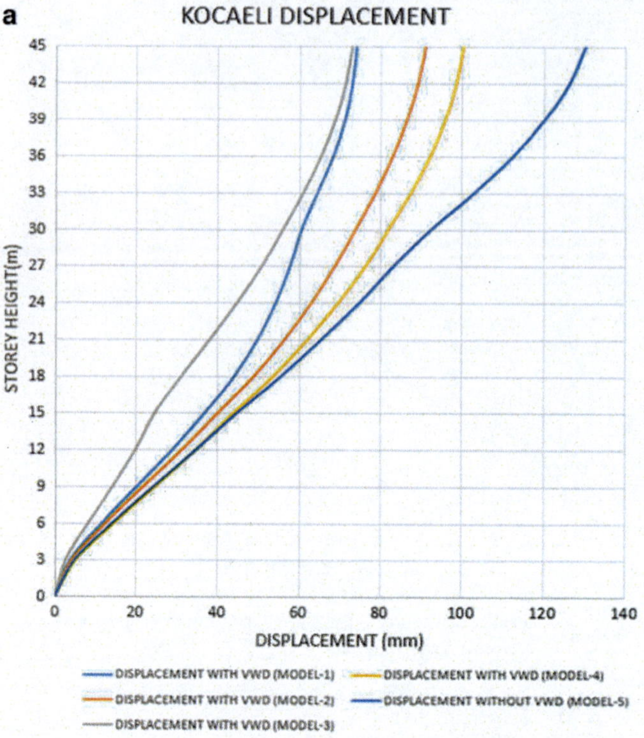


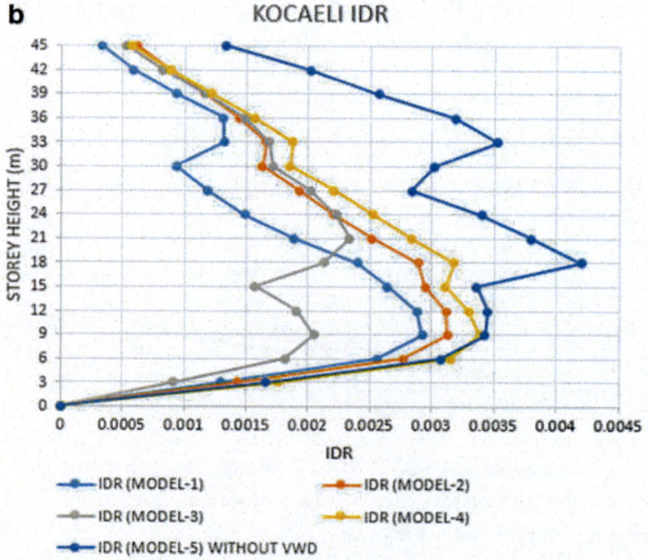


**Fig. 7** **a**. Displacement for KOCAELI earthquake for all the four models with VWD and model without VWD. **b**. IDR for KOCAELI earthquake for all the four models with VWD and model without VWD

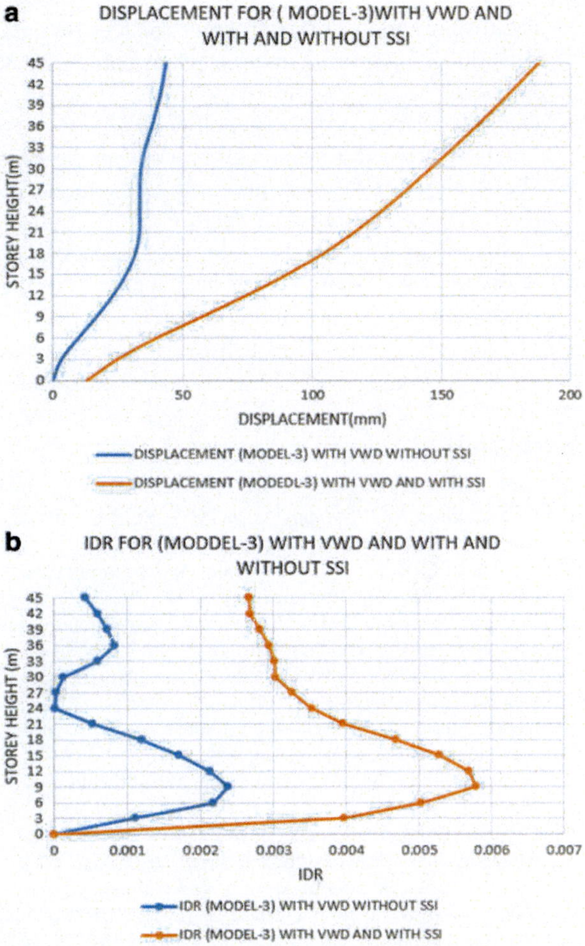


**Fig. 8** **a**. Displacement for Imperial earthquake with VWD positioned as per Model-3 with considering SSI and without considering SSI. **b**. IDR for Imperial earthquake with VWD positioned as per Model-3 with considering SSI and without considering SSI

So, by comparing the displacement and IDR of all the four models with viscous wall dampers, the model-3 Fig. 3c performs well in reducing the displacement up to 65% and also the IDR is reduced well in the model-3. So, the soil structure interaction is done for model-3 and the result are as follows (Fig. 8).

## 7 Conclusion

With fixed base the Displacement reduces up to 60%. And the IDR value also reduces by 65% using VWD and of all the four models with VWD the Model-3 perform well. So, it is taken for soil structure interaction.

- The displacement of the structure reduces by 65% while using dampers considering soil structure interaction when compared with the structure without wall dampers.
- The Inter-storey drift reduces by 60% while using dampers considering soil structure interaction when compared with the structure without wall dampers.
- The acceleration response at the top storey increases about 3 times the acceleration given at the bottom of the structure.
- By the use of dampers the axial force in the column increases.
- So, by the use of dampers the displacement, IDR can be greatly reduced, but the acceleration of the structure increases.

## References


1. Constantinou MC, Symans MD (1993) Seismic response of structures with supplemental damping. Struct Des Tall Buildings 2:77–92
2. Kaloop MR, Hu JW (2017) Seismic response prediction of buildings with base isolation using advanced soft computing approaches. Adv Mater Sci Eng. Angela De Bonis (eds)
3. Wang S, Mahin SA (2018) High-performance computer-aided optimization o viscous dampers for improving the seismic performance of a tall building. Soil Dyn Earthq Eng 113:454–461
4. Takewaki I (1997) Optimal damper placement for minimum transfer functions. Earthquake Eng Struct Dynam 26:1113–1124
5. Lu X, Zhou Y, Yan F (2008) Shake table test and numerical analysis of RC frames with viscous wall dampers. J Struct Eng 134(1)
6. Wong HL, Luco JE (1991) Structural control including soil-structure interaction effects. J Eng Mech 117(10)
7. Dicleli M, Albhaisi S, Mansour MY (2005) Static soil-structure interaction effects in seismic-isolated bridges. Pract Periodical Struct Des Constr 10(1)
8. Dynamic isolation system, viscous wall dampers guidelines for modelling, (2015)
9. Siddiq STM, Visuvasam J, Hejazi F (2019) Effect of viscous wall dampers on response of reinforced concrete structures subjected to seismic excitation. In: International conference on industry 4.0 and advancement in civil engineering. ISBN 978-93-87862-53-1, March 2019
10. Estabragh AR, Javadi AA (2014) Effect of soil density and suction on the elastic and plastic parameters of unsaturated silty soil. Int J Geomech
11. Saha R, Dutta SC, Haldar S (2012) Seismic response of soil-pile raft-structure system. J Civil Eng Manage
12. Gazetas G (1991) Formulas and charts for impedances of surface and embedded foundations. J Geotech Eng 117

# Response of Multistoried Building Considering Soil-Structural Interaction Under Lateral Loading

**Snehal Kaushik, Tabassum Nahar Saikia, Syed Maroof Hassan Syed, Suhail Jafri, and Banashree Baruah**

**Abstract** Conventionally soil-structure interaction (SSI) is neglected while designing any structure assuming it is situated on hard strata. For light structures on relative stiff soil like low rise buildings, neglecting the SSI can be affordable. But analyzing a multistoried building on relative soft soil without considering SSI may be relatively dangerous from the safety and stability point of view. In 1995 Kobe earthquake the seismic behavior of a structure was highly influenced by the response of the foundation and the ground [1]. By keeping these criteria in mind, it is essential to analyze any structure considering the effect of SSI. Total 12 different models of symmetric multistoried building, out of which, three models are simulated considering fixed at the base. Other models are simulated for three different soil properties and for three different heights of the buildings. The Influence of soil-structure interaction on the structure is investigated for 5-storey, 10-storey and 15-storey buildings by pushover analysis using the finite element program SAP2000. Displacement based pushover analysis is carried out and the structures are subjected to lateral displacement of 800 mm on the top node of the model. The capacity of the structures for varying building heights and soil types are obtained and thus the analysis helps us to get a better understanding of the interaction between the structure and the soil medium. The influence of the soil-structure interaction on the analysis of the structure is reflected in a decrease of the base shear values. It is observed that the capacity of the building with lesser height and hard soil type is more compared to other models.

**Keywords** Soil-structure interaction · Multistoried building · Pushover analysis

S. Kaushik (✉) · T. N. Saikia · S. M. H. Syed · S. Jafri · B. Baruah
Girijananda Chowdhury Institute of Management and Technology, Guwahati 781017, India
e-mail: snehal_ce@gimt-guwahati.ac.in
URL: http://www.gimt-guwahati.ac.in

 
T. G. Sitharam et al. (eds.), *Seismic Design and Performance*, Lecture Notes in Civil Engineering 120, https://doi.org/10.1007/978-981-33-4005-3_20

# 1 Introduction

In high seismic zone region, structures are usually considered as fixed at the base while analysis and design of the structure. However, the response of multistoried buildings under earthquake loads is greatly influenced by the interaction between the soil and the structure. Mostly the buildings are analyzed considering solid rock at the base and boundary condition is assumed as fixed, but in reality, the soil beneath the structure may have different properties which, need to be considered during the analysis. To study this effect, the linear or non-linear behavior of the soil beneath the structure should be considered for the analysis of multistoried buildings. The main objective of this study is to find out the effect of soil on the behavior of the building considering variation in the height.

In the past, various studies have been carried out to study the soil structure interaction which describes analysis procedures and system identification techniques for evaluating inertial SSI effects on seismic structural response. The analysis procedures are similar to provisions in some building codes but incorporate more rationally the influence of site conditions and the foundation embedment, flexibility, and shape on foundation impedance [2]. From the study, it is confirmed that the response of the structure depends not only on its dynamic characteristics but also on the interaction between the structure, the foundation and the soil [3]. Aydemir and Ekiz [4] studied the behavior of multistoried building with different aspect ratio. Inelastic displacement ratios and strength reduction factors are investigated for designed sample plane frames for 64 earthquake motions recorded on different site conditions such as rock, stiff soil, soft soil and very soft soil. It was observed that strength reduction factors of sample buildings considering soil–structure interaction are found to be smaller than design strength reduction factors, given the current seismic design codes. Luco and Contesse [5] studied a dynamic interaction, through the soil, between two parallel infinite shear walls placed on rigid foundations. The steady-state response of both structures for a vertically incident SH wave is obtained and compared with the corresponding values resulting from consideration of only one structure. Investigations of soil–structure interaction have shown that the dynamic response of a structure supported on flexible soil may differ significantly from the response of the same structure when supported on a rigid base [6–8].

# 2 Methodology

Analysis and design of a building is generally carried out considering solid rock at the base and boundary condition at the base of the building is assumed as fixed. But in real this may not be the situation; the soil beneath may not be rock. The soil should be considered during the analysis and design of the structures. In the present study, the effect of soil on the behavior of the building is evaluated considering variation in the soil type and height of the building.

A 5-storey, 10-storey and 15-storey reinforced concrete buildings are modeled for the purpose of the present study using the computer program SAP2000 [9]. The buildings are assumed to be located in Seismic zone *V* [10]. Earthquake load is resisted by the frame action of lateral load resisting members in the building. Only the frame structure of the building comprising of the beams, columns and slabs have been considered. The buildings have a rectangular shaped plan of dimension 19 m × 11 m. The heights of the buildings are 17.5 m, 35 m and 52.5 m respectively.

The soil types considered for soil-structure interaction analysis are loose sand, medium clay and hard, gravelly soil, which have been modelled in SAP2000. Figures 1 and 2 show the models considered for analysis with fixed base and with

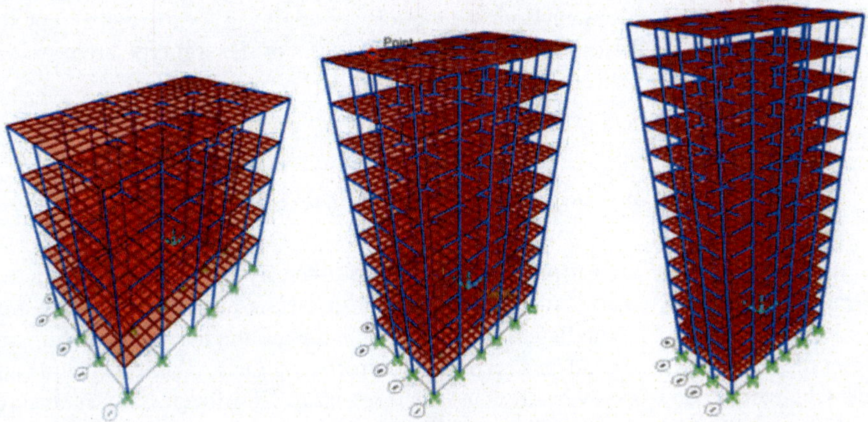

(a) (b) (c)

**Fig. 1** Buildings with rigid base: **a** 5-storey; **b** 10-storey, and **c** 15-storey

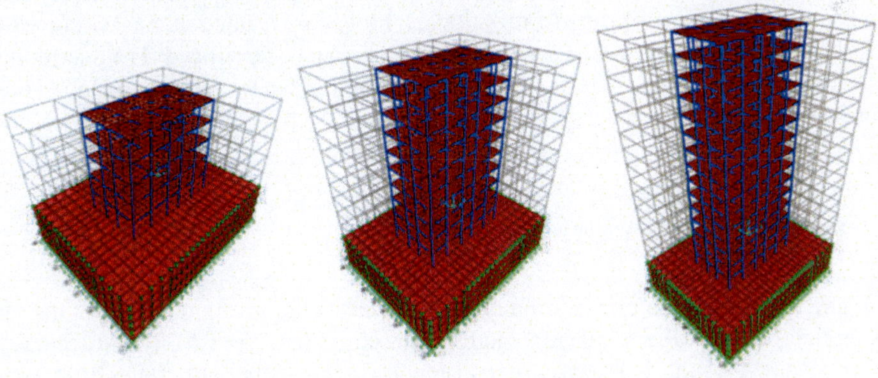

(a) (b) (c)

**Fig. 2** Buildings with soil-structural interaction: **a** 5-storey; **b** 10-storey, and **c** 15-storey

**Table 1** Details of the soil type considered for analysis

| Soil type | Modulus of elasticity (N/m$^2$) | Poisson's ratio | Unit weight (kN/m$^3$) |
|---|---|---|---|
| Loose sand (Soil 1) | 24,000 | 0.3 | 18 |
| Medium clay (Soil 2) | 15,000 | 0.45 | 20 |
| Hard/Gravel (Soil 3) | 65,000 | 0.3 | 18 |

soil-structural interaction respectively. The soil is modelled beneath the structures to a depth of 8 m from the base of the structure and extending 8 m on all sides from the edges of the structure. The modelled soil is then subdivided into smaller sections or meshed so that uniformity is maintained and the results obtained are more precise. In case of soil-structural interaction, the models are analyzed for three different soil types obtained from previous research [11], the details of the soil type are provided in Table 1.

## 3 Introduction to Non-linear Static Analysis

In pushover analysis, the seismic demand is computed by subjecting the structure to monotonically increasing lateral loads with an invariant height-wise distribution until a predetermined target displacement is reached. Both the force distribution and target displacement are based on a response controlled by the fundamental mode and an unchanged mode shape even after the structure yields. Non-linear static analysis or Pushover analysis is a static nonlinear procedure to analyze the seismic performance of a building where the computer model of the structure is laterally pushed until a specified displacement is attained or a collapse mechanism has occurred. The loading is increased in increments with a specific predefined pattern such as uniform or inverted triangular pattern and the gravity load is kept as a constant during the analysis. The structure is pushed until sufficient hinges are formed such that a curve of base shear versus corresponding roof displacement can be developed. The maximum base shear the structure can resist and its corresponding lateral drift can be found out from the Pushover curve.

## 4 Validation of Analytical Model

For any detailed finite element study, the material parameters and the modeling methodology need to be validated for judging the consistency, reliability and accuracy of the obtained results. In the past, several analytical and numerical studies have been carried out on the behavior of buildings considering the effect of soil. However, very few experimental works are available on this topic. Goktepe et al. [12] studied a

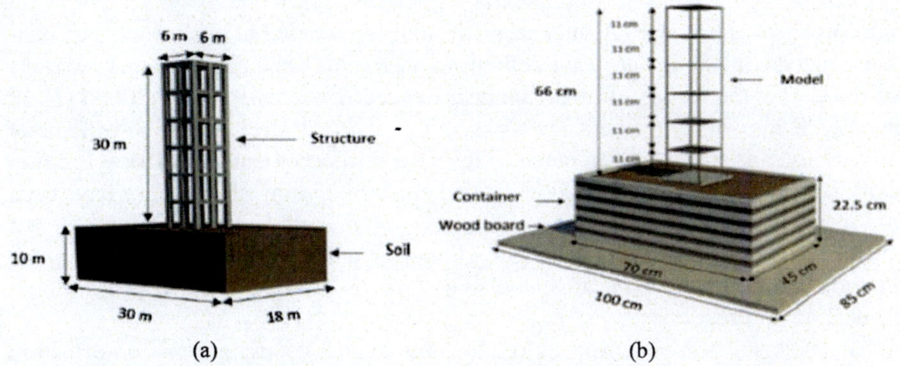


**Fig. 3** **a** 3D prototype building of the soil-structure; **b** 3D model of soil-structure interaction [12]

scaled soil structure interaction model for shake table testing. The prototype of soil-structure interaction system with frame-type building, resting on a sandy soil has been selected. The test model, consisting of six-story steel frame-type building, natural soil and laminar soil container, is designed and constructed to realistically simulate the seismic soil-structure interaction effects of building structures in shaking table tests. The prototype of the experimental tests is a soil structure system with dimensional characteristics illustrated in Fig. 3a and the experimental test model of SSI is shown in Fig. 3b.

From the study it is proposed that numerical soil-structure model is a valid and qualified method of simulation for numerical investigations of dynamic soil-structure interactions. Hence, similar modeling of the structure and soil is considered in the present study to study the behavior of the high-rise building interacting with soil.

## 5 Structural Modeling

Two different cases are studied, in the first case, the building is assumed to be founded on rocky stratum and soil-structure interaction effects are not considered in the study. Thus, the translational and rotational degrees of freedom are restrained at the bottom nodes. In the second case, the soil structure interaction is considered to observe the behavior under lateral loading. In addition, the effect of unreinforced brick masonry infill walls was not considered in this analysis.

The beams and columns are modeled using two noded frame elements and floor slab are modeled using four noded shell elements. Nonlinear static analysis (pushover analysis) is carried out under design load combinations, as prescribed in IS: 1893 (Part 1)—2016 [10]. The members are designed against the critical forces and moments obtained through the load combinations. The buildings have beams of cross-section 250 mm × 300 mm and columns of cross-section 400 mm × 400 mm.

Reinforced concrete slab of thickness 150 mm is modeled at each floor. For non-linear analysis and hinge property definitions, designing of beams, columns and slabs are done from the results of linear analysis in accordance with IS 456:2000 [13]. In beams, 20 mm diameter bars are used as longitudinal steel and 8 mm diameter bars are used as shear reinforcement. The shear reinforcement is provided 150 mm apart. Beams have been provided with clear cover of 25 mm and columns have been provided with clear cover of 40 mm. M25 grade of concrete and Fe415 reinforcement steel is considered for the analysis. For all models, a live load of 3 $KN/m^2$ is applied on floor slabs. Seismic loads are calculated as per IS 1893:2016-Part 1 [10] for both $X$ and $Y$ plan directions.

Adequate seismic performance implies that available strength and deformation capacities of a structure must be greater than the demands imposed by a design earthquake. In nonlinear static analysis, any structure is pushed under a pre-specified lateral force or lateral displacement [14] with the monitoring of possible nonlinear behaviour at various locations. Finally, the total base shear is plotted against a reference displacement to obtain the global capacity curve for the structure. In the present study, displacement-controlled pushover analysis of the building was carried out considering material non-linearity only.

In the present study, lumped plasticity is considered in the frame members by assigning flexural plastic hinges at the ends of beams and columns of all the stories in the building. The length of the plastic hinge for the beams and columns was assumed as 10% of the length of the member. Three different plastic hinges are modelled, namely (a) axial force—bending moment (P-M) hinge, (b) moment—rotation ($M$-$\theta$) hinge, and (c) shear force—shear deformation ($V$-$\Delta$) hinge. The $P$-$M$ and $M$-$\theta$ hinges were specified using the auto hinge option in SAP2000, using Tables 6-7 and 6-8 of FEMA 356 [14]. The plastic hinge rotation capacities can be estimated by calculating moment curvature relationship using realistic material stress strain curves. Base shear-displacement relationships for different models with and without soil structure interaction are studied for different models. Different levels of inelasticity are developed at the end regions of the beams and columns for analyses along the directions of width and length of the building.

## 6 Nonlinear Analysis: Important Outcomes

Two different cases are considered to study the response of a multistoried building. In the first case, the building is assumed to be situated on hard strata and base of the structure is considered as fixed. The peak responses are investigated that is, maximum base shear and maximum peak displacement. In the second case, the effect of soil structure interaction is studied using three different types of soil, that is, loose sand, medium clay and hard strata/gravel. These four different types of soil conditions are applied to three buildings with different heights, that is, 5 storey, 10 storey and 15 storey. Displacement controlled pushover analysis was carried out to a maximum displacement of 800 mm. For all the 12 cases, pushover analysis is carried

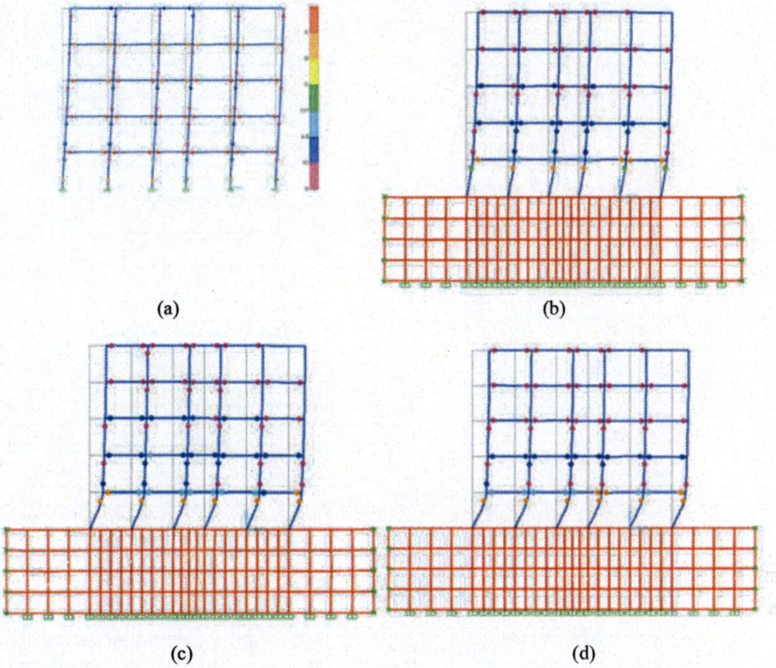


**Fig. 4** Formation of plastic hinges in the final stage of pushover analyses along the length of the 5 storey building **a** rigid base, **b** Soil type 1, **c** Soil type 2 and **d** Soil type 3

out considering different load cases and combinations along the length and width of the building.

Figure 4 shows the plastic hinge formation for 5 storey building when pushed along the length of the building. As expected the top displacement is less for the building situated on hard strata than the cases when soil-structure interaction is considered. The formation of hinges is in similar fashion for all the models when soil-structure interaction is considered. The hinges are first developed in the beams and then in columns. It is observed that the building behaves in a flexible manner, and observed that less hinge failure, when soil structure interaction is considered.

Figures 5 and 6 show the formation of hinges in the final stage of analysis for 10 storey and 15 storey buildings respectively. In both the cases the hinges are formed in the beams first and then in column following the strong column weak beam concept. Hinge failure is observed in beam for the fixed base case (Fig. 5a) while they are not reached the failure stage in case of soil structure interaction. However more hinges are observed in columns considering soil-structure interaction. As the height of the building increases, the failure of hinges decreases. However, the failure in beams is more in case of soil type 1 irrespective of the aspect ratio of the building. From the hinge formation it is observed that the effect of soil-structure interaction is less in case of 5 storey building than the higher stories. It can be concluded that the soil structure interaction must be consider while analyzing and designing high-rise buildings.

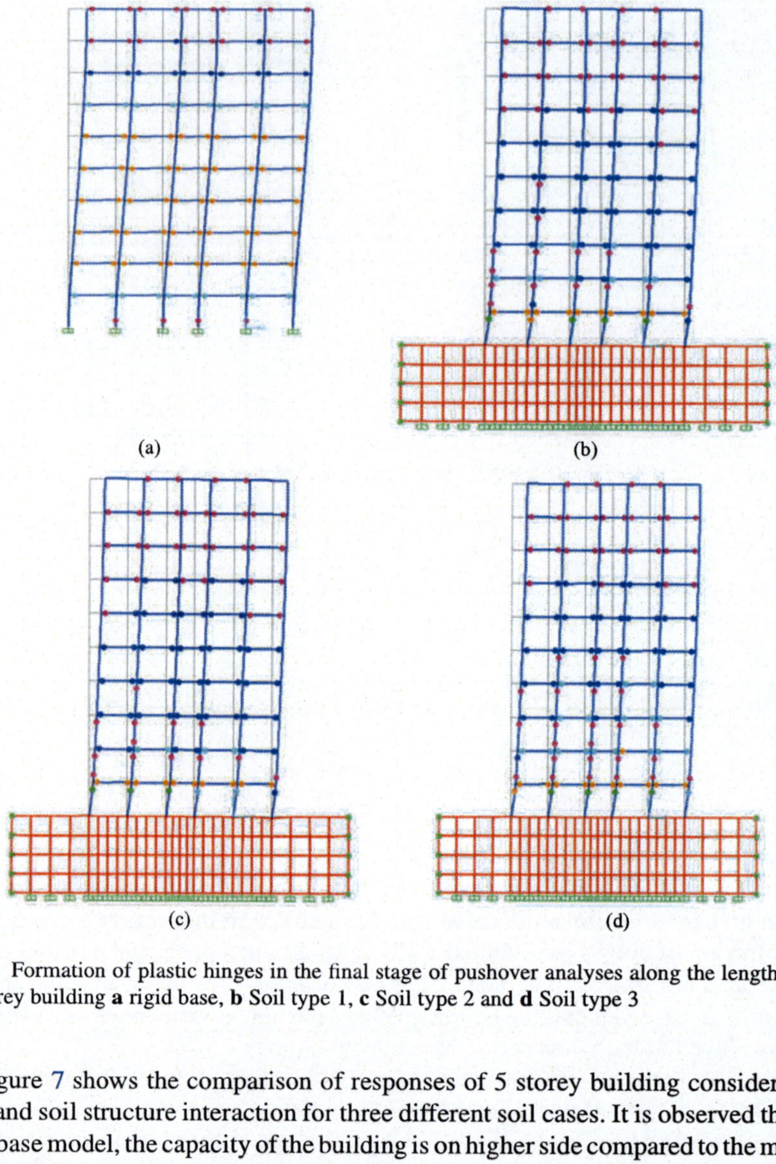


**Fig. 5** Formation of plastic hinges in the final stage of pushover analyses along the length of the 10 storey building **a** rigid base, **b** Soil type 1, **c** Soil type 2 and **d** Soil type 3

Figure 7 shows the comparison of responses of 5 storey building consider rigid base and soil structure interaction for three different soil cases. It is observed that for rigid base model, the capacity of the building is on higher side compared to the models with soil-structure interaction and the deflections are quite high in buildings with SSI. Conventionally, the analysis of the building is carried out considering the rigid base without accounting for soil modeling. This is because this type of modeling gives the overestimated result. However, considering the soil structure interaction the base shear capacity of the building reduces. In reality the soil is always present beneath the structure, so in spite of higher capacity of the rigid base model, soil structure interaction should be considered.

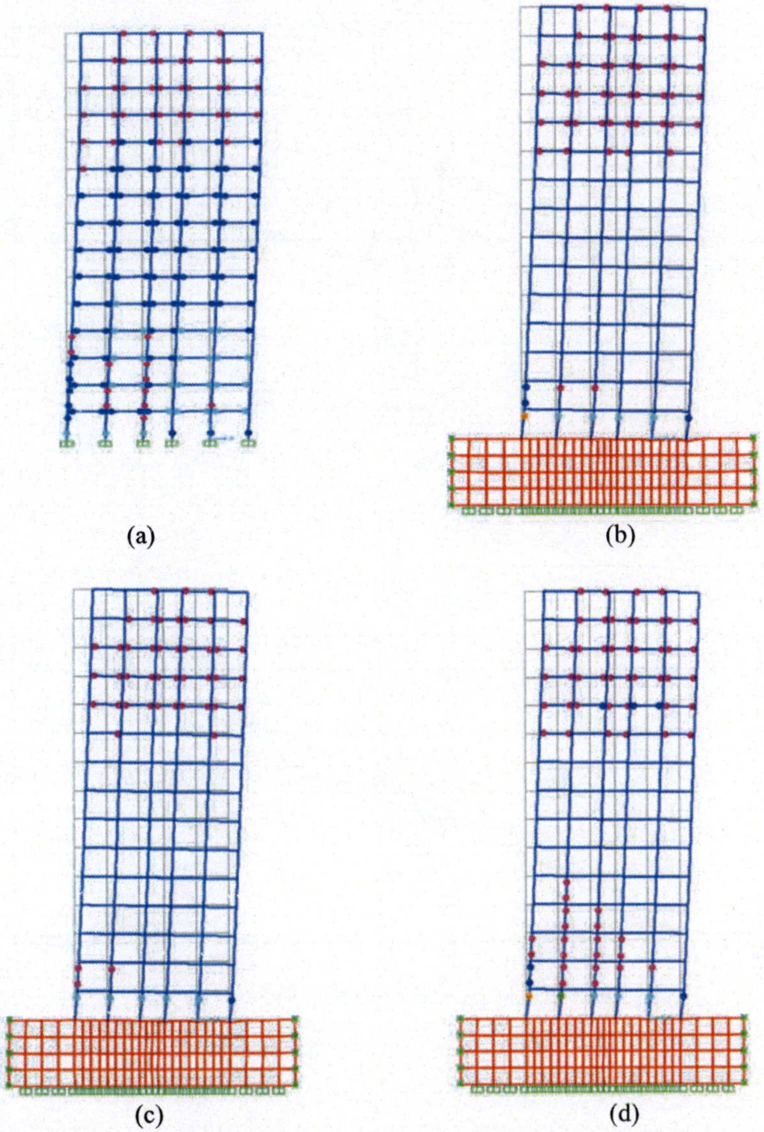


**Fig. 6** Formation of plastic hinges in the final stage of pushover analyses along the length of the 15 storey building **a** rigid base, **b** Soil type 1, **c** Soil type 2 and **d** Soil type 3

It is also observed that for the lowest stiffness soil (Soil type 2), the capacity of the building gets reduced as the height of the building increases (Figs. 8 and 9). The capacity of the structure decreases with the decrease in the stiffness of the soil (Table 2). Also, at certain displacement for a type of soil, the capacity reduces as the height of the building increases.

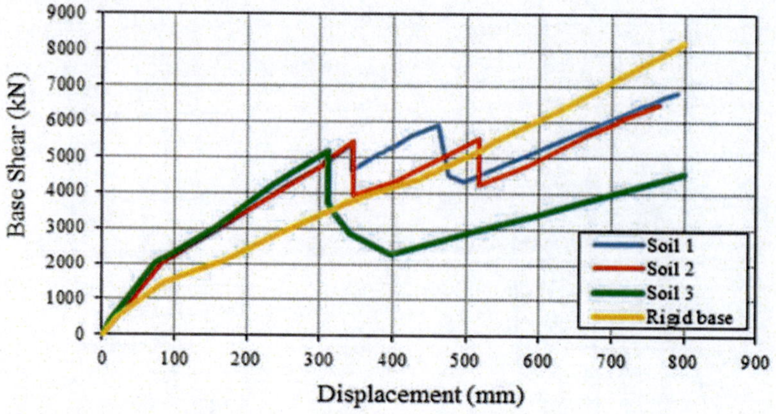


**Fig. 7** Pushover Curve for 5 storey building interacting with different soil types for earthquake along the length of the building

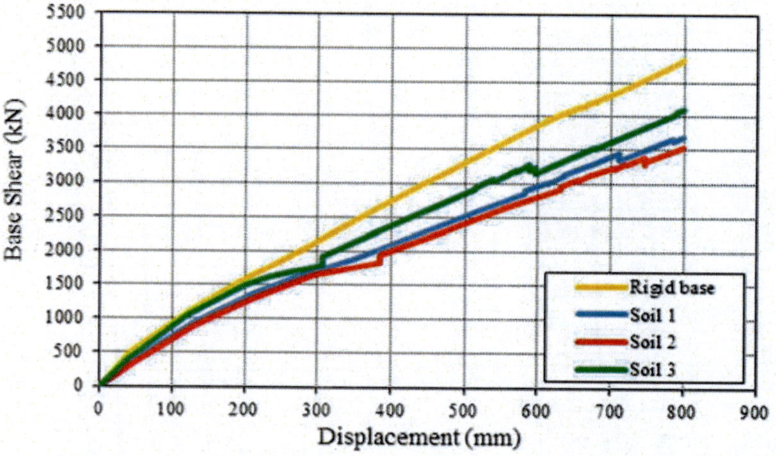


**Fig. 8** Pushover curve for 10 storey building interacting with different soil types for earthquake along the length of the building

# 7 Conclusions

Three buildings with different heights are model with and without SSI to study the effect of the soil on the overall capacity of the structure. The following conclusions are drawn from the present study:

1. It is observed that the base shear capacity of the structure gets reduced with the increase in height of building considering the effect of soil-structure interaction.
2. With the increase in the soil stiffness value, the capacity of the building increases.

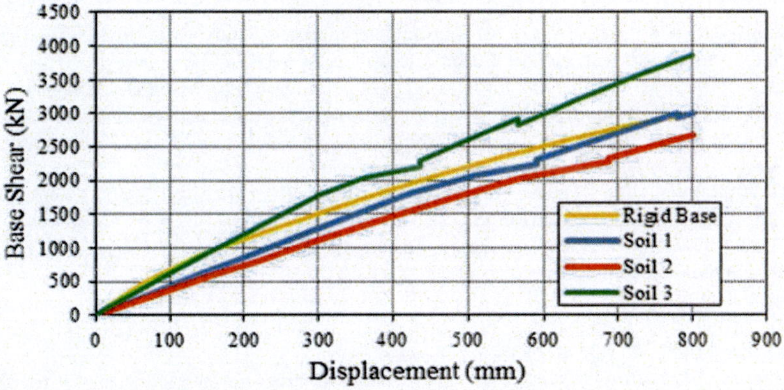


**Fig. 9** Pushover curve for 15 storey building interacting with different soil types for earthquake along the length of the building

**Table 2** Comparison of base shear for different type of buildings

| Soil type | Rigid base | Loose sand (Soil 1) | Medium clay (Soil 2) | Hard/Gravel (Soil 3) |
|---|---|---|---|---|
| Height (m) | Base shear (kN) | | | |
| 5 | 8000 | 5500 | 5400 | 5250 |
| 10 | 4800 | 3640 | 3500 | 4100 |
| 15 | 2900 | 2800 | 2600 | 3800 |

3. The formation of plastic hinges in the structure and the overall capacity of the building reduce with the reduction in the stiffness of soil beneath the building.
4. It can also be concluded that response of the structure is over-estimated without considering the soil-structure interaction. Hence, for analysis and design of a structure, soil-structure interaction should be considered.

## References


1. Mylonakis G, Gazetas G, Nikolaou S, Michaelides O (2000) The role of soil on the collapse of 18 piers of the hanshin expressway in the kobe earthquake. In: Proceedings of 12th world conference on earthquake engineering, New Zealand. Paper No. 1074
2. Stewart JP, Fenves GL, Seed RB (1999) Seismic soil-structure interaction in buildings. I: analytical methods. J Geotech Geoenvironmental Eng 125(1)
3. Mekki M, Elachachi SM, Breysse D, Nedjar D, Zoutat M (2014) Soil-structure interaction effects on RC structures within a performance-based earthquake engineering framework. Eur J Environ Civil Eng 18(8):945–962
4. Aydemir ME, Ekiz I (2013) Soil–structure interaction effects on seismic behaviour of multistorey structures. Eur J Environ Civil Eng 17(8):635–653
5. Luco EJ, Contesse L (1973) Dynamic structure-soil-structure interaction. Bull Seismol Soc Am 63(4):1289–1303

6. Chopra AK, Gutierrez JA (1974) Earthquake response analysis of multistory buildings including foundation interaction. Earthq Eng Struct Dyn 3:65–77
7. Bielak J (1976) Modal analysis for building-soil interaction. Eng Mech Div ASCE 771–786
8. Iguchi M (1978) Dynamic interaction of soil–structure with elastic rectangular foundation. In: Proceeding of the fifth Japanese earthquake engineering symposium. Tokoyo, Japan, pp 457–64
9. SAP-2000 Integrated software for Structural Analysis and Design (2013) Computers and Structures Inc., Berkeley, California.CSI
10. IS 1893 (Part1):2016 (2016) Criteria for earthquake resistant design of structures. Bureau of Indian Standards, New Delhi
11. Dhanalakshmi P, Ramesh BM, Manjunatha K, Priya M (2017) Seismic vulnerability of plan irregular buildings with soil-structure interaction. Int Res J Eng Technol (IRJET), Bangalore, Karnataka, India, 4 (10):322–328
12. Goktepea F, Omidb AJ, Celebi E (2017) Scaled soil-structure interaction model for shaking table testing. Acta Phys Pol, A 132(3):588–590
13. IS 456:2000 (2016) Plain and reinforced concrete—code of practice, Bureau of Indian Standards, New Delhi
14. FEMA 356 (2000) Prestandard and commentary for the seismic rehabilitation of buildings. Federal Emergency Management Agency, USA

# Seismic Response of Buildings Resting on Sloping Ground Considering SSI Effect—An Experimental Study

**R. M. Thejaswini, L. Govindaraju, V. Devaraj, and R. S. Preethi**

**Abstract** Due to the rapid urbanization and infrastructure developments, there is a demand of construction of high-rise buildings. Due to this extended demand, many high-rise buildings have been invariably built on sloping grounds and on hill slopes. These buildings which are built on sloping ground fail vulnerably when subjected to seismic forces. Structures on slopes differ from those on level ground because they are irregular horizontally as well as vertically. The chances of failure further increase with the effect of sub soil conditions where the buildings are founded. This paper highlights the results of seismic response analysis of multistory buildings resting on sloping ground with soil-structure- interaction effects. The prototype buildings are transformed into model buildings by using appropriate scaling laws. Further, a comparison has been made for the seismic response of buildings on sloping ground with buildings resting on level ground.

**Keywords** Prototype · Scaled model · Soil structure interaction · Resonance

## 1 Introduction

Insufficiency of level ground has led to increase in construction on the sloping ground The buildings constructed on sloping ground can be categorized into two types namely step back and step back—set back buildings as shown in Fig. 1.

Due to the effect of sloping ground these buildings fail vulnerably under seismic action. The chances of failure further increase with the effect of soil where the buildings are founded as shown in Fig. 2.

In the seismic analysis of a buildings founded on ground, the ground motion passes to the base of buildings and then loads on buildings. The response of the foundation system affects the response of the structure and vice versa, which is called dynamic soil-structure interaction (SSI). Soil structure interaction plays an important role in

R. M. Thejaswini (✉) · L. Govindaraju · V. Devaraj · R. S. Preethi
Department of Civil Engineering, UVCE, Bangalore University, Bengaluru, Karnataka, India
e-mail: thejaswini08@gmail.com

 
T. G. Sitharam et al. (eds.), *Seismic Design and Performance*, Lecture Notes in Civil Engineering 120, https://doi.org/10.1007/978-981-33-4005-3_21

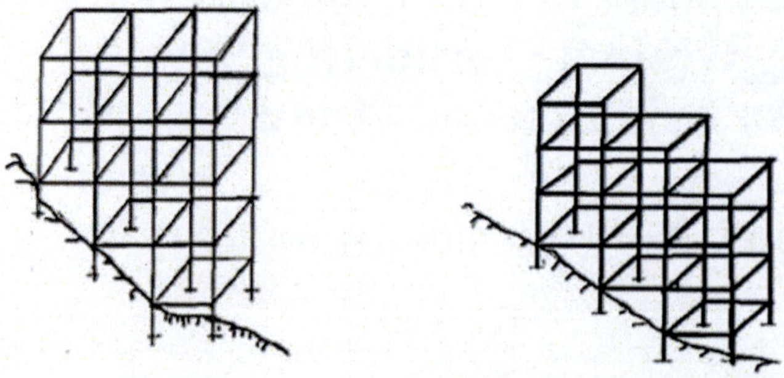

(a) Step back Building (b) Step back-Set back Building

**Fig. 1** Set back and step back building configurations

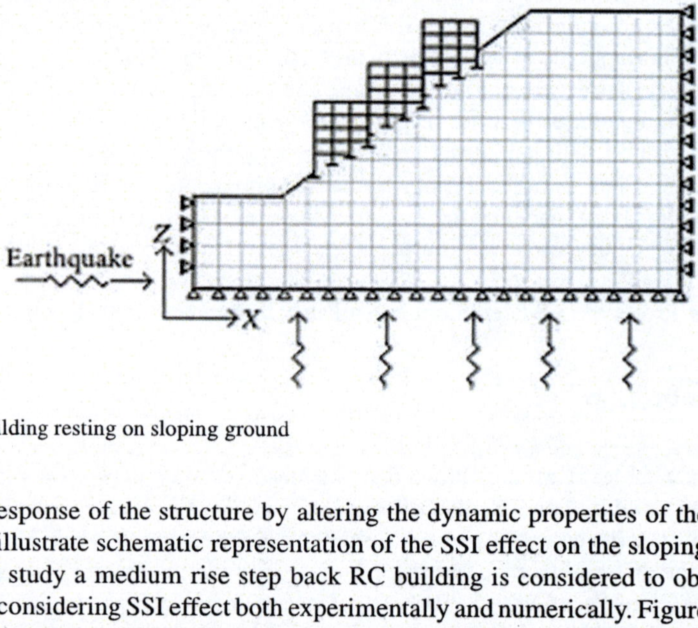


**Fig. 2** Building resting on sloping ground

seismic response of the structure by altering the dynamic properties of the system. Figure 3 illustrate schematic representation of the SSI effect on the sloping ground.

In this study a medium rise step back RC building is considered to observe the behavior considering SSI effect both experimentally and numerically. Figure 4 shows the elevation and three-dimensional view of the building considered in the present study. The plan at base as shown in Fig. 5. This irregular building has a uniform storey height of 3 m. A sloping angle of 27° has been considered because which is neither too slope nor too flat [1].

The RC building shown in Fig. 4 is designed according to IS: 1893(Part 1)-2016 and IS: 456-2000. While designing it has been considered that the first mode of vibration is obtained along longitudinal direction ($X$-axis). Live load of 3 kN/m$^2$ and floor finish of 1 kN/m$^2$ is considered at each floor. At roof, live load of 1.5 kN/m$^2$ and floor finish of 2 kN/m$^2$ is considered in design of this building. The size of the beam and columns are 300 mm × 400 mm and 250 mm × 600 mm respectively. The thickness of the slab is 150 mm. The analysis has been carried out considering

**Fig. 3** SSI effect on sloped building

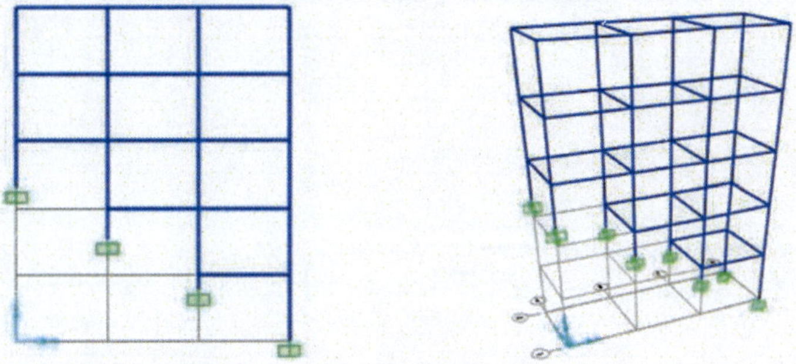

(a) Elevation (b) 3- Dimensional View

**Fig. 4** Three-dimensional view of the step back building (27° slope)

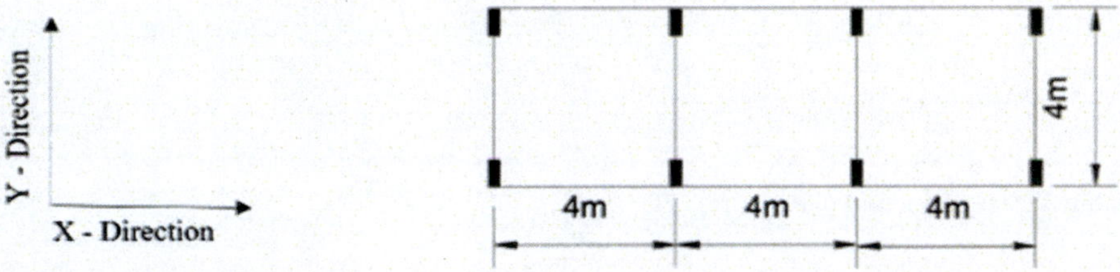


**Fig. 5** Plan of the building

**Table 1** Properties of the soil (Hokmabadi et al. 2014)

| Sl. No | Contents | Description |
|---|---|---|
| 1 | Young's modulus | 25 MPa |
| 2 | Poisson's ratio | 0.4 |
| 3 | Density of soil | 1470 kg/m$^3$ |
| 4 | Shear wave velocity | 200 m/sec |

special moment resisting frame structure (SMRF) located on seismic zone V with an importance factor of one.

The soil medium beneath the structures is a clayey soil whose properties are presented in Table 1. The peak load that can be carried by the pile or at which the pile continues to sink without further increase in load is known to be as ultimate bearing capacity of pile or ultimate bearing resistance of pile. Safe load that can be carried by the pile is called as its allowable load and it is obtained by dividing ultimate bearing capacity of pile by factor of safety of 2.5.

A friction type of pile group is adopted in this study which is having a pile group of 1 × 2 as shown in Fig. 6. Table 2 shows the details of the piles and pile cap.

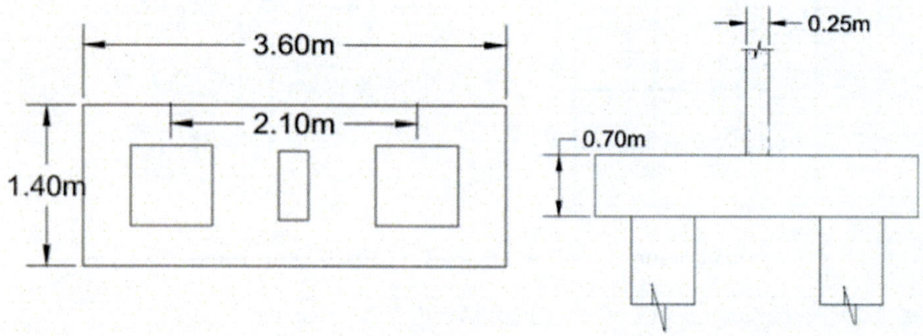


(a) Plan (b) Elevation

**Fig. 6** Details of pile cap

**Table 2** Size of pile and pile cap

| Sl. No | Contents | Description |
|---|---|---|
| 1 | Pile size | 0.7 × 0.7 m |
| 2 | Pile length | 9.6 m |
| 3 | Pile spacing | 3 times the pile size |
| 4 | Pile cap size | 1.4 × 3.6 × 0.7 m |
| 5 | Material | M25 grade concrete |

Continuum modelling i.e., prototype has been modelled using FEM based software SAP 2016. For experimental study the structure, substructure and soil have been scaled down to 1:30 as per scaling laws. This model is termed as scaled model.

## 2 Scaling of the Structure

To develop an experimental model which is able to represent prototype with less de- gree of distortion is the crucial part of the study. So geometric scaling, dynamic scal- ing and material scaling have been chosen properly in the present study. Table 3 shows the parameters for scaling along with scale factor.

A geometric scale factor of 30 is adopted in the present study. The appropriate modulus of elasticity of concrete has been adopted and also taken care that it has to be help full in fabrication of the model. Table 4 shows the geometric and material properties of the scaled model along with thickness of slabs and column dimensions.

**Table 3** Scaling relationships in terms of geometric scaling factor (Hokmabadi et al. 2014)

| Sl. No | Parameters | Scale factor (S) |
|---|---|---|
| 1 | Mass density | 1 |
| 2 | Stiffness | S2 |
| 3 | Force | S3 |
| 4 | Modulus | S |
| 5 | Acceleration | 1 |
| 6 | Frequency | S-1/2 |
| 7 | Time | S1/2 |
| 8 | Shear wave velocity | S1/2 |
| 9 | Length | S |
| 10 | Stress | S |
| 11 | Strain | 1 |
| 12 | EI | S5 |

**Table 4** Geometric and material properties of scaled model

| Sl. No | Parameters | Description |
|---|---|---|
| 1 | No. of story's | 5 |
| 2 | Storey height | 0.1 m |
| 3 | Bay width (*X*-axis) | 0.133 m |
| 4 | Bay width (*Y*-axis) | 0.133 m |
| 5 | Slab thickness | 11 mm |
| 6 | Size of columns | 2 mm × 12 mm |
| 7 | Material | Aluminium |
| 8 | Youngs moduls of aluminum material | 69 Gpa |

**Table 5** Mass by volume ratios of model

| Sl. No | Description | Setback building ($kg/m^3$) |
|---|---|---|
| 1 | Scaled model | 335.74 |
| 2 | Prototype | 344.15 |

The mass by volume ratios of the prototype to scaled model are considered such that dynamic similarity is achieved and these are presented in Table 5. According to scaling relations (Table 3). The frequency of the scaled model ($f_n$) and prototype ($f_p$) has been matched with the relation given in Eq. (1).

$$f_m/f_p = S^{1/2} \tag{1}$$

For connection between slabs to columns, bolts of 6 mm diameter are used. 4 numbers of exterior columns are connected with 6 mm diameter bolts which are driven into the plates. At the junction i.e., at the connection of the plates with the interior columns small angle sections with the bolts are employed.

## 2.1 *Scaling of Soil*

The scaling of soil has been carried out as per Aslan (2014) which consisted of 60% kaolinite clay, 20% bentonite, 10% Class F fly ash and 10% lime and 100% water. Accordingly, the soil density on the second day was determined to be 1470 $kg/m^3$.

## 2.2 *Scaling of Pile*

Adopting acrylic material, dimensional scaling has been adopted for scaling of pile and pile cap. The scaled dimensions of the pile group is as shown in Table 6.

**Table 6** Size of scaled pile and pile cap

| Sl. No | Contents | Description |
|---|---|---|
| 1 | Pile size | 17 mm x 17 mm |
| 2 | Pile length | 320 mm |
| 3 | Pile spacing | 70 mm |
| 4 | Pile cap size | 120 mm x 50 mm x 4.7 mm |

## 3 Experimental Study Using Shake Table

The Shake Table at the Department of Civil Engineering, UVCE, Bangalore, is an uniaxially driven having table size 1 m × 1 m with maximum payload capacity of 100 kg. The table has an operating frequency range of 0.05–25 Hz. In the present study, the objective is to evaluate the change in the dynamic characteristics of scaled models such as natural frequency or time period for flexible base condition.

In order to get natural frequency of scaled model, the model was subjected to a gradually increasing uni directional harmonic excitation (sine sweep wave) with sweep rate in the range of frequency 0.05–15 Hz. The response parameter such as displacements, accelerations and resonant frequences were recorded by Data Acquisition System (DAQ). Figure 7 shows the experimental setup of the step back building resting on soil.

In the experimental study, frequency has been sweeped from 0.05 to 13 Hz as shown in Fig. 8. From Fig. 8 it is observed that the resonant frequency is about 10.15 Hz. The damping of the soil structure model was determined using Half power band width adopting Eq. (2). From Fig. 8 damping ratio of 6.89% is obtained. Table 7 shows the storey displacements of th scaled model at resonance.

$$\xi = \left(\frac{f_2 - f_1}{2 f_n}\right) \tag{2}$$

Numerical prototype studies have been carried out considering SSI with continuum modelling using FEM based software SAP 2016. Figure 9 shows the elevation and three-dimensional view of the prototype structure.

The frequency obtained in the numerical model is 1.77 Hz. Using scaling laws (1:30) the fundamental frequency of prototype of the scaled experimental model was deter- mined to be 1.85 Hz. It can be observed that the predominant frequency of structure are nearly the same (with a deviation of 4.4%). Table 8 gives the summery

(a) Soft soil with pile and pile cap

(b) Stepback model resting on soil

**Fig. 7** Experimental setup of step back building

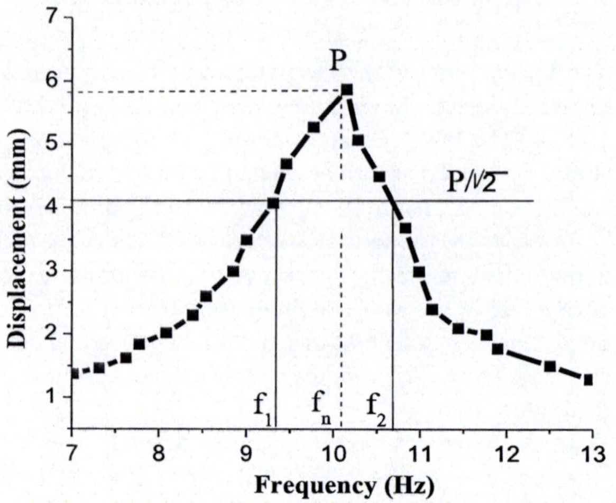


**Fig. 8** Displacement versus frequency of step back scaled model

**Table 7** Storey displacement of scaled model

| Storey No's | Displacement of stepback scaled model (mm) |
|---|---|
| 1 | 0.12 |
| 2 | 1.65 |
| 3 | 3.18 |
| 4 | 4.18 |
| 5 | 5.17 |

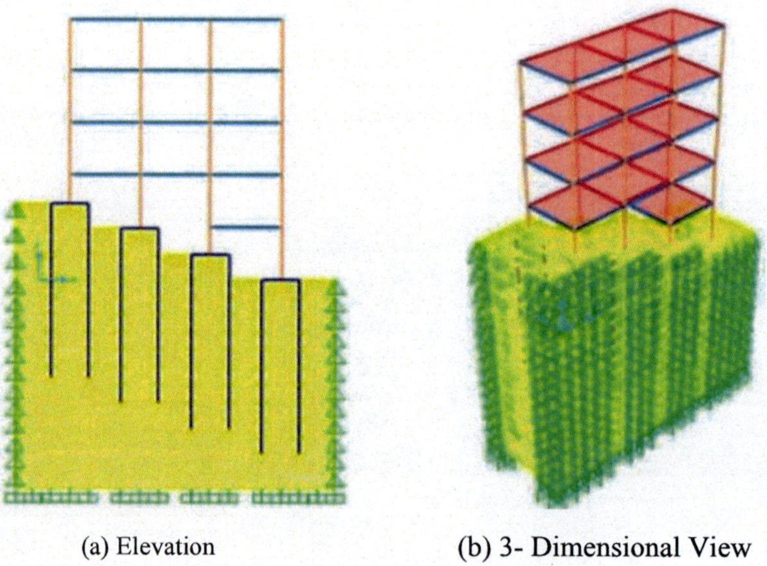


**Fig. 9** Three-dimensional view of the step back building (27° slope)

**Table 8** Displacement of scaled model and prototype

| Storey No's | Displacement of scaled model (mm) | Displacement of scaled model (mm) *30 | Displacement of prototype (mm) |
|---|---|---|---|
| 1 | 0.12 | 3.6 | 8.97 |
| 2 | 1.65 | 31.65 | 35.98 |
| 3 | 3.18 | 95.40 | 91.66 |
| 4 | 4.18 | 125.4 | 136.02 |
| 5 | 5.17 | 155.1 | 161.03 |

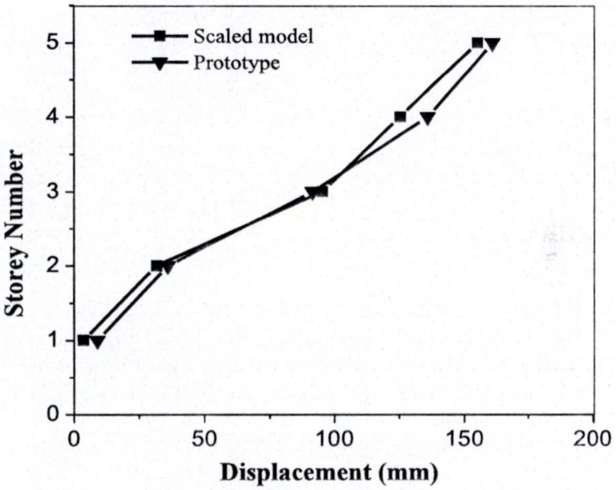


**Fig. 10** Storey displacements of scaled model and prototype

of storey displacements both for model and prototype structures. The storey displacements of scaled model and prototype structure are graphically presented as in Fig. 10. Figure 11 shows the storey drifts of the of scaled model and prototype structure.

## 4 Conclusions

Both experimental and numerical studies have been carriedout on the stepback building considering SSI. The response parameters such as storey displacements, storey drifts and resonant frequencies are determined. Based on the study the following conclusions can be drawn.

- The resonant frequencies of the scaled model and prototype are in close agreement using appropriate scaling laws.
- The displacement at each storey both for scaled model and prototype show almost comparative results.
- The storey drifts are well within the limits as per IS: 1893-2016 provisions.

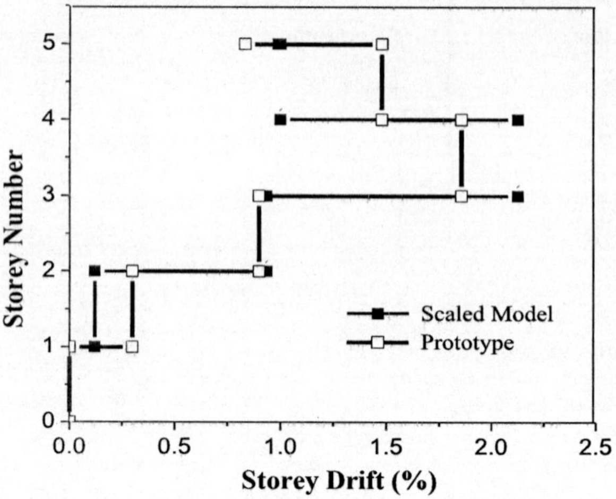


**Fig. 11** Storey drift of scaled model and prototype

# References

1. Birajdar BG, Nalawade SS (2004) Seismic analysis of buildings resting on sloping ground. In: 13th world conference on earthquake engineering. Vancouver, B.C., Canada, Paper No. 1472
2. Hokmabadi AS, Fatahi B, Samali B (2014) Physical modeling of seismic soil pile-structure interaction for buildings on soft soils. Int J Geomech ASCE. ISSN 1532-3641

# A Study on Dynamic Response of Hyperbolic Cooling Tower for Fixed Base Condition

**C. L. Mahesh Kumar, B. C. Shanthappa, and K. Manjunatha**

**Abstract** Hyperbolic Cooling towers are the portraying area characteristics of power station. Hyperbolic cooling towers rank among the largest reinforced concrete thin shell structures. They contribute both to an effective vitality yield and to a cautious offset with our condition. These structures are most effective measures for cooling of thermal power plants by limiting the need of water and maintaining a strategic distance from warm contamination of water bodies. The present day cooling towers are exceptional structures in view their sheer size and complexities. Present paper deals with study of Dynamic response that is modal analysis, seismic analysis of the two different cooling towers varying the $H/D$ ratio and thicknesses with fixity at the base boundary condition. Results show that there is a significant effect on structural elements for the earthquake behavior of Hyperbolic Cooling towers.

**Keywords** Cooling tower · Shell element · Modal analysis · Maximum principal stress

## 1 Introduction

Hyperbolic cooling towers are enormous, slight shell strengthened solid structures which contribute to the power generation efficiency, reliability and to environmental

C. L. M. Kumar (✉)
Department of Civil Engineering, Nitte Meenakshi Institute of Technology, Yelahanka, Bangalore 560064, India
e-mail: Maheshkumar.cl@nmit.ac.in

Department of Civil Engineering-VTU RRC, Jnana Sangama, Belagavi 590018, India

B. C. Shanthappa
Principal, SJM Institute of Technology, NH-4 Bypass, PB No 73, Chitradurga 577501, India
e-mail: Shanthappa2204@yahoo.com

K. Manjunatha
Department of Civil Engineering, UBDT College of Engineering, Hadadi Road, Davangere 577004, India
e-mail: drkmanjunatha@yahoo.com

 
T. G. Sitharam et al. (eds.), *Seismic Design and Performance*, Lecture Notes in Civil Engineering 120, https://doi.org/10.1007/978-981-33-4005-3_22

protection. Hyperbolic cooling towers is also called as Natural draft Cooling Tower. A Natural draft cooling tower is an enclosed device where hot water gets cooled under direct contact with air. Natural draft cooling towers rank among the largest reinforced concrete thin shell structures. Natural draft cooling towers are present in many thermal and nuclear power stations. The thin outer shell of the tall natural draft cooling tower can be said to be the greatest structural innovations. With thickness as little as 150 mm for a height of 75 m, the shell is a remarkable structure and its analysis and design demands considerable precision and care. The present day Natural draft cooling towers are exceptional structures in view their sheer size and complexities [5, 6]. It works on the principle of temperature contrast between the air inside and outside the tower. Hyperbolic state of cooling tower is generally favored because of its quality and soundness and bigger accessible territory at the base. Hyperbolic fortified solid cooling towers are viably utilized for cooling enormous amount of water in thermal power stations, treatment facilities, nuclear power plants, steel plants, cooling and other mechanical plants. The hyperbolic cooling tower is very important and essential component in thermal and nuclear power plants [2, 4]. They contribute both to a good yield and to a cautious offset to site condition. These shell structures are exposed to environmental loads, for example, Seismic and thermal gradients that is stochastic in nature.

The Present Paper manages the investigation of static and dynamic examination of hyperbolic cooling towers (for example self-weight, seismic load). The existing cooling towers dimensions are taken from following plants Pingwei phase II project, Haishen power station in Wuhai and Power station in Shouguang, Shandong province [1], The boundary conditions considered are Top end free and Bottom end is fixed. Busch et al. [3] demonstrated the optimization of a 200 m height natural draught cooling tower by varying the height of throat and inclination of meridian in reducing the stress due to wind load.

## 2 Objectives of Study

- To compare the Natural Frequencies of different Cooling towers with different $(H/D)$ ratio.
- To compare the Natural Frequencies of Cooling towers with different Thickness.
- To compare the Maximum Displacement of cooling towers subjected to Seismic force (zone IV) with different $(H/D)$ ratio with fixed base.

# 3 Methodology

## 3.1 Description and Geometry of Cooling Towers

The Total Height of Cooling Tower(CT1) is 150 m. The tower has base, throat and top radii of 54.2 m, 33 m and 36 m respectively, with throat located at 119 m from base, with varying thickness of 200, 300 and 500 mm.

The Total Height of tower is 190 m. The tower has base, throat and top radii of 65.25 m, 42 m and 43.45 m respectively, with throat located at 119 m from base, with varying thickness of 200, 300 and 500 mm. The geometrical details of cooling towers is shown in Table 1.

The geometry of the Hyperboloid revolution is

$$\frac{R_0^2}{a_0^2} - \frac{Y^2}{b^2} = 1 \tag{1}$$

where $R_0$ is the horizontal radius at any vertical coordinate, $Y$ with the origin of coordinates being defined by the center of the tower throat, $a_0$ is the radius of the throat, and $b$ is some characteristic dimension of the hyperboloid. The Typical geometry of cooling towers is shown in Fig. 1.

In Uttar Pradesh, Natural Draft Cooling Towers are less. Hence, in this research two Models of Natural Draft Cooling Towers are to be modeled using ansys FEM package. Response spectrum analysis was carried out according to Codal provisions.

# 4 Results and Discussions

In Uttar Pradesh, Natural Draft Cooling Towers are less. Hence, in this research two Models of Natural Draft Cooling Towers are to be modeled using ansys FEM package. Models are analyzed for seismic loads and Results are predicted in terms of Lateral Displacement, Principal stresses for Zone-IV.

**Table 1** Geometrical description of CT1, CT2

| Sl. No | Description | Symbols | Parametric values of CT1 in m and CT2 in m | |
|---|---|---|---|---|
| 1 | Total height | $H$ | 150 | 190 |
| 2 | Height of throat | $H_{\text{thr}}$ | 119 | 142.5 |
| 3 | Diameter at top | $D_t$ | 72 | 86.9 |
| 4 | Diameter at bottom | $D_b$ | 108.4 | 130.5 |
| 5 | Diameter at throat level | $D_{\text{thr}}$ | 66 | 84 |

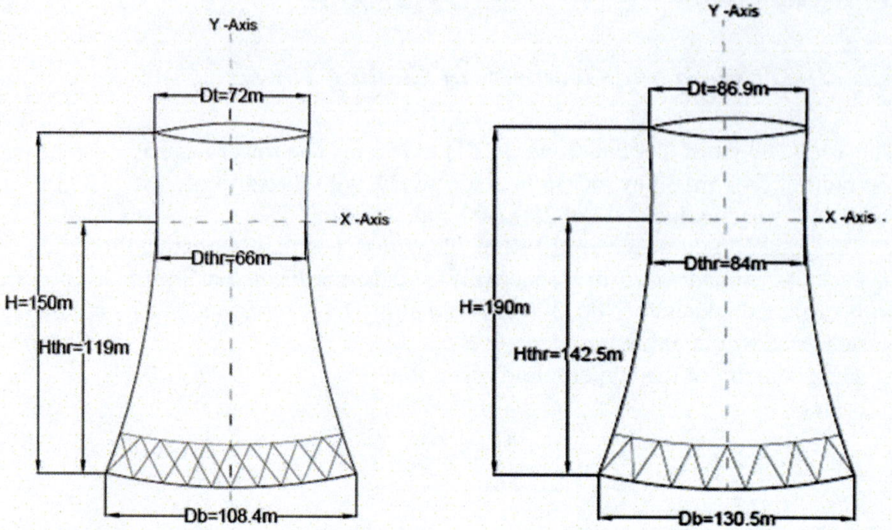


**Fig. 1** Dimensions of typical CT1 and CT2

For the generation of the Design Spectra, the following factors are considered.
Zone factor: For Zone IV, $Z = 0.24$ [8]
Importance factor, $I = 1.50$
Response reduction factor, $R = 3.00$
Average response acceleration coefficient $Sa/g$ = Soft soil site condition.

## 4.1 *Modal Analysis Results*

| Model | *H*/*D* | Thcikness (mm) | Natural frequencies (Hz) |
|---|---|---|---|
| CT-1 | 1.384 | 200 | 0.863 |
| CT-1 | 1.384 | 300 | 0.969 |
| CT-1 | 1.384 | 500 | 1.112 |
| CT-2 | 1.456 | 200 | 0.604 |
| CT-2 | 1.456 | 300 | 0.709 |
| CT-2 | 1.456 | 500 | 0.855 |

**Table 2** Natural frequency values for CT-1

| Mode | Thickness (mm) | Natural frequencies (Hz) |
|---|---|---|
| 1 | 200 | 0.863 |
| 2 | 200 | 0.863 |
| 1 | 300 | 0.969 |
| 2 | 300 | 0.969 |
| 1 | 500 | 1.112 |
| 2 | 500 | 1.112 |

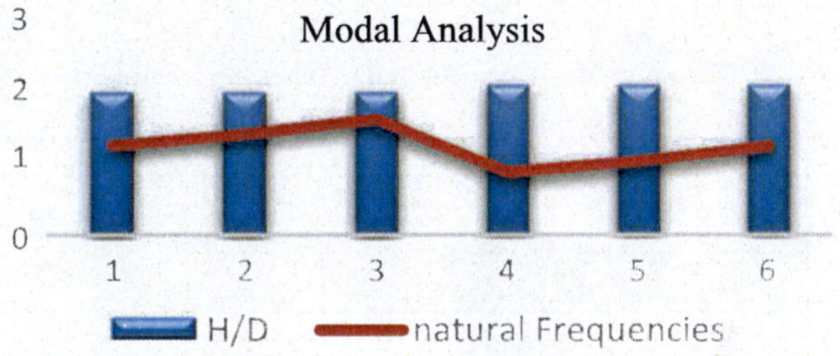


**Fig. 2** *H/D* versus Natural Frequency graph

The natural frequency values obtained for CT-1 is shown in Table 2 and also the graph of H/D ratio versus natural frequency is drawin and shown in Fig. 2. From the graph it can be seen that natural frequency values are increasing. Similarly mode shapes of CT-1 for 300 mm and CT-2 for 300 mm extracted from ansys are shown in Figs. 3 and 5 also the maximum displacement and principal stress plots is shown in Figs. 4 and 6. From the IS Code Book [7] H/D Should be 1.2 to 1.55.

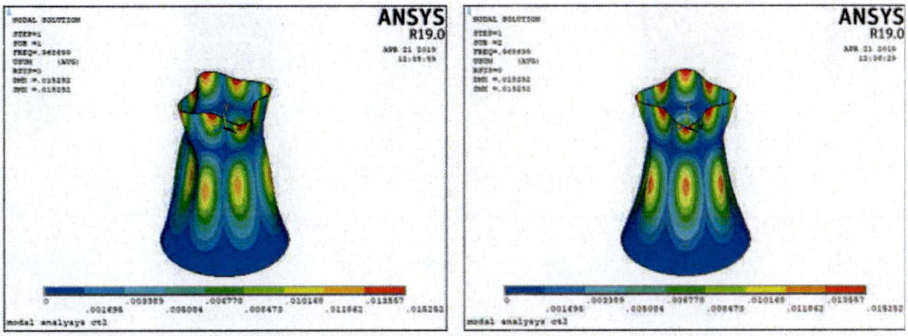


**Fig. 3** Mode shapes for CT-1 300 mm thickness

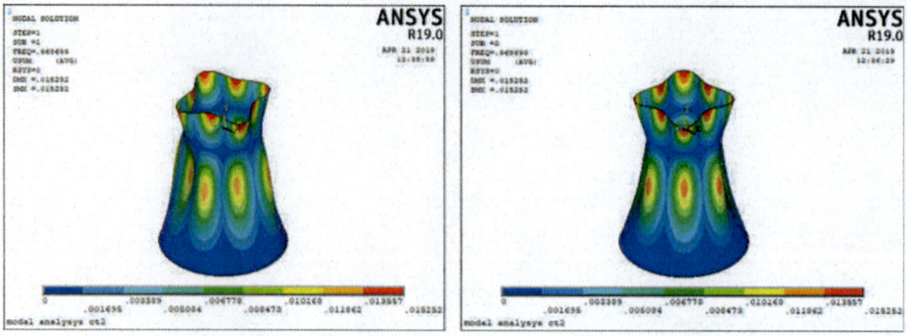


**Fig. 4** Maximum displacement and maximum principal stress for CT-1 300 mm thickness

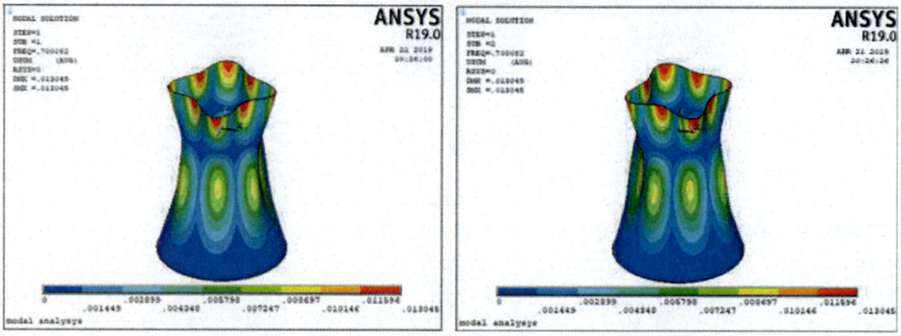


**Fig. 5** Mode shapes for CT-2 300 mm thickness

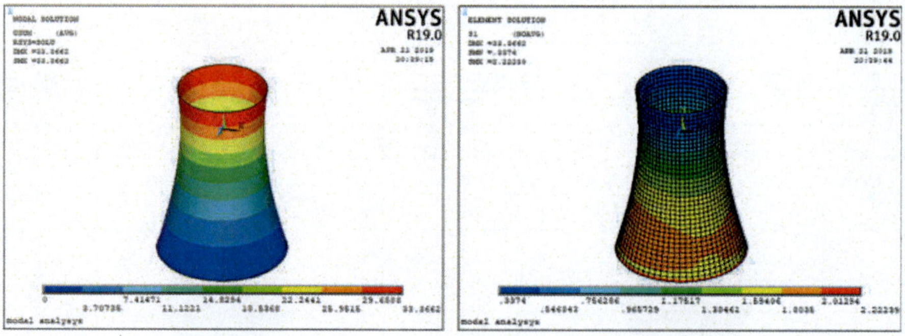


**Fig. 6** Maximum displacement and maximum principal stress for CT-2 300 mm thickness

## 5 Conclusions

From the above discussions the following conclusions can be arrived.

- The displacement of the Cooling tower model-1(CT-1) and Cooling tower model-2(CT-2) for 200 mm thickness due to response spectrum analysis is 18.93 and 33.31 mm. So the maximum displacement is 33.3 mm for CT-2. This is Because as height varies displacement also varies.
- The displacement of the Cooling tower model-1(CT-1) and Cooling tower model-2(CT-2) for 300 mm thickness due to response spectrum analysis is 18.90 and 33.36 mm.
- The displacement of the Cooling tower model-1(CT-1) and Cooling tower model-2(CT-2) for 500 mm thickness due to response spectrum analysis is 18.95 and 33.39 mm. The Displacement values are increased for 5% for change in thickness of cooling towers.
- The Maximum Principal Stress of the Cooling tower model-1(CT-1) and Cooling tower model-2(CT-2) for 200 mm thickness due to response spectrum analysis is 1.57 and 2.12 Mpa.
- The Maximum Principal Stress of the Cooling tower model-1(CT-1) and Cooling tower model-2(CT-2) for 300 mm thickness due to response spectrum analysis is 1.58 and 2.22 Mpa.
- The Maximum Principal Stress of the Cooling tower model-1(CT-1) and Cooling tower model-2(CT-2) for 500 mm thickness due to response spectrum analysis is 1.72 and 2.20 Mpa. This Shows that Maximum principal stress is increasing as thickness of shell increases.

## References

1. Ke S, Yu W, Zhua P,Ge Y, Hou X (2018) Full-scale measurements and damping ratio properties of cooling towers with typical heights and configurations. Thin-Walled Struct 124:437–448
2. Naghshineh A, Alavi E, Rezaee MR (2012) Seismic behavior of column-supported and innovative fixed-base cooling towers with ring beam. In: Proceedings of the 15th world conference on earthquake engineering. NITK Surathkal
3. Busch D, Harte R, Kratzig WB, Montag U (2002) New natural draft cooling tower of 200 m of height. Eng Struct 24(12):1509–1521
4. Murali G, Vivek Vardhan CM, Prasanth Kumar Reddy BV (2012) Response of cooling towers to wind loads. ARPN J Eng Appl Sci 7(1):114–120
5. Prabhakar N (1990) Structural design aspects of hyperbolic cooling towers. Technical Session IV, Paper no 9, pp 65-72, National Seminar on Cooling towers. Chennai
6. Gaikwad TG, Gore NG, Sayagavi, Kiran Madhavi VG, Sandeep (2014) Effect of wind loading on natural draught cooling tower. IJEAT 4(1):34–39
7. IS:11504(1985) (1985) Criteria for structural design of reinforced concrete natural draught cooling towers. Bureau of Indian Standards, New Delhi
8. IS 1893 (Part 1)—2016, Criteria for earthquake resistant design of structures. Bureau of Indian Standards, New Delhi

# Numerical Simulation of Soil-Tunnel Interaction Under Surface Blast Loading

**Jagriti Mandal, A. K. Agarwal, and M. D. Goel**

**Abstract** In the present investigation, interaction of soil with the tunnel, subjected to surface blast load, has been numerically simulated using commercially available software LS-DYNA®. In this simulation, three different contact approaches, namely shared nodes, Desai thin-layer element, and contact friction method, are employed to understand their effect on soil-tunnel interaction. In shared node method, adjacent nodes of soil mass and tunnel are merged together, thus generating a fully jointed interface without allowing any slip between the interfaces, whereas in Desai thin-layer method, a thin layer of solid elements is introduced in between soil and the tunnel. In this approach, Mohr–Coulomb material model is used to define the behavior of the thin layer. Further, contact friction method is based on Coulomb friction model, wherein, interaction is defined using coefficient of friction. This approach allows to model slippage, separation, rotation, and rebound, thus representing the real physical behavior. FE models have been prepared using Multi-Material Lagrangian Eulerian (MM-ALE) formulation, wherein air, soil, and explosive are modeled using Eulerian formulation, and tunnel is modeled using Lagrangian formulation. FE analysis is conducted, wherein the effect of these three approaches is compared based on downward displacement of the tunnel crown. Based on this analysis, it is observed that the maximum displacement for the contact friction method is significantly lower in comparison with the other two methods used herein. Thus indicating the importance of proper understanding of the interaction modeling under such extreme loading.

**Keywords** Soil-structure interaction · LS-DYNA · Numerical simulation · MM-ALE · Contact

---

J. Mandal (✉) · A. K. Agarwal
Department of Mining Engineering, Visvesvaraya National Institute of Technology, 440010, Nagpur, India
e-mail: jagritimandal@students.vnit.ac.in

M. D. Goel
Department of Applied Mechanics, Visvesvaraya National Institute of Technology, 440010, Nagpur, India

 
T. G. Sitharam et al. (eds.), *Seismic Design and Performance*, Lecture Notes in Civil Engineering 120, https://doi.org/10.1007/978-981-33-4005-3_23

## 1 Introduction

Blast resistant analysis of underground structures has become an active topic of research in recent years. Conducting experimental blast test on buried structures is not feasible, and the ones conducted by military sector are not open for civilian research. However, with advancement made in numerical analysis, it has become possible to simulate the blast phenomenon with very high accuracy and precision. Further, it involves less risk and saves up a lot of resources and time. Few factors play major roles when simulating any structure buried in soil subjected blast loading such as dynamic properties of the media and the structure along with the properties used to define the interface between them. There are quite a few methods available to model soil-structure interface under various type of loadings. Studies have been done to compare the efficiency in simulating the real behavior of the interface and load transfer [1]. However, the same has not been done considering blast as the loading condition. The blast response of buried structure is significantly affected by the soil medium in which it is buried, and the interaction between the medium and structure especially when the explosion is external.

The objective of present study is to simulate the interface between soil and tunnel subjected to surface blast load using three different contact approaches. The approaches considered herein are shared nodes, Desai thin-layer element, and contact friction methods. The concept behind each of these methods is discussed in subsequent sections. Multi-Material Arbitrary Lagrangian Eulerian (MM-ALE) method is used to model the entire problem. The comparison is made based on the downward displacement of the tunnel crown.

## 2 Numerical Methods for Soil-Structure Interface

When modeling the soil-structure interface in Finite Element (FE) special care has to be taken in simulating the relative translation and rotation of the surfaces in contact. Furthermore, realistic load transfer across the contact has to be ensured. However, available contact methods do not fulfill all of the above mentioned requirements when considering extreme loading such as blast. Among all the contact methods used in FE analysis, three methods are used commonly, namely shared nodes, Desai thin-layer element, and contact friction methods.

In shared nodes method, the adjacent soil and structure nodes are merged together making the contact rigid in nature as shown in Fig. 1(a). The contact is further defined using CONSTRAINED_LAGRANGE_IN_SOLIDS option offered by LS-DYNA®, where tunnel, which has been modeled using Lagrangian formulation, is set as slave, whereas soil, which is modeled using Eulerian formulation, is set as master. This option enables the interaction between Lagrangian and Eulerian elements, where the slave nodes follow the master nodes [2].

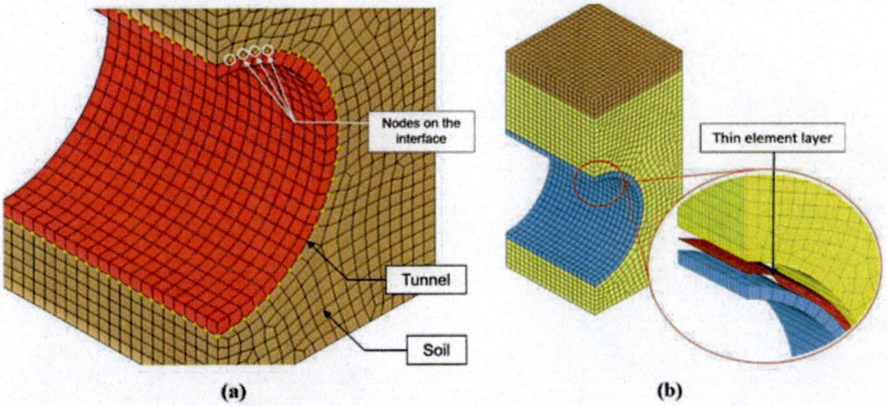


**Fig. 1** FE model showing (**a**) Shared nodes, (**b**) Desai thin-layer elements in the interface between soil and tunnel lining

In the second method, the interface is modeled as a layer of solid elements of thickness 0.01–0.1 times the longest dimensions. The properties of this one element thick layer are kept similar to the soil surrounding the structure. Generally, the strength of the thin interface is less than the soil strength, while the normal stiffness is greater than that of the soil. The shear modulus may be similar to that of soil [3]. For the present investigation, thickness of the element is taken as 5 mm as illustrated in Fig. 1(b).

Contact friction method is based on Coulomb friction model, wherein interaction is defined using coefficient of friction. This approach allows to model slippage, separation, rotation, and rebound, thus representing the real physical behavior. This method is applied by using AUTOMATIC_SURFACE_TO_SURFACE option which requires the input of static coefficient of friction. Further, to avoid penetration of elements, pinball segment-based contact constraint is used.

## 3 Geometry and Finite Element Modeling

A tunnel with circular cross section of diameter 5 m buried 3 m below the ground level, as shown in Fig. 2, is considered in the present investigation. The tunnel lining is assumed to have a thickness of 300 mm and made of reinforced concrete. The explosive is located at the interface of top soil and air centrally above the tunnel which is worst case scenario in case of surface explosion. TNT equivalent is used as the explosive. Since the main focus of the study is to compare the contact methods, the model is kept simple, wherein the element size is taken as 300 mm. The model is further cut along symmetrical planes, and the translation of nodes perpendicular to these planes along the cut sides is restricted. This saves a lot of computational time

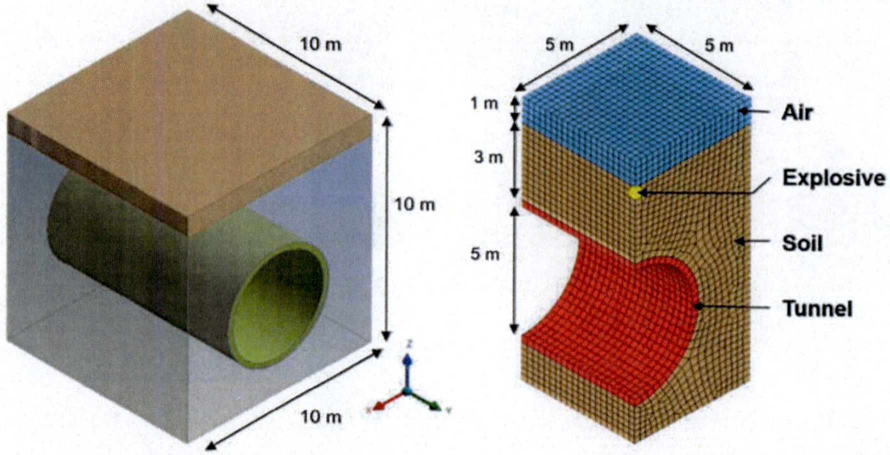


**Fig. 2** Geometrical details of buried tunnel

and memory. On the other, two lateral sides and bottom non-reflecting boundary conditions are applied to avoid reflection of blast waves.

The tunnel lining is usually made of reinforced concrete which has both non-linear and strain rate-dependent behavior. However, in the present investigation, strain rate effects are ignored. Modeling the concrete and steel separately increases the complexity of the model; hence, to keep the model simple, principle of equivalent stiffness is applied, where both concrete and steel are assumed to be as one part. The material properties used to define the behavior of reinforced concrete are given in Table 1 [4].

Federal Highway Association (FHWA) material model is used to model the soil. This model is generally used for simulating soil with varying saturation. It was developed by Lewis in 2004 using modified Mohr–Coulomb plasticity surface, and it, further, incorporates strain rate effects, strain softening, and kinematic hardening. Table 2 enlists all the parameters which require to model the soil [5].

**Table 1** Material properties of reinforced concrete [4]

| Parameters | Magnitude |
|---|---|
| Density (kg/m$^3$) | 2650 |
| Young's modulus (GPa) | 39.1 |
| Poisson's value | 0.25 |
| Yield stress (MPa) | 100 |
| Tangent modulus (MPa) | 4000 |
| Hardening parameter | 0.5 |
| Effective plastic strain | 0.8 |

**Table 2** Material properties for soil [5]

| Parameters | Magnitude |
|---|---|
| Peak friction angle | 35° |
| Residual friction angle | 30° |
| Cohesion | 6.5 kPa |
| Void ratio | 1.0 |
| Saturated bulk density | 18 kN/m$^3$ |
| Bulk modulus | 32.4 MPa |
| Shear modulus | 19.4 MPa |
| Specific gravity | 2.684 |

**Table 3** Material and equation of state parameters for air [4]

| ρ (kg/m$^3$) | $C_0$ | $C_1$ | $C_2$ | $C_3$ | $C_4$ | $C_5$ | $C_6$ | $E_0$ (MPa) | $\rho_0$ (kg/m$^3$) |
|---|---|---|---|---|---|---|---|---|---|
| 1.29 | 0 | 0 | 0 | 0 | 0.4 | 0.4 | 0 | 0.25 | 1 |

**Table 4** Material and JWL parameters for TNT [4]

| ρ (kg/m$^3$) | $\upsilon_D$ (m/s) | $P_{cut}$ (MPa) | $A$ (MPa) | $B$ (MPa) |
|---|---|---|---|---|
| 1630 | 6930 | $2.1 \times 10^4$ | $3.738 \times 10^5$ | $3.747 \times 10^3$ |
| $R_1$ | $R_2$ | ω | $V_0$ | $E_0$ (MPa) |
| 4.15 | 0.9 | 0.35 | 1 | 6000 |

Air and TNT are modeled using null and high explosive burn material models along with linear polynomial and Jone-Wilkins-Lee (JWL) equations of state, respectively. Their respective material parameters are listed in Tables 3 and 4, where $C_0$, $C_1$, $C_2$, $C_3$, $C_4$, $C_5$, and $C_6$ are constants, ρ is density, $E_0$ is initial internal energy per volume, $V_0$ is initial relative volume, $\upsilon_D$ is detonation speed, $P_{cut}$ is Chapman-Jouget pressure, and $A$, $B$, $R_1$, $R_2$, and ω are JWL parameters [4].

Mohr–Coulomb elastoplastic model is employed to simulate the interface behavior in Desai thin-layer element contact method. The material paramters used for the layer are: density = 1226.67 kg/m$^3$; elastic shear modulus = 14.55 MPa; Poisson's ratio = 0.4; coefficient of internal friction = 28.65 °; and cohesion value = 6.9 kPa [5].

## 4 Validation

The FE model is validated before proceeding toward the analysis to gain confidence over the accuracy of the obtained results. The validation is carried out on basis of the size of the crater formed on the ground surface due to detonation of explosive. For

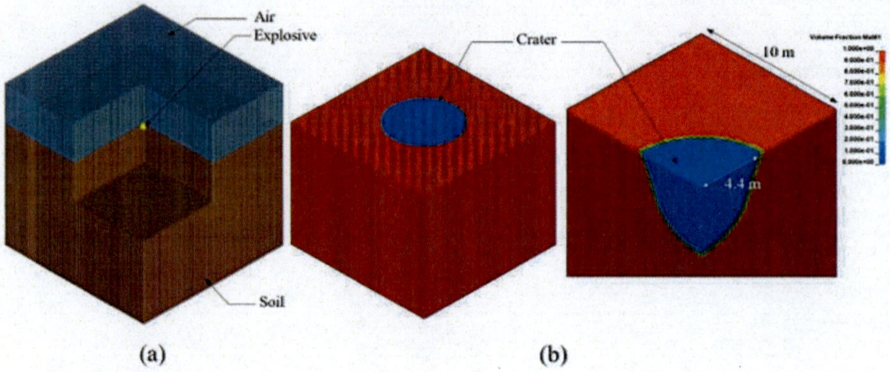


**Fig. 3** FE model showing **a** location of explosive with respect to soil, **b** formation and size of crater for TNT charge weight of 250 kg

the same, three explosive charge weights are considered, 150, 200, and 250 kg. A soil domain of 20 × 20 × 20 m is considered for this purpose, where the explosive is placed such that its top surface coincides with the top surface of the soil, shown in Fig. 3a. The numerically obtained radius of the crater is compared with the values obtained from an empirical formula based on theory of model similarity which is stated as,

$$R_{vd} = k_{vd}^{*}\sqrt[3]{W} \tag{1}$$

where $R_{\mathrm{vd}}$ is the radius of the crater formed, $k_{\mathrm{vd}}$ is the coefficient of proportionality which is taken as 0.694, and $W$ is the explosive charge weight [6]. Figure 3b shows the formation of crater on the soil and its radius on detonation of TNT with charge weight of 250 kg. Table 5 shows the comparison of the numerically and empirically obtained values of crater radius. The difference between these values is very small, thus validating the FE model.

**Table 5** Comparison of crater sizes obtained from numerical simulation and empirical formula

| TNT weight (kg) | Crater size obtained from numerical simulation (m) | Crater size obtained from the empirical formula [6] (m) | Percentage difference (%) |
|---|---|---|---|
| 150 | 3.43 | 3.69 | 7.04 |
| 200 | 4.18 | 4.06 | 2.95 |
| 250 | 4.44 | 4.37 | 1.60 |

## 5 Results and Discussion

The comparison between each of the contact methods considered herein is done on basis of displacement time histories at the crown of tunnel. As it can be observed in Fig. 4, the downward displacement curves for shared nodes and Desai thin-layer element methods coincide well till a time instant of 0.085 s, approximately. However, their displacement time curves vary greatly during the rebound stage. In the case of shared nodes method, rebound displacement is significantly more than that of Desai thin-layer element method. This can be attributed to the fact that in the former case, contact is rigid resulting in greater rebound reaction, whereas in the second case, incorporation of thin layer of elements provides some extend of flexibility in the interface. On arrival of blast wave, the thin element layer expands which simulates the opening of the interface and thus transferring the load to the tunnel lining causing its downward deflection. As the blast wave dissipates, tunnel lining tries to rebound back to its original position. This is accommodated by compression of the thin element layer which simulates the closing of interface. This phenomenon can be clearly seen in Fig. 5. The correlation between positions of top and bottom nodes of the thin element layer along z-axis right above the tunnel crown, and the visuals at various time instants give a better insight of the phenomenon.

On the other hand, the displacement obtained using contact friction method is significantly less than both the above discussed methods. A peak displacement of 1 mm, approximately, is reached before the tunnel crown rebounds back to its original

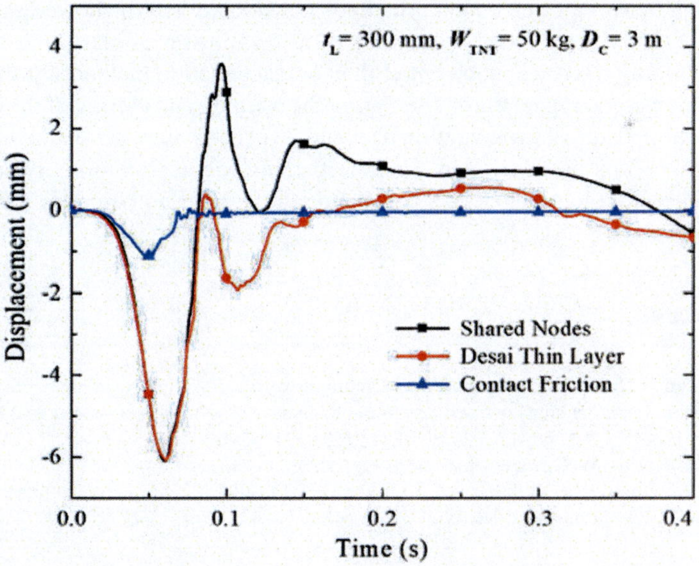


**Fig. 4** Comparison of displacement time histories of three contact methods: shared nodes, Desai thin layer, and contact friction methods

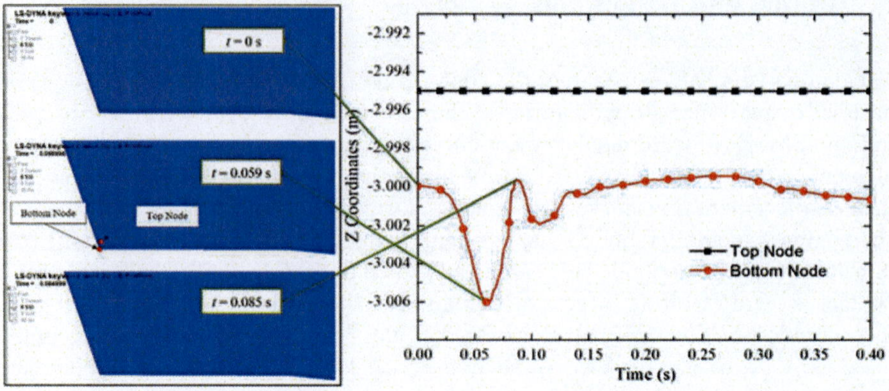


**Fig. 5** Opening and closing of interface shown as expanding and compression of Desai thin element layer

position and stabilizes. This states that the blast load is not transferred completely from soil to tunnel lining.

## 6 Conclusions

Comparison between three contact methods considered herein showed that shared nodes and Desai thin element layer methods allow transfer of blast load across the soil-tunnel lining interface much better than contact friction method. Further, Desai thin element method is capable of simulating the opening and closing of the interface. However, more detailed investigation needs to be carried out with an aim to develop a numerical technique which can simulate the entire complex mechanism of soil-structure interface including bond stage, bond failure stage, and friction sliding stage.

## References

1. Qian X, Yuan H, Li Q, Zhang B (2013) Comparative study on interface elements, thin-layer elements, and contact analysis methods in the analysis of high concrete-faced rockfill dams. J Appl Math 2013:1–11. https://doi.org/10.1155/2013/320890
2. LSTC (2018) LS-DYNA theory manual
3. Desai CS, Zaman MM, Lightner JG, Siriwardane HJ (1984) Thin-layer element for interfaces and joints. Int J Numer Anal Methods Geomech 8:19–43. https://doi.org/10.1002/nag.1610080103
4. Yang Y, Xie X, Wang R (2010) Numerical simulation of dynamic response of operating metro tunnel induced by ground explosion. J Rock Mech Geotech Eng 2:373–384. https://doi.org/10.3724/SP.J.1235.2010.00373

5. Han Y (2014) Mechanism of saturated ground-tunnel interaction under medium blast loading. The City College of New York
6. Henrych J (1979) The dynamics of explosion and its use. Elsevier Scientific Publishing Company

# Dynamic Analysis of Deep Beams on Vlasov Foundation

**Ashis Kumar Dutta and Jagat Jyoti Mandal**

**Abstract** In the present study, higher-order shear-deformable theories have been used for the dynamic analysis of deep beams on Vlasov foundation. For modeling the subsoil, the modified Vlasov model is used. It considers shear strain in the soil and gives a dish-shaped deflection for a uniformly distributed load on the beam. The values of the modulus of the sub-grade reaction, k, and the shear parameter, 2t, are computed using the properties of the soil. The modulus of elasticity and Poisson's ratio of the soil is assumed constant from top to bottom i.e. from ground surface to top of the assumed rigid base. A Matlab code is developed and convergence study is carried out and then some realistic cases of loaded beam on elastic foundations are solved to determine the dynamic response like displacements, velocity and acceleration. The results, thus obtained, are compared, with the available results obtained by other researchers.

**Keywords** Soil-structure interaction · Vlasov foundation · Shear modulus · Deep beams

## 1 Introduction

Dynamic analyses of beams on elastic foundations have very common applications in civil engineering. Many researchers consider Euler Bernoulli beam theory & Timoshenko beam model. Thin beam theory under estimate the deflection and overestimate the natural frequency. The limitations of Euler Bernoulli theory of beam and Timoshenko theory led to the development of higher order shear deformation theories. Developing more realistic foundation models and simplified methods to solve this

A. K. Dutta (✉)
Junior Engineer (Civil), Water Resources Investigation and Development Department, Government of West Bengal, Kolkata, India
e-mail: dutta.ashis2013@gmail.com

J. J. Mandal
Department of Civil Engineering, National Institute of Technical Teachers' Training and Research, Kolkata 700106, India
e-mail: jjm_civil03@yahoo.co.in

 
T. G. Sitharam et al. (eds.), *Seismic Design and Performance*, Lecture Notes in Civil Engineering 120, https://doi.org/10.1007/978-981-33-4005-3_24

complex soil-structure interaction problem are very important for safe and economical design. Majority of the problems cannot be solved by theoretical approach. In solving this type of problems; one can use the numerical techniques.

Numerous studies have been performed to estimate the dynamic response of the beams resting on Vlasov foundations. Several researchers had tried to improve the Winkler model by considering the shear strain energy in the soil in addition to the strain energy for normal strains as used in the Winkler model. Two-parameter foundation models account for the displacement continuity of the foundation which is the major defect of the Winkler foundation by the introduction of a second parameter. The two-parameter foundation models derived by Filonenko Borodich [1], Hetenyi [2] and Pasternak [3] provide for the displacement continuity of the soil medium by the adding of a second spring which interacts with the first spring of the Winkler model. The two-parameter foundation models improve by Vlasov and Leontiev [4] who made simplifying assumptions to the formulation of elastic continuum foundations by introducing functions for the distribution of displacements in the soil medium. Of all models, a two-parameter model by Vlasov using a variational method has attracted the attention of many engineers. The Vlasov model accounts for the effect of the neglected shear-strain energy in the soil and the shear forces on the beam edges that come from the surrounding soil. To measure the value of the vertical deformation parameter within the subsoil, Vallabhan and Das [5–7] developed an iterative technique to solve problems of beams on elastic foundations by introducing a modified Vlasov model. Ayvaz and Ozqan [8] considered the modified Vlasov model to analyze the free vibration of beams resting on elastic foundations. Kim and Kim [9] have considered vibration of beams with general restrained boundary conditions. Friswell et al. [10] have considered vibration of beams, with a nonlocal viscoelastic foundation model using the finite element method. Senalp et al. [11], have considered dynamic response of Euler–bernoulli beam on linear and nonlinear viscoelastic Winkler foundations to a concentrated moving force. Hizal and Çatal [12] study the dynamic analysis of axially loaded beams on modified Vlasov foundation. Thambiratnam and Zhuge [1] have formulated a basic model to study the dynamic behavior of beam on a flexible foundation under moving load condition.

Dynamic analysis of deep beams on Vlasov foundation is vast topic. In this study, an efficient method is introduced for the analysis of the free and as well as forced vibration behavior of deep beams on an elastic foundation.

## 2 Methodology

### 2.1 Higher Order Displacement Variations Written as

$$u(x,z) = z\theta_x(x) + z^3\theta_x{}^*(x) \text{ and } w(x,z) = w(x) + z^2 w^*(x)$$

$$\varepsilon_{xx} = \frac{\partial u}{\partial x} = z\frac{\partial \theta}{\partial x} + z^3\frac{\partial \theta_x{}^*}{\partial x} = z\chi_x = z^3\chi_x{}^*;$$

$$\varepsilon_{zz} = \frac{\partial w}{\partial z} = 2zw^*(x) = z\chi_z$$

$$\gamma_{xz} = \frac{\partial u}{\partial z} + \frac{\partial w}{\partial x} = \theta_x + 3z^2\theta_x{}^* + \frac{\partial w}{\partial x} + z^2\frac{\partial w^*}{\partial x};$$

$$\gamma_{xz} = \left(\theta_x + \frac{\partial w}{\partial x}\right) + z^2\left(3\theta_x{}^* + \frac{\partial w^*}{\partial x}\right);$$

$$\gamma_{xz} = \varphi_x + z^2\varphi_x{}^*;$$

$$Z_b = \begin{bmatrix} z & 0 & z^3 \\ 0 & z & 0 \end{bmatrix} \text{ and } Z_s = \begin{bmatrix} 1 & z^2 \end{bmatrix};\ C_{11} = \frac{E}{(1-\nu^2)} \text{ and } G = \frac{E}{2(1+\nu)};$$

$$C_{22} = C_{11};\ C_{12} = \nu C_{11};\ C_{21} = C_{12};\ C_{44} = G;$$

$$C_b = \begin{bmatrix} C_{11} & C_{12} \\ C_{21} & C_{22} \end{bmatrix} \text{ and } C_s = G;\ D_b = \int_{-\frac{h}{2}}^{\frac{h}{2}} [Z_b]^T[C_b][Z_b]\mathrm{d}z$$

$$\text{and } D_s = \int_{-\frac{h}{2}}^{\frac{h}{2}} [Z_s]^T[C_s][Z_s]\mathrm{d}z;$$

$$\begin{Bmatrix} \varepsilon_{xx} \\ \varepsilon_{zz} \end{Bmatrix} = \begin{bmatrix} z & 0 & z^3 \\ 0 & z & 0 \end{bmatrix}\begin{Bmatrix} \chi_x \\ \chi_z \\ \chi_x^* \end{Bmatrix} \text{ and } \gamma_{xz} = \begin{bmatrix} 1 & z^2 \end{bmatrix}\begin{Bmatrix} \varphi_x \\ \varphi_x^* \end{Bmatrix}$$

$$[B_{bi}] = \begin{bmatrix} 0 & \frac{\partial N}{\partial x} & 0 & 0 \\ 0 & 0 & 2N & 0 \\ 0 & 0 & 0 & \frac{\partial N}{\partial x} \end{bmatrix};\ [B_{si}] = \begin{bmatrix} \frac{\partial N}{\partial x} & N & 0 & 0 \\ 0 & 0 & \frac{\partial N}{\partial x} & 3N \end{bmatrix}$$

The strain energy $U_r$ in a linear spring is given by Eq. $= \frac{1}{2}Kw^2$ (Fig. 1).

$$\text{Strain energy } U_r = \frac{1}{2}\int Kw^2\mathrm{d}A \quad K \text{ is the soil parameter.}$$

As '$w$' is a scalar, so $w^2 = w^{\mathrm{T}}w = \{d\}^{\mathrm{T}}[N_f]^{\mathrm{T}}[N_f]\{d\}$

$$\text{Strain energy } U_r = \frac{1}{2}\int K\{d\}^{\mathrm{T}}[N_f]^{\mathrm{T}}[N_f]\{d\}\mathrm{d}A$$
$$= \frac{1}{2}\{d\}^{\mathrm{T}}[K_f]\{d\}$$

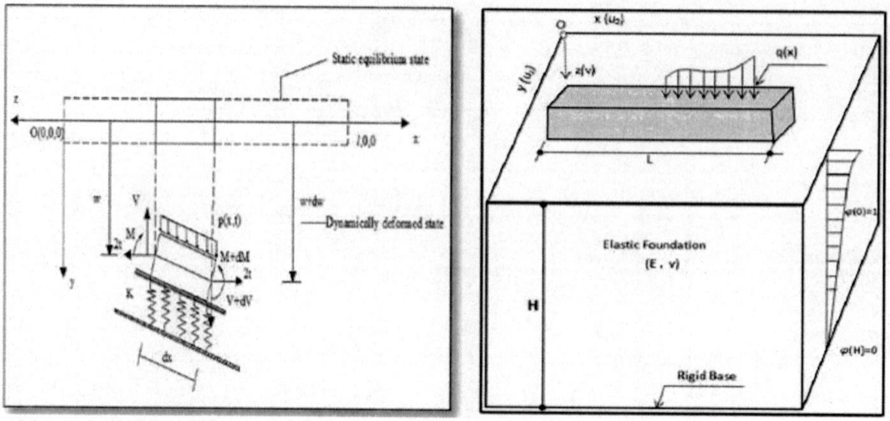


**Fig. 1** Deformed beam on two parameter Vlasov foundation

In which the foundation stiffness matrix for the element is, $[K_f] = \int K[N_f]^{\mathrm{T}}[N_f]\mathrm{d}A$. If the problem deals with a beam on elastic foundation, $[N_f]$ is identical to the shape function matrix $[N]$ of the beam.

Strain energy stored by foundation parameter '2*t*' is given by

$$U = \frac{1}{2}\int_0^l 2t\left(\frac{\partial w}{\partial x}\right)^2 \mathrm{d}x = \frac{1}{2}\int_0^l 2t\{d\}^{\mathrm{T}}[N_{f'}]^{\mathrm{T}}[N_{f'}]\{d\}\mathrm{d}x$$

$$= \frac{1}{2}\{d\}^{\mathrm{T}}[K_e]\{d\}$$

where $[K_e] = \int_0^l 2t[N_{f'}]^{\mathrm{T}}[N_{f'}]\mathrm{d}x$

$$\text{Kinetic energy, } T = \frac{1}{2}\rho A\int_0^l \left(\frac{\partial w}{\partial t}\right)^2 \mathrm{d}x$$

$$= \frac{1}{2}\{d'\}^{\mathrm{T}}\rho A\int_0^l [N]^{\mathrm{T}}[N]\mathrm{d}x\{d'\}$$

For mass matrix

$$[m] = \rho A\int_0^l [N]^{\mathrm{T}}[N]\mathrm{d}x$$

### 2.2 Parameter

$$k = \frac{E_0 b}{(1 - \nu_0^2)} \int_0^H \left(\frac{d\varphi}{dz}\right)^2 dz; \; 2t = \frac{E_0 b}{2(1 + \nu_0)} \int_0^H \varphi^2(z) dz;$$

$$m_0 = \rho_s b \int_0^H \varphi^2(z) dz; \; E_0 = \frac{E}{(1 - \nu^2)} \; \& \; \nu_0 = \frac{\nu}{(1 - \nu)}$$

$E$, $\nu$ is the Young's modulus of elasticity and poison's ratio of soil

where $\rho_s$ is the mass cdensity of soil.

$E_0$, $\nu_0$ is the effective modulus of elasticity and effective poison's ratio of soil.

$H$ is the effective finite depth of the foundation that is dynamically activated.

The parameter given by Vlasov [4] and Vallabhan and Das [6] has been shown that

$$\varphi(z) = \frac{\sin h\gamma \left(1 - \frac{z}{H}\right)}{\sin h\gamma}$$

$\gamma$ is the parameter denotes the vertical deformation within subsoil.

Now how obtained the $\gamma$ parameter

$$\left(\frac{\gamma}{H}\right)^2 = \frac{1 - 2\nu}{2(1 - \nu)} \frac{\int_{-\infty}^{\infty} \left(\frac{dw}{dx}\right)^2 dx}{\int_{-\infty}^{\infty} (w)^2 dx}$$

Solution of the governing equation of beams on elastic foundation is dependent on the vertical deformation profile $\varphi(z)$, which in turn depends upon vertical deformation parameter within the subsoil.

## 3 Results and Discussion

Several numerical examples are provided in this section with some comparisons with similar studies done by other researchers are made. The present method can be applied to analyze the dynamic response of a beam with various boundary conditions, including free-free and subjected to all types of loadings.

An example has been chosen from the study done by Friswell et al. [10] beam on Winkler foundation. The material properties for the beam, foundation and load are presented in Table 1

**Table 1** The material properties for the beam, foundation and load

| | |
|---|---|
| Length of the beam | 6.096 m |
| I of the beam | 0.001439 $m^4$ |
| E of the beam | $24.82 \times 10^6$ $KN/m^2$ |
| Mass per meter of the beam | 446.3 kg/m |
| Foundation parameter, k | 16550$KN/m^2$ |
| Foundation parameter, Gb/2t | 0 |

**Table 2** Comparison of natural frequencies (Hz) of vibration for the Simple beam

| Mode No | Analytical | M. I. Friswell | Present study |
|---|---|---|---|
| 1 | 32.898 | 32.898 | 32.810 |
| 2 | 56.808 | 56.812 | 56.313 |
| 3 | 111.90 | 111.95 | 110.021 |
| 4 | 193.76 | 194.08 | 188.748 |

The natural frequencies (Hz) of vibration for the Simple beam, modeled with 20 finite elements, on an elastic foundation. Poison's ratio of beam material taken as 0.3 for all problem below (Table 2).

The above problem has been chosen for forced vibration and dynamic external load was applied to this beam midpoint as shown in Fig. 2. Newmark's average acceleration method is applied for all problems below (Table 3).

**Fig. 2** Forcing function

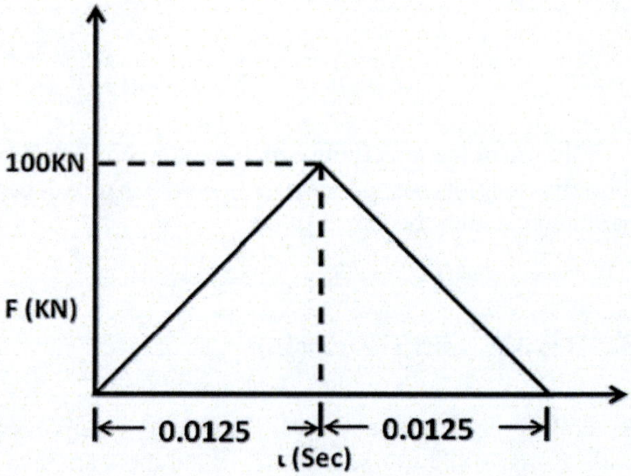


**Table 3** Comparison of displacement (mm)

| In mm | Sapountzakis [13] | Hizal and Catal [12] | Present study |
|---|---|---|---|
| Max | 2.630 | 2.635 | 2.659 |
| Min | −2.500 | −2.482 | −2.499 |

**Table 4** The properties of foundation

| | |
|---|---|
| The vertical deformation parameter within the subsoil, $\gamma$ | 1 |
| E of the soil | 20000KN/m$^2$ |
| Mass density of the soil | 1700 kg/m$^3$ |
| Poison's ratio of the soil | 0.25 |
| Depth of rigid base | 5 m |

**Table 5** Comparison of natural frequencies (Hz) of vibration for the Simple beam

| Mode No | Analytical | Present study |
|---|---|---|
| 1 | 10.0760 | 10.0445 |
| 2 | 29.5349 | 29.1758 |
| 3 | 63.5815 | 61.8656 |
| 4 | 111.5388 | 106.3513 |

**Table 6** Mid-point displacement response (mm)

| In mm | Present study |
|---|---|
| Max | 4.55 |
| Min | 0 |

The above problem has been chosen on Vlasov foundation with following modification (Table 4).

The natural frequencies (Hz) of vibration for the Simple beam, modeled with 20 finite elements, on an elastic foundation (Table 5).

Mid-point displacement response (Table 6).

The above problem has been chosen on Vlasov foundation with a full UDL of 50 KN/m and free-free boundary condition.

Mid-point displacement response (Table 7).

A finite simply supported beam is considered.

The material properties for the beam, foundation and load are presented in Table 8.

The results are presented below for the above beam

Max$^{m}$ static deflection = 1.9733 mm.
Max$^{m}$ Dynamic deflection = 4.12 mm.
Max$^{m}$ Dynamic Velocity = 0.39 m/s.
Max$^{m}$ Dynamic Acceleration = 1143.7 m/s$^2$ (Fig. 3).

**Table 7** Mid-point displacement response (mm)

| In mm | Present study |
|---|---|
| Max | 43.13 |
| Min | 0 |

**Table 8** The material properties for the beam, foundation and load

| | |
|---|---|
| Length of the beam | 5.0 m |
| Width of beam | 0.5 m |
| Depth of beam | 0.4 m |
| Where response is required from left end | 2.50 m |
| E of the beam | $2.1 \times 10^8$KN/m$^2$ |
| Poison's ratio of the beam material | 0.3 |
| Damping ratio | 0.05 |
| Initial displacement | 0.0 m |
| Initial velocity | 0.0 m/s |
| Mass density of beam material | 7850 kg/m$^3$ |
| Concentrated load | 500.00 KN |
| Position of load from left end | 2.5 m |
| The vertical deformation parameter within the subsoil, $\gamma$ | 1.988 |
| E of the soil | 20000KN/m$^2$ |
| Mass density of the soil | 1700 kg/m$^3$ |
| Poison's ratio of the soil | 0.25 |
| Depth of rigid base from top surface | 5 m |

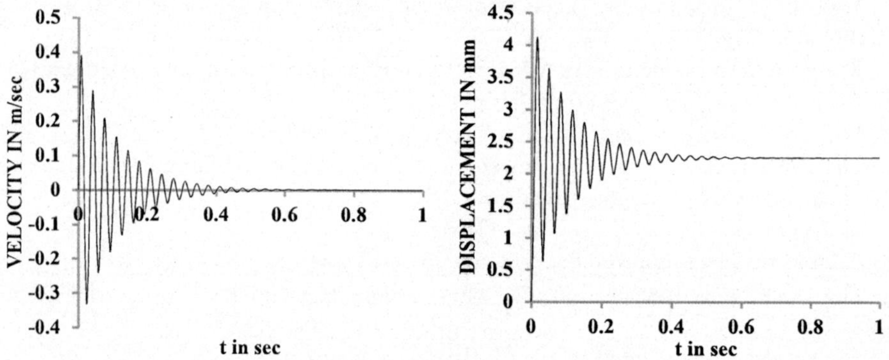


**Fig. 3** Time response of beam midpoint (velocity and Displacement)

The above problem has been chosen on Vlasov foundation with following modification.

Support condition is free-free considered. The vertical deformation parameter within the subsoil, $\gamma = 0.875$.

The results are presented below for the above beam

Max$^{\text{m}}$ static deflection = 26.025 mm.

Max$^{\text{m}}$ Dynamic deflection = 48.16 mm.

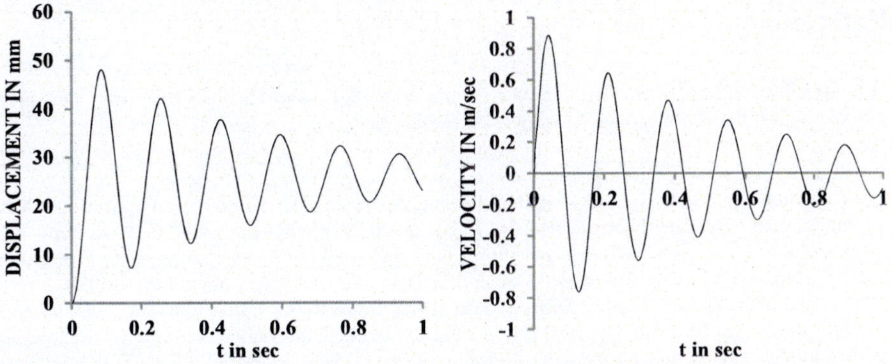


**Fig. 4** Time response of beam midpoint (Displacement and velocity)

**Table 9** Mid-point displacement response (mm)

| In mm | Present study D = 1.0 m | Present study D = 2.0 m |
|---|---|---|
| Max | 46.92 | 50.51 |
| Min | 0 | 0 |

$\text{Max}^{\text{m}}$ Dynamic Velocity = 0.89 m/s (Fig. 4).

The above problem has been chosen with the depth of beam modified as 1.0 and 2.0 m.

Mid-point displacement response (Table 9).

## 4 Conclusions

The following conclusions can be drawn:

- The maximum dynamic deflection of a deep beam resting on two-parameter Vlasov foundation is smaller than that of the beam on the Winkler foundation.
- It is also found that the dynamic deflection of the deep beam on a two-parameter Vlasov foundation decreases with the increased sub-grade shear modulus.
- The responses obtained by using HSDT is on the higher side and natural frequency on the lower side, compared to thin and moderate thick beam theory as expected.

## References

1. Filonenko-Borodich MM (1945) A very simple model of an elastic foundation capable of spreading the load. Shornite Moskovkovo Elektro Instituta
2. Hetenyi M (1946) Beams on elastic foundation: theory with applications in the fields of civil and mechanical engineering. The University of Michigan Press, Ann Arbor, Michigan
3. Pasternak PL (1954) On a new method for analysis of an elastic foundation by means of two foundation constants (in Russian) Gosud. Izd. Lit. po Stroitelstvu Arkhitekture, Moscow
4. Vlasov VZ, Leontiev NN (1966) Beams, plates and shells on elastic foundations. Translated from Russian by Israel Program for Scientific Translations Accession No N67–14238
5. Vallabhan CVG, Das YC (1989) Beams on elastic foundations: a new approach. In: Proceedings of the American society of civil engineers conference on foundation engineering: current principles and practices, Evanston, Illinois
6. Vallabhan CVG, Das YC (1988) Parametric study of beams on elastic foundations. J Eng Mech Division 114:2072–2082
7. Vallabhan CVG, Das YC (1991) Modified Vlasov model for beams on elastic foundations. J Geotech Eng 117:956–966
8. Ayvaz Y, Ozqan K (2002) Application of modified Vlasov model to free vibration analysis of beams resting on elastic foundations. J Sound Vib 255:111–127
9. Kim HK, Kim MS (2001) Vibration of beams with generally restrained boundary conditions using fourier series. J Sound Vib 245(5):771–784
10. Friswell MI, Adhikari S, Lei Y (2007) Vibration analysis of beams with non-local foundations using the finite element method. Int J Numer Meth Eng 71:1365–1386
11. Senalp AD, Arikoglu A, Ozkol I, Dogan VZ (2010) Dynamic response of a finite length Euler-Bernoulli beam on linear and nonlinear viscoelastic foundations to a concentrated moving force. J Mech Sci Technol 24(10):1957–1961
12. Hizal Ç, Çatal HH (2016) Comparative dynamic analysis of axially loaded beams on modified Vlasov foundation. Struct Eng Mech 57(6):969–988
13. Sapountzakis EJ, Kampitsis AE (2010) Nonlinear dynamic analysis of Timoshenko beam-columns partially supported on tensionless Winkler foundation. Comput Struct 88(21–22):1206–1219
14. Thambiratnam DP, Zhuge Y (1996) Dynamics analysis of beams on an elastic foundation subjected to moving loads. J Sound Vib 198(2):149–169

# Seismic Analysis of Soil-Pile-Structure System in Stratified Ash-Soil Deposit

**Amit Kumar Ram and Supriya Mohanty**

**Abstract** The present study deals with the seismic analysis of soil-pile-structure system in a stratified soil deposit subjected to EI Centro (Mw: 6.9), Northridge Santa Monica (Mw: 6.7), and San Fernando (Mw: 6.5) earthquakes. The influence of pile diameter (0.5 and 1.0 m) and shape has been considered for the analysis. The soil profile having fly ash as top layer followed by weathered soil and weathered rock of thickness 8 m, 10 m, and 10 m, respectively, was considered for the study. The response analysis has been done using finite element software SoilWorks. The results are presented in the form of displacement, velocity, and acceleration time histories. Pile responses have also been recorded in form of axial force, shear force, and bending moment for better design of pile foundation. Amplification in acceleration value has been noticed within the fly ash layer as compared to soil for the soil-pile-structure system under different earthquakes. This amplification in acceleration is due to low modulus of elasticity of fly ash layer, and the acceleration in the case of pile diameter 1.0 m is high as compared to pile diameter 0.5 m. The acceleration and velocity response show high magnitude for Northridge earthquake when moving away from footing center. Bending moment shows its maximum peak at the interface of fly ash and weathered soil. The results of the study are very much useful for safe design of pile foundation in ash-soil deposit.

**Keywords** Seismic response · Pile foundation · Soilworks · Seismic excitation · Ash-soil deposit

A. K. Ram (✉) · S. Mohanty
Department of Civil Engineering, Indian Institute of Technology (BHU), Uttar Pradesh, Varanasi 221005, India
e-mail: amitkumarram.rs.civ18@itbhu.ac.in

S. Mohanty
e-mail: supriya.civ@iitbhu.ac.in

 
T. G. Sitharam et al. (eds.), *Seismic Design and Performance*, Lecture Notes in Civil Engineering 120, https://doi.org/10.1007/978-981-33-4005-3_25

# 1 Introduction

The seismic response analysis of soil structure system has significant contribution for the safe design of earthquake resistant structure. The earthquake generated waves have devastating effect on sub-surface structure mostly pile foundation, and it results in the failure of the superstructure resting on it. Seismic activity is a massive challenge in front of design engineers from the serviceability point of view for the civil engineering structures. The majority of failures of the structure occur due to lack of knowledge of how earthquake's wave affects existing building and due to the trashy constructions. When earthquake occurs, the bases of the foundation have a tendency to move but the superstructure would be at rest because of the inertia forces. These unbalanced forces develop shearing of the structure resulting to the failure or collapse of the structure. The inertia force will be more for those structures which have higher mass; this is the reason why lighter structure sustains the earthquake movement in a better way. For the better load transfer mechanism of any structure, there is a need of sound understanding of static as well as seismic loading. The plethora of studies has been conducted on the seismic response of pile foundation subjected to vertical as well as lateral load [1–4]. Numerical analysis of foundations resting on different types of homogeneous or stratified soil/coal ash subjected to varying input motion has been studied by several investigators [5–12]. Most of the simplified approaches are available for the analysis of single pile and group of pile subjected to seismic excitation [13, 14], and these methods gave results in good agreement with mathematical models [15, 16]. Castelli and Maugeri [3] have made a numerical model for the determination of internal response of pile subjected to earthquake loading based on "p-y" sub-grade reaction method. For the investigation of soil-pile-structure interaction, many researchers have performed shaking table model test and dynamic model test [17–19]. Gazetas and Dobry [1] have considered the flexible pipe inside the stratified soil system for the development of realistic procedure to estimate lateral dynamic stiffness and damping. Maheswari et al. [20] presented seismic response of structure resting on pile foundation with the help of successive coupling incremental solution scheme, and seismic response was found considering the material nonlinearity. Sen et al. [2] have used the fundamental solution derived by Kobayashi and Nishimura [21] and develop an efficient algorithm which was applicable to combination of horizontal, vertical, and moment loading for the analysis of piles encapsulated in homogeneous soil. Phanikanth et al. [22] analyzed the behavior of single pile subjected to earthquake embedded in liquefiable deposits, and with the help of seismic deformation method, both kinematic and inertial interaction were evaluated. Rayhani and Naggar [23] established a numerical model for the seismic response of a rigid foundation in soft soil based on finite difference program. Haldar and Babu [24] used finite difference technique considering plain strain analysis to determine the parameters responsible for the failure mechanism of pile foundations in liquefiable soil. The equivalent linear method was found to be good for horizontal layered soil and unsuitable for deep soft soil in response analysis [25]. Liang et al. [26] considered 3D soil column in OpenSees for non-linear ground response analysis and verified the accuracy with the help of dynamic centrifuge model test. Li et al. [27] employed equivalent multi-degree of freedom system and compared it with existing discrete model, which exactly presents natural frequency and damping of soil. Volpini and Douglas [28] proposed an equation, which was used to distinguish between one-dimensional and two-dimensional analysis of quasi-horizontal layered deposits.

Numerous studies were carried out considering shallow foundation resting on homogeneous/non-homogeneous soil and pile foundation subjected to vertical and lateral loads. But, a very few study has been carried out taking pile as foundation system subjected to seismic excitation placed on different industrial waste as subgrade. An attempt has been made in this study to find out the response of the pile embedded in three layered soil-ash system under seismic excitations (EL Centro: Mw–6.9, Northridge: Mw–6.7: San Fernando: Mw–6.6 earthquakes). The three layer soil consists of fly ash as top most deposit underlain by weathered soil followed by weathered rock. The outcomes of the analysis are represented in the form of displacement, velocity, acceleration and axial force, shear force, and bending moment of soil-pile system at different locations.

## 2 Input Parameters

The input parameters contain time history record of considered earthquake, footing/pile foundations properties, and geotechnical properties of soil-ash deposit, which are explained in the following sections.

### *2.1 Footing-Pile Parameters*

The footing is resting on the surface of fly ash layer having width of 12.5 m and depth of 2.5 m. Both the pile and footing are made up of concrete which has Young's modulus of elasticity of 20E + 6 (kN/m$^2$), unit weight of 24.5 (kN/m$^2$), and Poisson's ratio of 0.19. Pile has length of 20 m and diameter of 1 and 0.5 m.

### *2.2 Properties of Layered Deposit*

The stratified soil profile considered is of dimension 40 m x 28 m (length x depth) with pile group installed at the middle of the domain. Three layered soil deposit consists of fly ash as top layer having thickness of 8 m, weathered soil as middle layer with thickness of 10 m, which is sandwiched between fly ash and weathered rock shown in Fig. 1. Similarly, weathered rock was considered as hard strata of 10 m, which

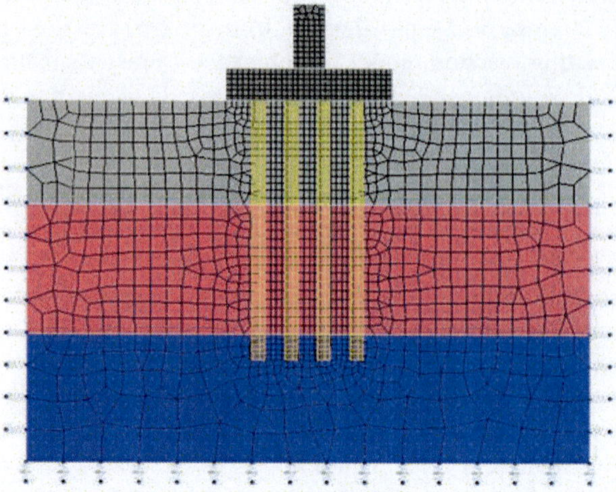

Fig. 1 Typical representation of generated mesh of soil-ash-foundation system

was supporting fly ash and weathered soil deposit. The detailed properties required for this study have been represented in Table 1.

### 2.3 *Seismic Input Motions*

The present study specially focuses on the seismic response analysis of soil-pile-structure system in stratified ash-soil deposit under different earthquake loadings. Three earthquake motions were considered as seismic input: El Centro earthquake (1940) (Mw–6.9), Northridge earthquake (1994) (Mw–6.7), San Fernando earthquake (1971) (Mw–6.6), respectively. The peak values of acceleration, velocity, and displacement for these three earthquakes are shown in Table 2. The acceleration time history plot of input motions is shown in Fig. 2.

## 3 Finite Element Modeling

The seismic analysis of soil-pile-structure system has been done using finite element software SoilWorks. The geometry of soil-pile-structure system has been prepared in Auto-CAD and imported into the SoilWorks for the modeling. Here, three layers of strata, i.e., fly ash, weathered soil, and weathered rock have been considered with thickness 8 m, 10 m, and 10 m, respectively. The whole soil domain was discretized using the very fine mesh system into 1857 number of elements and 2172 number of nodes.

**Table 1** Material property of layered deposit and footing

| S. no. | Ground type | Modulus of elasticity (kN/m$^2$) | Poisson's ratio | Unit weight (kN/m$^2$) | Saturated unit weight (kN/m$^2$) | Cohesion (kN/m$^2$) | Friction angle (°) | Cp (kN-sec/m$^2$) | Cs (kN-sec/m$^2$) |
|---|---|---|---|---|---|---|---|---|---|
| 1 | Fly ash | 18,279.88 | 0.3 | 15.89 | 16.87 | 50.39 | 39.3 | 1461.12 | 721.609 |
| 2 | Weathered soil | 530,000 | 0.4 | 18 | 19 | 17.5 | 30 | 252.506 | 134.970 |
| 3 | Weathered rock | 2,550,000 | 0.35 | 20 | 21 | 31 | 32 | 524.319 | 251.875 |
| 4 | Footing | 20,000,000 | 0.19 | 24.5 | 24.5 | – | – | – | – |

**Table 2** Peak response of all seismic input motion taken for analysis

| Earthquake | Peak acceleration (cm/s/s) | Peak velocity (cm/s) | Peak displacement (cm) |
|---|---|---|---|
| EL Centro | 350.119 | 33.449 | 10.866 |
| Northridge | 865.965 | 41.751 | 14.316 |
| San Fernando | 309.405 | 17.236 | 4.293 |

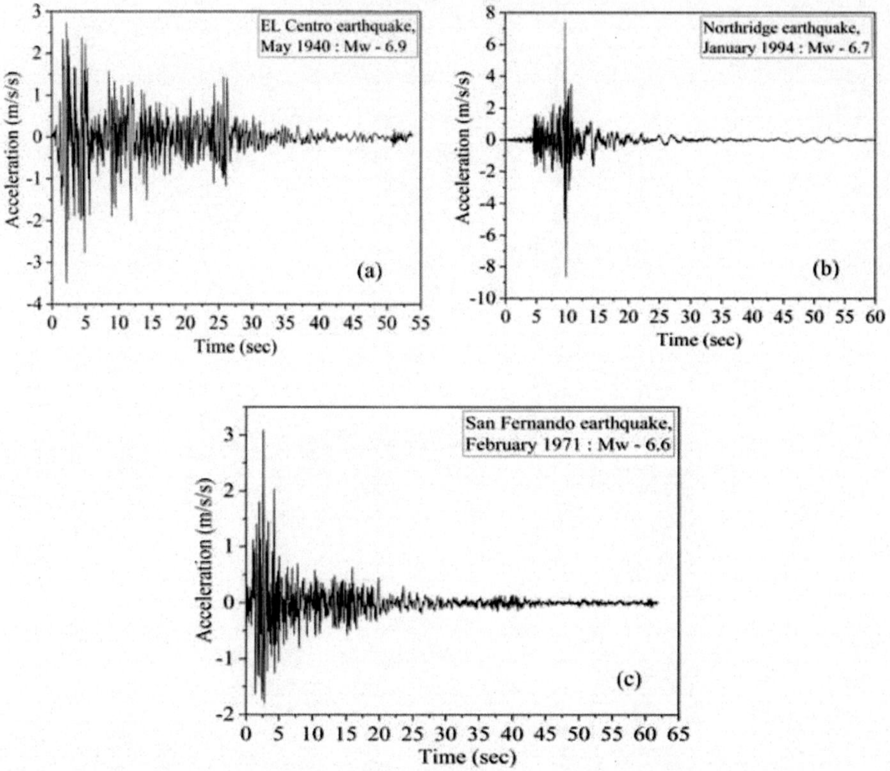


**Fig. 2** Acceleration time history plot of **a** El Centro earthquake, **b** Northridge earthquake, and **c** San Fernando earthquake

The analysis of the soil structure system can be done in following way such as: eigenvalue analysis, response spectrum analysis, time history analysis, and equivalent linear analysis; among these, time history analysis has been considered in the present study for better evaluation of responses of soil-pile-structure system in ash-soil deposit under seismic condition. The boundary condition was applied using the surface spring having different damping constant for different layer. Moreover, it was considered as viscous boundary, which will absorb the incoming wave energy. The base of the bottom layer was taken as fixed. The modeling of soil-ash-foundation system had been done by defining the basic material properties of layered deposit, and the pile embedded in this deposit was subjected to both vertical and seismic loads. The vertical load was evaluated considering base resistance and side friction resistance. The allowable load calculated for diameters 1 m and 0.5 m were 31,792.57 kN and 10,019.36 kN, respectively. The interface behavior between soil and pile system has taken care by employing Coulomb's friction model. The seismic responses were recorded at center, 0.5B and 1.22B (B: width of the footing).

## 4 Results and Discussion

The outcomes of seismic analysis of soil-ash foundation system have been discussed in terms of acceleration, velocity, displacement, and forces in pile along the depth of deposit. The seismic responses are explained in the following sections.

### *4.1 Acceleration Responses*

The acceleration has been extracted from the outcome of soil-pile foundation system subjected to different earthquake at different location below the top surface. The locations considered are (i) at center of footing (ii) at the edge of the footing or (0.5B), and (iii) at 9 m right from the edge of footing or (1.22B). Typical variation of acceleration along the depth for 1 m and 0.5 m diameter of pile is represented in Figs. 3 and 4, respectively. According to the Campbell's [29], empirical relation developed after regression analysis of worldwide earthquake, the horizontal and vertical attenuation of acceleration depends on magnitude of earthquake, depth of basement rock. Both the figures show the higher value of acceleration at the top surface/layer, and it decays as move from top to bottom. The reason behind this amplification of acceleration is due to the presence of low unit weight of top layer

**Fig. 3** Acceleration response of soil-ash foundation system subjected to different earthquake (pile dia. = 1 m)

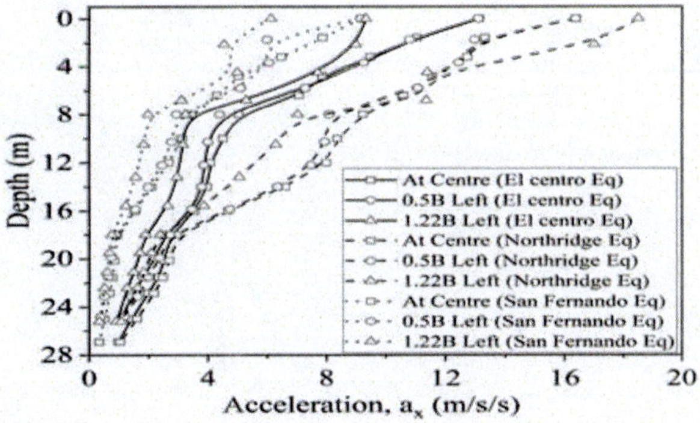

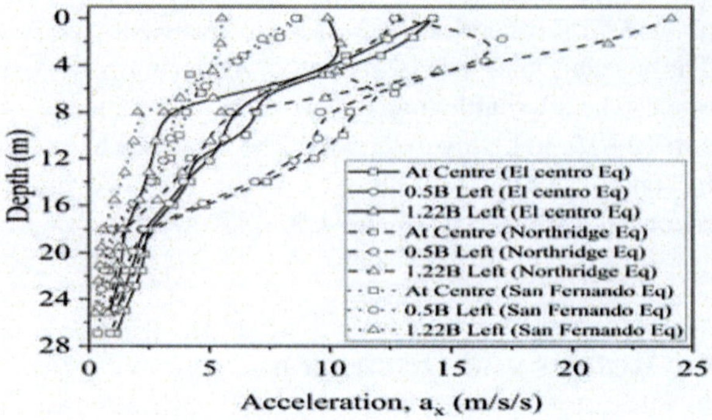


**Fig. 4** Acceleration response of soil-ash foundation system subjected to different earthquake (pile dia. = 0.5 m)

compared to the bottom two layers. Also, the variation of acceleration decreases when move away from center of the footing, and it can be seen in Fig. 3.

Among the three earthquakes considered, the Northridge earthquake shows higher acceleration than that of El Centro and San Fernando; this is because of having higher peak ground acceleration (PGA—8.66 m/s/s). The peak acceleration value of 16.39 m/s/s has been observed at 10.71 s for the case having pile diameter of 1 m.

## 4.2 *Velocity Responses*

The velocity response of soil-ash foundation system subjected to three different seismic excitation has discussed at center of footing, 0.5B and 1.22B, respectively. The variations of velocity response along the depth of stratified soil profile are demonstrated in Figs. 5 and 6, respectively.

The peak values of velocities are 1.05 and 1.25 m/s at 3.19 s for both the pile diameter, i.e., 1 and 0.5 m under the excitation of El Centro earthquake. The velocity variation also shows the similar trend of decreasing amplitude when moving away from the center of footing as already seen in acceleration responses. In both the

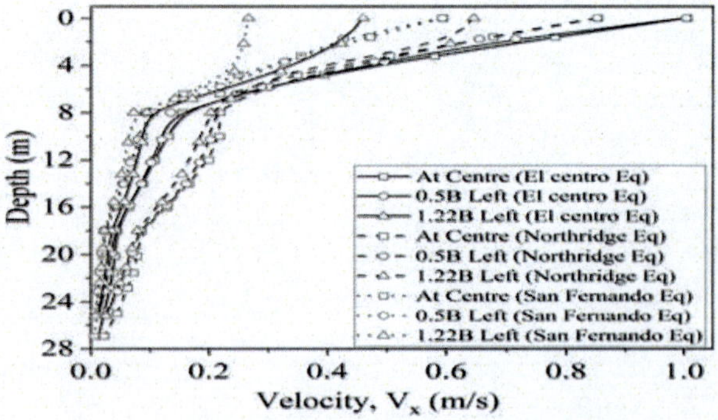


**Fig. 5** Velocity response of soil-ash foundation system subjected to different earthquake (pile dia. = 1 m)

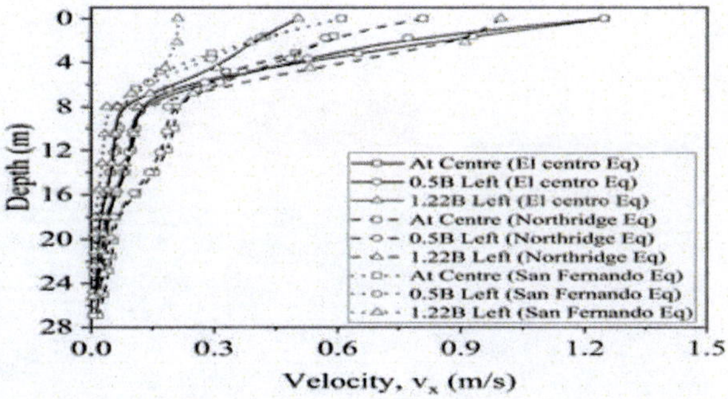


**Fig. 6** Velocity response of soil-ash foundation system subjected to different earthquake (pile dia. = 0.5 m)

cases, El Centro earthquake has higher velocity response than that of Northridge and San Fernando earthquakes. Campbell [29] recommended empirical relationship for finding the attenuation for free-field horizontal and vertical component of peak ground velocity; according to Campbell, it is a function of many regression factors; out of which magnitude of earthquake and depth of basement rock is the influencing factors of velocity responses.

## *4.3 Displacement Responses*

The horizontal displacement response of soil-ash foundation system subjected with El Centro, Northridge, and San Fernando earthquake was analyzed considering different pile diameter (1 and 0.5 m). The horizontal displacement response under the effect of El Centro earthquake has shown in Fig. 7. From Fig. 7, it can be seen that the horizontal displacement shows higher magnitude at the center of the footing. And this is also decreasing when moving away from the center. The displacement is directly dependent on magnitude of earthquake and shows higher value for higher magnitude of earthquake in the present study.

## *4.4 Pile Responses*

The responses of the pile include axial force, shear force, and bending moment within pile member generated during the seismic excitation. The representation of pile member forces has carried out considering the outermost pile, which was away from center of footing along the depth of pile. The variation of axial force, shear force, and bending moment along the depth of the pile is shown in Figs. 8, 9, and 10. The axial force response is higher in magnitude within the zone of weathered soil, whereas shear force and bending moment have higher magnitude near the junction between fly ash and weathered soil as represented in figure below.

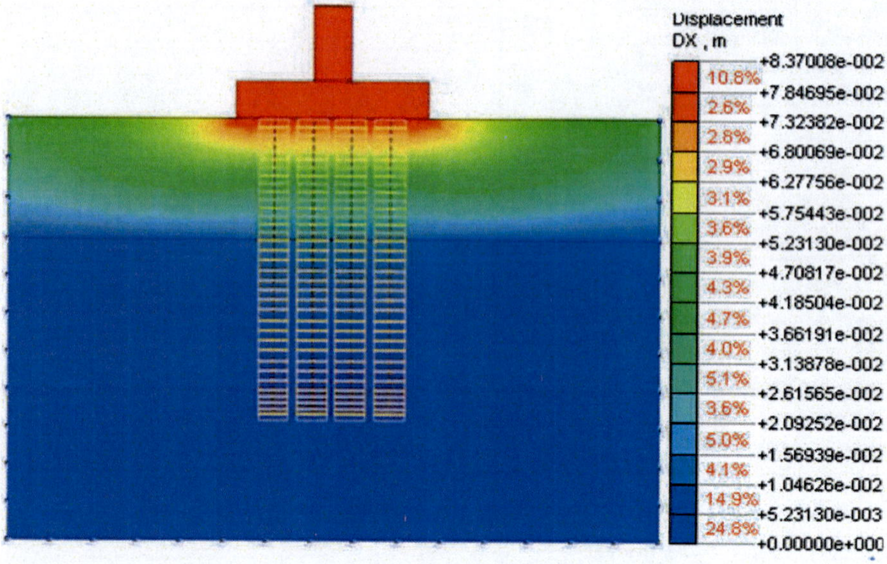


**Fig. 7** Typical variation of horizontal displacement of soil-ash foundation system under the excitation of El Centro earthquake

**Fig. 8** Variation of axial force of pile along the depth under different earthquake loadings

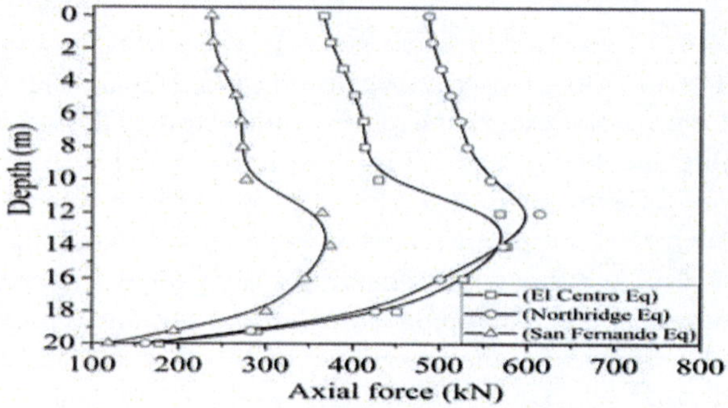


**Fig. 9** Variation of shear force of pile along the depth under different earthquake loadings

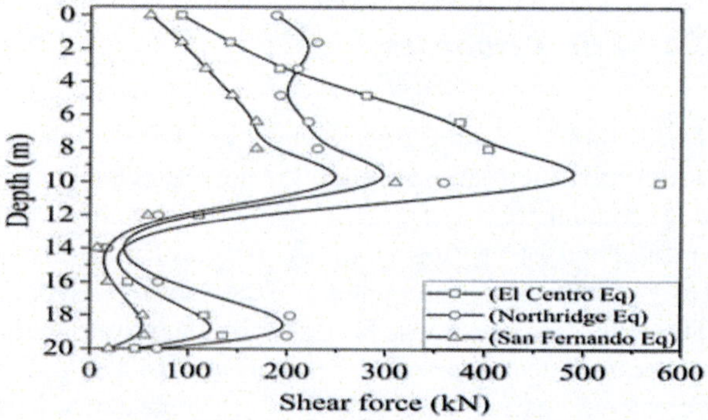

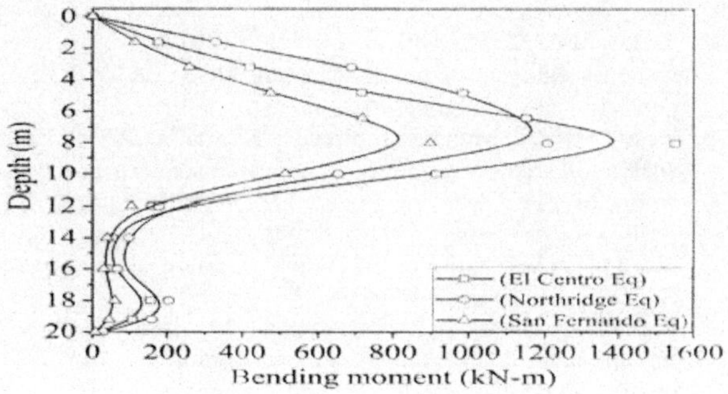


**Fig. 10** Variation of bending moment of pile along the depth under different earthquake loadings

## 5 Conclusions

In the present study, the seismic responses of soil-pile-structure system subjected with seismic disturbance have been investigated considering pile of different diameter. The result discussed above shows maximum value for acceleration, velocity, and displacement for Northridge earthquake. This is because of higher PGA of Northridge as compared with the El Centro and San Fernando earthquake. From figures, it can be seen that the horizontal acceleration is high at the top portion of the deposit. This amplification in acceleration is due to low modulus of elasticity of fly ash layer and also the acceleration in the case of pile of diameter 1.0 m is high as compared with acceleration of pile of diameter 0.5 m. The results of the study are very much useful for safe design of pile foundation in ash-soil deposit. To increase the life of any structure, the designer must study the seismic response analysis of ground and sub-surface structure, which will result in safe design of structure.

## References

1. Gazetas G, Dobry R (1984) Horizontal response of piles in layered soils. J Geotech Eng 110(1):20–40
2. Sen R, Davies TG, Banerjee PK (1985) Dynamic analysis of piles and pile groups embedded in homogeneous soils. Earthq Eng Struct Dynam 13(1):53–65
3. Castelli F, Maugeri M (2009) Simplified approach for the seismic response of a pile foundation. J Geotech Geo Environ Eng 135(10):1440–1451
4. Chatterjee K, Choudhury D, Poulos HG (2015) Seismic analysis of laterally loaded pile under influence of vertical loading using finite element method. Comput Geotech 67:172–186
5. Yoshimi Y, Tokimatsu K (1977) Settlement of buildings on saturated sand during earthquakes. Soils Found 17(1):23–38
6. Liu L, Dobry R (1997) Seismic response of shallow foundation on liquefiable sand. J Geotech Geoenviron Eng ASCE 123(6):557–567
7. Mohanty S, Reddy MVRK (2016) Seismic analysis of shallow foundation on layered soil deposits. In: International conference on advances in civil engineering and sustainable construction, vol 1, pp 532–537. (Published in RILEM proceedings) (e-ISBN: 978-2-35158-161-2, 2017, proc103, 130–135)

8. Reddy MVRK, Mohanty S, Rehana S (2018) Comparative study of 1D, 2D and 3D ground response analysis of pond ash from odisha under different earthquake motions. In: Indian geotechnical conference-2018, pp 1–16
9. Reddy MVRK, Mohanty S, Pradeep Kumar R (2019) Comparative study of dynamic response analysis of shallow foundation on layered soils. In: Soil dynamics and earthquake geotechnical engineering, Lecture notes in civil engineering. Springer, Berlin
10. Ram AK, Mohanty S (2018) Behavior of sheet pile wall with pond ash as backfill material under seismic condition. In: 16th symposium on earthquake engineering, IIT Roorkee, Paper No. 91, pp 1–10
11. Ram AK, Mohanty S (2019a) Numerical analysis of layered soil-ash- foundation system. In: 7th Indian young geotechnical engineers conference, 7IYGEC 2019, NIT Silchar, Assam, India, pp 298–305
12. Ram AK, Mohanty S (2019b) Nonlinear analysis of laterally loaded pile group in layered soil-ash deposit. In: Indian geotechnical society, IGC 2019, SVNIT Surat, Gujarat, India
13. Liyanapathirana DS, Poulos HG (2005) Pseudostatic approach for seismic analysis of piles in liquefying soil. J Geotech Geoenviron Eng 131(12):1480–1487
14. Poulos HG (2007) Ground movements–a hidden source of loading on deep foundations. DFI J J Deep Found Inst 1(1):37–53
15. Novak M (1974) Dynamic stiffness and damping of piles. Can Geotech J 11(4):574–598
16. Tabesh A, Poulos HG (2001) Pseudostatic approach for seismic analysis of single piles. J Geotech Geoenviron Eng 127(9):757–765
17. Matsuda T (1994) Shaking table test and dynamic effective stress analysis of light structure under saturated sandy ground. In: Proceedings of 8th international conference on computer method and advance in geomechanics, Morgantown, Balkema, pp 935–941
18. Taji Y, Sato M, Kagawa T, Minowa C (1997) Investigation on soil-pile-structure interaction in liquefying sand by a large scale shaking-table test and centrifuge test. In: Proceedings of the 24th symposium of Japan earthquake engineering, pp 277–280
19. Murono T (1997) Model test on a group-pile foundation considering the vibration of ground in seismic evaluation. In: Proceedings of the 24th symposium of japan earthquake engineering, pp 625–628
20. Maheshwari BK, Truman KZ, El Naggar MH, Gould PL (2004) Three-dimensional nonlinear analysis for seismic soil–pile-structure interaction. Soil Dyn Earthq Eng 24(4):343–356
21. Kobayashi S, Nishimura N (1980) Green's tensors for elastic half-spaces, an application of boundary integral equation method. Memoir fac. Engineering Kyoto University, XLII, Part 2
22. Phanikanth VS, Choudhury D, Reddy GR (2012) Behavior of single pile in liquefied deposits during earthquakes. Int J Geomech 13(4):454–462
23. Rayhani MH, El Naggar MH (2008) Numerical modeling of seismic response of rigid foundation on soft soil. Int J Geomech 8(6):336–346
24. Haldar S, Babu GS (2010) Failure mechanisms of pile foundations in liquefiable soil: parametric study. Int J Geomech 10(2):74–84
25. Hashash Y, Phillips C, Groholski DR (2010) Recent advances in non-linear site response analysis
26. Liang F, Chen H, Huang M (2017) Accuracy of three-dimensional seismic ground response analysis in time domain using nonlinear numerical simulations. Earthq Eng Eng Vibration 16(3):487–498
27. Li C, Yuan J, Yu H, Yuan Y (2018) Mode-based equivalent multi-degree-of-freedom system for one-dimensional visco-elastic response analysis of layered soil deposit. Earthq Eng Eng Vibration 17(1):103–124
28. Volpini C, Douglas J (2019) An accessible approach for the site response analysis of quasi-horizontal layered deposits. Bull Earthq Eng 17(3):1163–1183
29. Campbell KW (1997) Empirical near-source attenuation relationships for horizontal and vertical components of peak ground acceleration, peak ground velocity, and pseudo- absolute acceleration response spectra. Seismol Res Lett 68(1):154–179

# Dynamic Response of Circular Base Slabs (CBS) Attached to the Seismic Bearing Systems, Inserted for Base-Isolation Underneath the Buildings

**K. Kamal**

**Abstract** Base-isolation may be looked upon as state-of-the-art technique as it is being adopted now successfully in design of some seismic resistant buildings. In this technique, superstructure of the buildings are kept separated from substructure and which floats horizontally with limit barriers, safely over number of Circular Base Slabs attached with seismic bearings which are inserted underneath each columns of the framed super structure of such buildings. These circular base slabs need be so designed as to resist the entire static downward gravity loads of the superstructure resting over them and also these must safely respond to any dynamic loads that may get imposed onto the buildings. In the present study conducted here on these circular slabs attached to the seismic bearings provided with several shear layers of viso-elastic materials, all have been modelled as slabs that appear as resting on Winkler-Pasternack type foundations. In order to understand the static and dynamic behaviours, such base slabs are analysed then, using Chebyshev Polynomials since reported earlier as a rapidly converging and one of the best suited analytical technique. This study is another appendage to those several studies already conducted and published in the literature on circular elastic plates, resting on elastic foundations. Several of the numerical results so generated on the design parameters were then plotted graphically and are found to have offered certain interesting observations that are believed to be the useful inputs for deigning such circular base slabs for attaching them to seismic bearings systems that may safely be deployed in base-isolation process for buildings.

**Keywords** Bearing slabs · Chebyshev-polynomials · Seismic base-isolation

## Abbreviations

| | |
|---|---|
| $D$ | $Eh^3/\,12(1-\upsilon^2)$ |

K. Kamal (✉)
Ex Head, Civil Engineering Research Centre of VTU at JSSATE-B Campus, Uttarahalli-Kengeri Road, Bengaluru 560 060, India
e-mail: drkkamal.jss@gmail.com

 
T. G. Sitharam et al. (eds.), *Seismic Design and Performance*, Lecture Notes in Civil Engineering 120, https://doi.org/10.1007/978-981-33-4005-3_26

| | |
|---|---|
| $E, \upsilon$ | Young's Modulus & Poisson's ratio of the material |
| $q$ | Distributed load per unit area, positive load downwards |
| $\gamma$ | Mass density of the slab material |
| $\tau$ | Time variable $\sqrt{(D/ha^4)}.t$ |
| $a$ | Radius of circular slab |
| $h$ | Thickness of circular slab |
| $r, \theta$ | Polar coordinate |
| $g$ | Gravitational acceleration |
| $N, M$ | Number of terms in Chebyshev series |
| $Mr$ | Radial bending moment |
| $Q$ | $qa^4/Eh^4$ |
| $P$ | Non dimensional load $q \cdot a^3/D$ |
| $T$ | Time in seconds |
| $\nabla^2$ | $\left\{\frac{\partial^2}{\partial r^2} + \frac{1}{r}\frac{\partial}{\partial r}\right\}$ |
| $\nabla^4$ | $\nabla^2 \bullet \nabla^2$ |
| $W$ | Displacement in transverse direction |
| $\overline{w}$ | $W/a$ |
| $W_r$ | Coefficient of Chebyshev series for $W$ |
| $\rho$ | $r/a$ Non dimensional radius |
| $T_r(\rho)$ | The rth Chebyshev polynomial in the range $-1 \leq \rho \leq 1$ |
| $G$ | Pesternack Foundation coefficient |
| $K$ | Winkler foundation modulus |
| $Z$ | Wnikler foundation factor $Ka^4/D$ |
| $V$ | Pesternack Foundation Factor $-Ga^2/D$ |
| $f(z), g(\rho)$ | Arbitrary functions in the range $-1 \leq \rho \leq 1$ |
| Superscript ' | First term of expansion to be halved |
| Subscript $_j$ | Step of marching variable |
| Superscript $^{(k)}$ | Order of derivative |

## 1 Introduction

Most practiced method in design of any seismic resistant buildings, the world over has been, to increase ductility and/or increase elastic strength of the superstructures. Reported studies in literature on various applicable methods however also include a method of seismic base isolation that may be created at the top of substructure in the buildings [1]. The entire superstructure in this process is kept separated horizontally from its substructure and the entire building thus floats safely over number of suitably designed Base Slabs attached to the Seismic Bearings system which are inserted underneath each column footings. Literature reveals that on deployment of such base isolation it is found to facilitate non development of any structural damages in the superstructure as well it also offers the required level of comforts to the occupants inside these buildings during any ground excitation caused by the earthquakes. Usually, in any base isolation technique, lubricated Ball Bearings/sliding Roller are, either inserted or some strong magnetic levitation using suitable means are created in between superstructure and substructure of the buildings. The type of Seismic Bearing system considered in the present study is therefore one which comprises of number of layers of synthetic/natural rubber layers, sandwiched between number of strong metallic, circular elastic plates of adequate thickness offering complete viso-elastic behaviour horizontally to absorb completely the earthquake energy. In this type of arrangement, the circular base slabs e.g., are attached to the seismic bearing systems, permitting a rigid body movement of the entire superstructure over the substructure in horizontal plane caused by earthquakes. Such base slabs attached to seismic bearing system, need be designed to offer the required hold in preventing any uplift of the entire superstructure of the building and at same time, it must also offer a desired flat surface that safely transfers the vertical loads of the building over it, onto the substructure. These attached base slabs at each of seismic bearings inserted underneath need therefore bear, without causing any elastic deformation of themselves, all the static and dynamic loads transferred downward of the floating superstructure of the buildings. For carrying out analysis and obtain the deflections and slopes at various points of these circular slabs attached to the seismic bearing systems comprise of the number of sandwiched visco-elastic layers and hence these base slabs need conveniently be modelled as circular plate resting on elastic foundations. Such modelling identically remains more so valid since it is in this way done in solving the problems of plates that are deployed in various other engineering applications successfully, for example; in water tanks, water hulls, under water vessels, in aircraft, rocket casings and cylindrical shells casings ete, by considering them all as resting on some elastic foundations.

In the published literature ample information is available to solve equations that govern the deformation of plates on elastic Foundations using various analytical and Numerical techniques. Analysis of circular plates for example, without any foundation but with edge support, was carried out by Way [2] using power series method. Nowinki [3] used Double Fourier Series method, Keller [4, 5] and Nash [6] worked with Finite Difference method and Furlong et al. [7] used the finite element Method. Some work on plates on elastic foundation was also found reported, as carried out by Sinha [8] wherein, he applied the Berger approximation [9] technique. Application of these methods in solving differential equations representing plates on foundations, as reported in the literature have been tedious and more time consuming. An analytical technique to solve differential equation as well as integral equation adopting Chebyshev Polynomials[10] has also been reported in the literature wherein the differential equations were then successfully solved using Chebyshev series in the range as $0 \leq \rho \leq 1$which were shown as rapidly converging and one of the best suited technique for analysis in the computer age. Use of Chebyshev Polynomials in the analysis, was then also continued further [11] and successfully demonstrated an adoption of Chybeshev series also in the range $-1 \leq \rho \leq 1$ and solved the governing differential equation for circular plate resting on elastic foundation. Several numerical

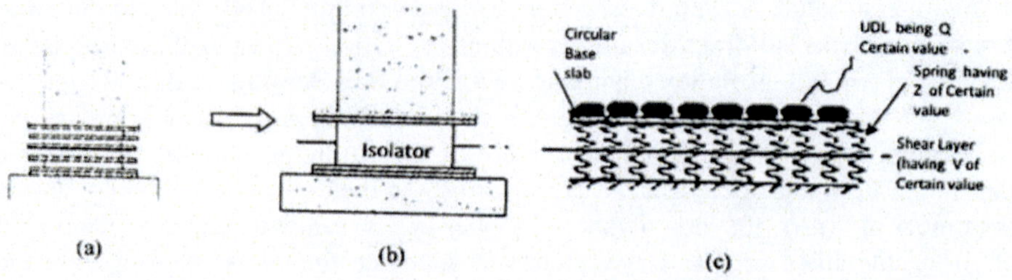


**Fig. 1** **a** Isolator as used, **b** Isolator as idealised, **c** Isolator analytical model

results were then computed for circular, elastic, isotropic and uniform in thickness plates with no foundation using simply supported and clamped edges respectively and were all found to tally well with those results available in the literature [12]. One of the foundation model considered then was Winkler [8] type having complete lack of continuity in the material of the foundation and for analytical purpose which was regarded as consisting of a series of independent springs capable of acting in tension and compression. Present study conducted and reported here is another appendage to those several studies on the behaviour of circular slabs on elastic foundations already reported earlier, but has been on the study of slabs that are attached to the seismic bearings inserted in buildings to convert existing or new buildings to be designed into the seismic resistant buildings. On simulating, the sandwich visco–elastic layers between number of metallic plates within the seismic bearings, this problem has been appropriately solved as circular base slabs resting on Winkler-Pasternak [13] type foundations so that it includes all the shearing effects created within any of the visco-elastic layers. The Winkler-Pesternack foundation model as depicted in Fig. 1, adopted for the present study is to obtain response of circular base slabs subjected to static as well as their response under any pulse loading that may be received from the building resting above them. In this study, as done earlier too, the dynamic terms in the differential equation, has been carried out using Houbolt timewise integration technique [14].

The governing differential equation for the base slabs with edge simply supported have been solved here and studied for the elastic modes and response to dynamic loads Influence of variation in values of foundation factors $Z$ and $V$ on elastic response mode of these slabs have been studied and reported. From many numerical results so obtained on personal computer are generated in the form of design charts and certain salient, useful features found are reported and discussed here in this paper.

## 2 Problem Definition and Methodology

As done in Engineering Mechanics, problem of analysis of the circular slabs resting on elastic foundations, subjected to any static and dynamic loadings, is defined as the differential equations which includes foundation parameters and inertia terms [12] and in polar coordinates it is

$$\begin{aligned}\nabla^4 W &= (\partial^2/\partial r^2 + 1/r \cdot \partial/\partial r + 1/r^2 \cdot \partial^2/\partial\theta^2) \\ &\quad \cdot (\partial^2 W/\partial r^2 + 1/r \cdot \partial W/\partial r + 1/r^2 \cdot \partial^2 W/\partial\theta^2) \\ &= \{q - (KW - G\nabla^2 W)\}/D + \{(h\gamma/_D) \cdot \partial^2 W/\partial t^2\}\end{aligned} \tag{1}$$

Solution as assumed for this differential equation thus comprise of those unknowns which are the function of one or several variables and contains not only the function themselves but also their derivatives. Reported in the literature are many solutions to choose from either Special function, Bessel function, Legendered polynomial, trigonometric series, or solutions can be obtained Numerically also from finite element method, finite difference, Dynamic Relaxtion method etc. Among all these solutions reported in the literature, the solution using Chybeshev polynomial approximation to a function $f(z)$ was found and reported as nearly minimaximal. In other words when the unknown constants in the Chybeshev series are determined by collocation with the zeros of the Chybeshv polynomials as the collocation points, the resulting series leads to a minimum of the maximum absolute error [15]. Also, out of all the ultra spherical polynomials, the chybeyshev polynomials reported to have the fasted rate of convergence. In the present analysis for solving the governing differential equations of circular base slabs resting on Visco-Elastic supports and subjected to Static and dynamic vertical loadings, solution adopted here uses these Chebyshev Polynomials, wherein,

An $r$th polynomial in a Chebyshev series is defined as

$$T_r(\rho) = \cos r\,\theta,\ \cos\theta = \rho\ ; \quad -1 \le \rho \le 1 \tag{2}$$

Chybeshev polynomials obey their properties of orthogonal functions as

$$\int_{-1}^{1} \left(1-\rho^2\right)^{1/2} T_m(\rho)\cdot T_n(\rho)\,d\rho = \begin{cases} 0, & m \ne n \\ \pi/2, & m = n \ne 0 \\ \pi & m = n = 0 \end{cases} \tag{3}$$

Then any given function can be expressed as a sum of such Chybeshev polynomials as

$$g(\rho) = a_{0/2} \cdot T_0(\rho) + \sum_{r=1}^{N} a_r T_r(\rho) \tag{4}$$

## 3 Mathematical Modelling and Analytical Solution

In order to analyse and obtain the solution of differential equation as in (1) above for Circular Base slabs attached with seismic bearings, the entire system, has to be considered as elastic, isotropic and of uniform thickness, resting on Visco- Elastic supports underneath i.e. Below the base slabs as if it is resting on Winkler-Pasternak type of foundation.

The solution to this problem was assumed in Chebyshev polynomials expressed here as

$$\bar{w} = \sum_{r=0}^{N,} \bar{w}_r \cdot \mathrm{T}_r(\rho) \tag{5}$$

and on substitution of this assumed solution (5) in Eq. (1), the slab differential equation becomes

$$\begin{aligned}
&\sum_{r=0}^{N-4,}\Bigg[\Bigg\{\overline{w_r}^{(4)} + 3/2\left(\overline{w_{r-1}}^{(4)} + \overline{w_{r+1}}^{(4)}\right) + 3/4\left(\overline{w_{r-2}}^{(4)} + 2\cdot\overline{w_{r+1}}^{(4)} + \overline{w_{r+2}}^{(4)}\right)\\
&+1/8\left(\overline{w_{r-3}}^{(4)} + 3\cdot\overline{w_{r+1}}^{(4)} + \overline{w_{r+3}}^{(4)}\right) + 2\left(\overline{w_{r-1}}^{(3)} + \overline{w_{r+1}}^{(3)} + \overline{w_{r+1}}^{(3)}\right)\\
&+1/2\left(\overline{w_{r-2}}^{(3)} + 2\cdot\overline{w_{r+1}}^{(3)} + \overline{w_{r+2}}^{(3)}\right)\Bigg\} - \left\{\overline{w_r}^{(2)} + 1/2\left(\overline{w_{r-1}}^{(2)} + \overline{w_{r+1}}^{(2)}\right)\right\} + \overline{w_r}^{(1)}\\
&-Z/16\Big\{\overline{w_r} + 3/2(\overline{w_{r+1}} + \overline{w_{r-1}})\\
&+3/4\left(\overline{w_{r-2}} + 2\cdot\overline{w_r} + \overline{w_{r+2}}\right) + 1/8\left(\overline{w_{r-3}} + 3\cdot\overline{w_{r-1}} + 3\cdot\overline{w_{r+1}} + \overline{w_{r+3}}\right)\Big\}\\
&+V/4\Big\{\overline{w_r}^{(2)} + 3/2\left(\overline{w_{r-1}}^{(2)} + \overline{w_{r+1}}^{(2)}\right) + 3/4\left(\overline{w_{r-2}}^{(2)} + 2\cdot\overline{w_r}^{(2)} + \overline{w_{r+2}}^{(2)}\right)\\
&+1/8\left(\overline{w_{r-3}}^{(2)} + 3\cdot\overline{w_{r-1}}^{(2)} + 3\cdot\overline{w_{r+1}}^{(2)} + \overline{w_{r+3}}^{(2)}\right) + \overline{w_r}^{(1)}\\
&+1/4\left(\overline{w_{r-2}}^{(1)} + 2\cdot\overline{w_r}^{(1)} + \overline{w_{r+2}}^{(1)}\right) + \left(\overline{w_{r-1}}^{(1)} + \overline{w_{r+1}}^{(1)}\right)\Big\}\Bigg] J\cdot T_r(\rho)\\
&= \frac{a^3}{16D}\cdot\Big[\sum_{r=0}^{N}\{q_r + 3/2\cdot(q_{r-1} + q_{r+1}) + 3/4\cdot(q_{r-2} + 2q_r + q_{r+2})\\
&+1/8(q_{r-3} + 3q_{r-1} + 3q_{r+1+}q_{r+3})\}\Big]_J \cdot T_r(\rho)\\
&+\sum_{r=0}^{N}\Big\{(5/32)\overline{w_{\tau,r}} + (15/8\times16)\overline{w_{\tau,r+1}} + (3/64)\overline{w_{\tau,r+2}}\\
&+(1/8\times16)\overline{w_{\tau,r+3}} + (1/8\times64)\overline{w_{\tau,r-3}} + (3/64)\overline{w_{\tau,2}} + (15/8\times10)\overline{w_{\tau,r-1}}\Big] J\cdot T_r(\rho)
\end{aligned} \tag{6}$$

This is the general governing equation of the base slab under dynamic loading and when it was solved, resulted into the solution of slab resting on winkler-Pasternack foundation, with the simple support edge conditions of slabs defined as

at edges $\rho = +1$

$$\sum_{r=0}^{N} \overline{w}_r T_r(\rho) = 0 \quad (7i)$$

$$-D\left[\sum_{r=U}^{N-2}\left\{\overline{w}_r^{(2)} + 1/2\left(\overline{w}_{r-1}^{(2)} + \overline{w}_{r+1}^{(2)}\right)\right\} + \upsilon \cdot \sum_{r=0}^{N-1}\left\{\overline{w}_r^{(1)}\right\}\right] = 0 \quad (7ii)$$

and at centre $\rho = -1$

$$\sum_{r=0}^{N-1} \overline{w}_r^{(1)} T_r(\rho) = 0, \quad (7iii)$$

$$\sum_{r=0}^{N-3} \overline{w}_r^{(3)} T_r(\rho) = 0 \quad (7iv)$$

In case of this isolator study considered here, the response of circular elastic slabs subjected to pulse type loading is solved including the dynamic part, for which the known Houbolt time wise integration technique has been used. Herein, the Inertia term, to start with the recurrence process, is expressed as

$$\left(d^2\overline{w}/d\tau^2\right)_{r,J} = 1/\Delta\tau^2\left\{\alpha\overline{w}_{r,J} + \beta\overline{w}_{r,J-1} + \delta\overline{w}_{r,J-2} + \eta\overline{w}_{r,J-3} + \zeta\right\} \quad (8)$$

where $\alpha$, $\beta$, $\delta$, $\eta$ and $\zeta$ are the constants for Houbolt Scheme and these are

When $J = 1$

$$\alpha = 6/\Delta\tau^2, \beta = 0, \delta = 0, \eta = 0 \text{ and } \zeta = -2P(\tau) \quad (9i)$$

When $J = 2$

$$\alpha = 2/\Delta\tau^2, \beta = -4\Delta\tau^2, \delta = 0, \eta = 0 \text{ and } \zeta = -2P(\tau) \quad (9ii)$$

When $J = 3$

$$\alpha = 2/\Delta\tau^2, \beta = -5\Delta\tau^2, \delta = 4\Delta\tau^2, \eta = 0 \text{ and } \zeta = 0 \quad (9iii)$$

When $J > 3$

$$\alpha = 2/\Delta\tau^2, \beta = -5\Delta\tau^2, \delta = 4\Delta\tau^2, \eta = -1\Delta\tau^2 \text{ and } \zeta = 0 \qquad (9iv)$$

## 4 Numerical Results and Discussions

Governing differential equation of circular slabs of any diameter of '2$a$', total thickness of '$h$', made of isotropic material, having edge simply supported all round, was initially subjected to static UDL of intensity '$q$' and was solved. Various numerical values were then assigned to the design parameters expressed only in the non-dimesionlised forms pertaining to geometry $\overline{w}$, $\rho$etc., the loads parameter $Q$, and the foundation factors $Z$ and $V$, starting with initial value as zero. The influence of values of these parameters on the responses of the slabs were studied under static and dynamic loadings. Numerical results so obtained correspond to variation of appropriate values of $K$ and $G$ respectively of the support system. These response of Base slabs as obtained has been plotted in the form of graphs for ready reference. Initially the response were obtained by assigning zero values to $Z$ and $V$, that represents the case of slabs with no foundation. Slabs of edge simply supported all round only are considered here. Deformation with varied $Q$ from zero to certain maximum, 10 for example, as shown in Fig. 2. As was expected, lateral deformations of various points in the slab about the neutral axis, when obtained and plotted, indicated a gradual increase in deformation values with increase in applied loads as shown in Fig. 2.

The centre point deformation of the base slabs obtained on varying values $Q$ representing the share of the portion of the entire building load by the respective column which is received by that base slab attached to bearing system considered in the study. It was varied from zero value to 10 for example, along with $Z$ value varied as 0–25, 30 and 35 respectively and initially assigning $V = 0$ only. The interesting results obtained show possibility of adopting suitable extrapolation and interpolation to obtain accurate values of $\overline{w}_{\max}$, corresponding to the desired values of $Q$ from these charts as furnished here in Fig. 3.

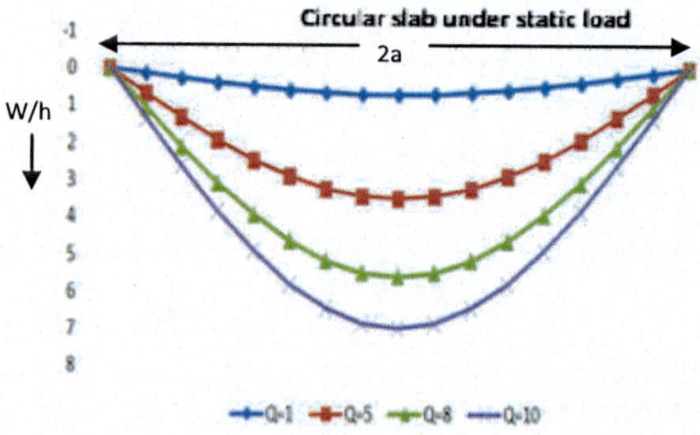


**Fig. 2** Lateral deformations of various points in the slab

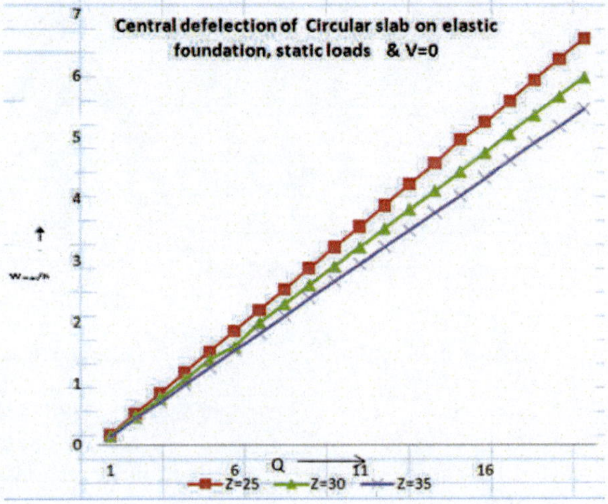


**Fig. 3** Central point deformation of the base slabs

Different values are assigned to $K$ representing various stiffness of the entire bearing system but keeping initially, $G = 0$ so as it does not include any of the visco-elastic layer therein. The response of central deformation point in the slab for example found to occur at the instant when the dynamic load remains imposed. It is clearly depicted as in Fig. 4, there remains drastic reduction in response and continues to reduce further, with increases in $Z$ values. Similarly, assigning different values then to $G$, represent those all present visco elastic shear layers and shears in the soil foundation underneath, but now keeping $Z = 0$, the case then resembles that of the circular slabs resting on the Pasternack type of foundation. On assigning values as $Z \neq 0$ and $V \neq 0$ the case, need not to clarify here as does ideally resemble of slabs on Wnkler-paternack foundation. The response of central deformation point on the slabs at the instant when the dynamic load was imposed, resulted with various values assigned of $Z$ and $V$ parameters. With a chosen value, for example $Z = 25$,

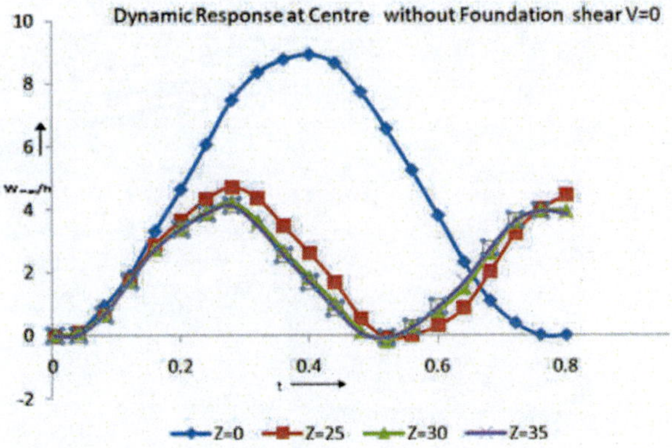


**Fig. 4** Slabs response without visco-elastic layer

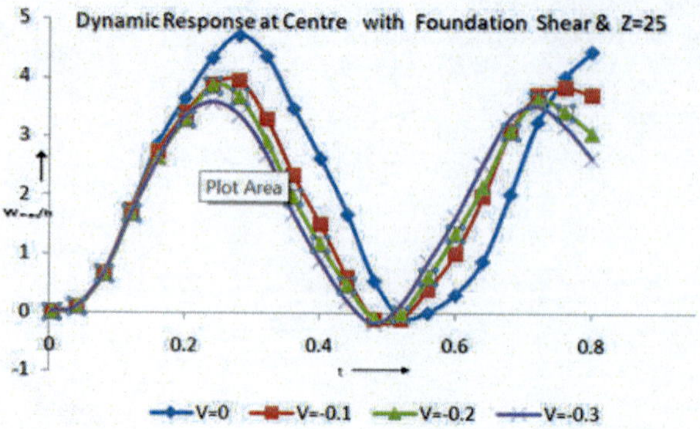


**Fig. 5** Slabs response with visco-elastic layer

and various value to $V$ as 0, −1, −2, −3 etc. for example, response was plotted, as depicted here in Fig. 5, shows a reduction in response still further with enhancing values of $V$.

It may be seen from these results as continuous reduction of deformation has been found with increase in values as $Z = 0, 25, 30, 35$ and $V = 0, -1, -2, -3$. Further, a clear shift in the occurrence of peak values of the response of central point in the time domain is also observed as depicted in Fig. 6. In other words, increased values of V effects in reduction of the wave lengths which otherwise observed was predominantly higher for both $V = 0$ and $Z = 0$ i.e. a case of slabs with no foundation. Therefore, use of suitably designed seismic bearings will certainly facilitate resisting (like dampens) the response to any dynamic loads if applied to these building, this has an additional feature of such bearings with base slabs, to act as an isolator for the entire superstructure from any damage loads caused by seismic effect on buildings.

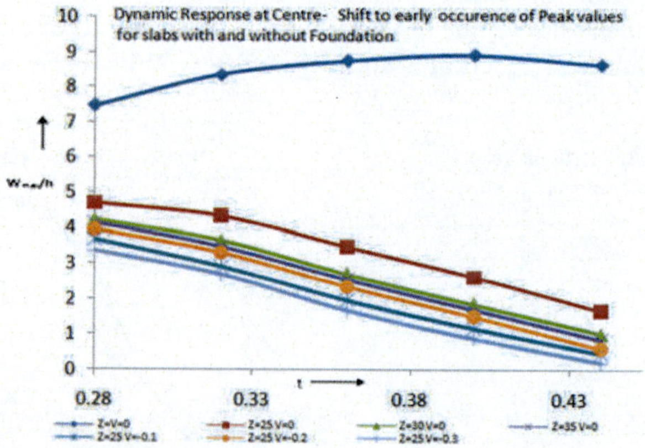


**Fig. 6** Depicting a shift in the occurrence of peak values

## 5 Conclusions

Use of appropriately designed bearings as isolators can turn medium-rise masonry (stone or brick) or reinforced concrete structures capable to withstanding earthquakes, protecting them from major damage under dynamic loads and also could offer high level of comforts to their occupants leading to no injury. The most appropriate simulation of bearings system as base slabs resting on Wnkler-paternack foundation is possible. On studying further, exact computation of values of $K$ and $G$ to be used in analyses, as equivalent spring-shear layers when used, response values may then be obtained safely either by extrapolation or intrapulation. Further this study did not cover the rigid body mode that will developed in horizontal direction and assumed to be damped out due to the presence of visco-elastic layers. which could however be obtained, using the same general equation for solution as presented here in his study, but considering the base slabs boundary conditions as freely supported at the edge of these slabs.

## 6 Inspiration

Efforts put in by Mr Ratish Jain, Chairman & Managing Director, Resistoflex Group in India who did execute a DST funded project as a study sanctioned to IITR earlier, through which practical deployment of such base Isolation bearing systems became possible it has been the inspiration to the Author to publish some work done earlier.

## References


1. Kircher CA (2006) Seismically isolated structure. NEHRP rcommended Provisions Design Example-academia.edu, Chapter 17 of ASCE/SEI-7-05 (the standard)
2. Way S (1934) Bending of circular plates with Large Deflection. Trans ASME 56:627–636
3. Nowinski J (1962) Nonlinear transverse vibration of circular elastic plates built-in-boundary. In: Proceedings, 4th US National Congress, Applied Mechanics, pp 325–334
4. Keller HB, Reiss EL (1958) Nonlinear bending and buckling of circular plates. In: Proc. Third congress applied mechanics, Providence R.I., pp 375–385
5. Keller HB, Reiss EL (1958) Iteative solution for the nonlinear bending and buckling of circular plates. Com Pue Appl Math, XI(3)
6. Nash WA, Kanematsu H (1969) Finite amplitude response of circular plates subjected to dynamic loading, IUTAM, Herenolb, pp 31–36
7. Furlong RW, Becker EB, Colville (973) J Struct Divsion. Proc ASCE, 99, ST3, 349
8. Sinha SN (1960) Analysis of large static and dynamic deflections of plates on elastic foundation, Thesis submitted for the degree of Doctor of Philosophy, Department of Mechanical Engineering , University of Colorado, pp 17 and pp 21–22
9. Nosh WA, Modeer JR (1960) Certain approximation analysis of nonlinear behaviour of plates and shallow shells. In: Proc theory of thin elastic shells, Netherlands, Int Soc, publish inc ny, pp 33–354

10. Alwar S, Nath Y (1976) Application of Chebyshev polynomials to the nonlinear analysis of circular plates. Int J Mech Sci 8:589–595
11. Kamal K (1978) Dynamic response of circular plates on elastic foundation. In: a disquisition submitted for the degree in civil engineering, MNREC, Allahabad
12. Timoshenko S, Woinowsky SK Theory of plates and shells. Mc GrawHill Book Company Inc., NY, pp 39, 260–265 and 558–561
13. Kerr AD (1964) Elastic and viscoelastic foundation models. J Appl Mech ASME 13:491–498
14. Houbolt JC (1950) A recurrence matrix solution for the dynamic response of elastic aircraft. J Aero Soc 7(9):540–550
15. Fox L, Parker IB (1968) Chebyshev polynomials in numerical analysis. Oxford University Press, London

# A Case Study on the Evaluation of Seismic Response of Earthen Dam

R. S. Preethi, L. Govindaraju, K. S. Swamy, and R. M. Thejaswini

**Abstract** The investigation of seismic behavior of existing earthen dams is one of the most challenging efforts and much important because its failure may prompt huge loss of lives and properties. Earthquakes develop transient forces that can cause the catastrophic failure and disastrous to living downstream of the dam. Hence, it is necessary to carry out the analysis of dams under dynamic condition prior to earthquake to mitigate the damages and for better performance during the earthquake. This paper focuses on the evaluation of stability of existing Chikkahole dam situated in Chamarajanagar District of Karnataka. Due to lack of recorded strong motion data in the study area, site-specific earthquake ground motions have been generated using RSPMatch [1] program based on wavelet technique approach. Pseudo-static and time history analyses have been carried out under different hydraulic boundary conditions employing finite element analysis using GEOSTUDIO 2012 [2]. In addition, the stability of dam is investigated for permeant deformation due to seismic force using yield acceleration concept.

**Keywords** Earth dam · Shear wave velocity · Synthetic earthquake · Stability

## 1 Introduction

Existing dams may prove difficult to analyze especially in those instances where the dam was designed before the development of modern design and construction technology or where adequate records are not available. There is a need to analyze and check the behavior of earth fill dam during and after the application of the seismic loading. Otherwise, a future quake might prove to be disastrous for those

R. S. Preethi (✉) · L. Govindaraju · R. M. Thejaswini
Department of Civil Engineering UVCE, Bangalore University, Bangalore, Karnataka, India
e-mail: suvarna.preethi72@mail.com

K. S. Swamy
Dy. Manager (ATES&SMI), AIMIL Limited, Bangalore, Karnataka, India



T. G. Sitharam et al. (eds.), *Seismic Design and Performance*, Lecture Notes in Civil Engineering 120, https://doi.org/10.1007/978-981-33-4005-3_27

living downstream of a dam; therefore, seismic slope stability analysis is becoming one of the most important criteria for safe design of earth dam so that adequate safety measures can be taken for better performance during earthquake.

## 2 Chikkahole Dam

The dam is built across Chikkahole River and is situated at Chamarajanagar, Sathyamangalam Road, Chamarajanagar District of Karnataka State.

Table 1 illustrates the technical details of the dam. Chikkahole dam built as stone masonry dam at Chamarajnagar taluk of Mysore in the year 1965, ruptured in December 1972 owing to severe flood of 40,000 cusecs and reconstructed in 1984 by encasing the existing dam by an earth fill.

**Table 1** Technical details of Chikkahole dam

| Sl. No. | Contents | Description |
|---|---|---|
| 1 | Length of dam | 755 m |
| 2 | Top width of dam | 6.10 m |
| 3 | Average river bed level | R.L. 733.30 M (2406.00Ft) |
| 4 | Lowest foundation level | R.L. 726.910 M (2385.00Ft) |
| 5 | Dead storage level | R.L. 736.970 M (2418.00Ft) |
| 6 | Full reservoir level | R.L. 754.040 M (2474.00Ft) |
| 7 | Maximum reservoir level | R.L. 755.59 M (2479.00Ft) |
| 8 | Minimum drawdown level | R.L. 738.490 M (2423.00Ft) |
| 9 | Top of dam level | R.L. 757.390 M (2485.00Ft) |
| 10 | Gross storage capacity | 10.64 Mcum (0.376TMC) |
| 11 | Dead storage capacity | 0.11 Mcum (0.004TMC) |
| 12 | Maximum water spread area | 400 acres at FRL |
| 13 | Type of spill way | Ogee |
| 14 | Number of gates | 4 Nos. |
| 15 | Discharging length | 39.0 m |
| 16 | Designed flood intensity | 85,000 Cusecs |
| 17 | Maximum flood lift | 45,000 Cusecs |

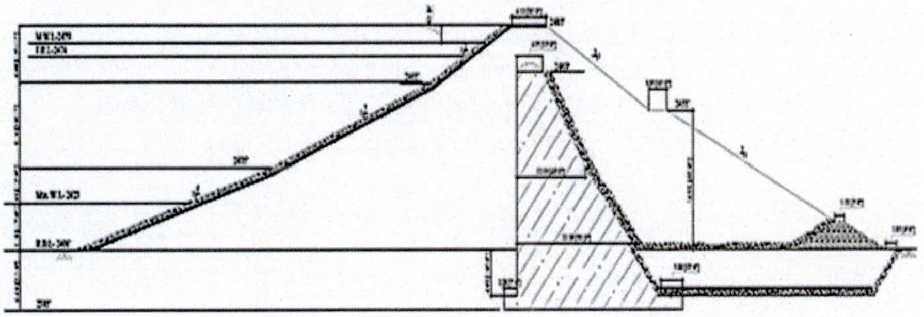

**Fig. 1** Dam cross section at existing spillway at chainage 1500ft (gorge portion)

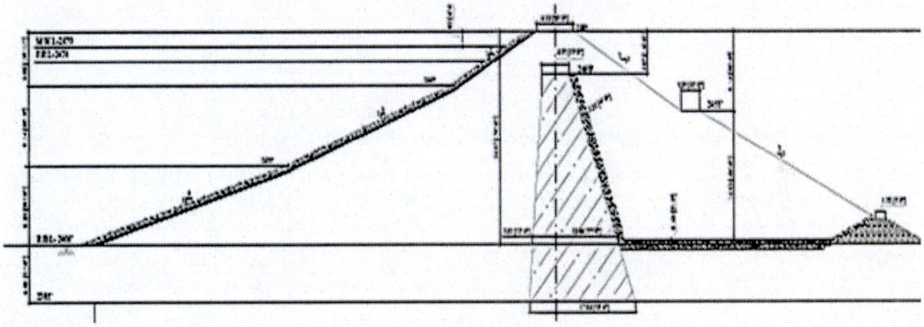

**Fig. 2** Dam cross section at existing dam section at chainage 1700ft (right bank)

Stability analyses have been carried out for the two sections (chainage) such as dam section for existing spillway at chainage 1500ft (gorge portion) and existing dam section at chainage 1700ft (right bank), cross section as shown in Figs. 1 and 2, respectively.

## 3 Site Characterization from Field Test Results

In the present study, extensive geotechnical site investigation data is collated and analyzed. These includes multi-channel analysis of surface wave (MASW) survey and geotechnical investigation including borehole data and laboratory test data providing information such as thickness of subsoil strata, index and engineering properties of Chikkahole dam. Tables 2 and 3 illustrate the dynamic properties of embankment and foundation soil, respectively, and Table 4 shows the geotechnical characteristics of dam section and foundation soils.

**Table 2** Dynamic properties of dam embankment layers using MAWS tests

| Depth (m) | Shear wave velocity (m/s) | Density (kg/m$^3$) | Shear modulus (MN/m$^2$) | Poisson ratio | Young's modulus (MN/m$^2$) |
|---|---|---|---|---|---|
| 0.0–1.8 | 1199 | 2000 | 2875 | 0.30 | 7475 |
| 1.8–4.0 | 911 | 1800 | 1493 | 0.30 | 3881 |
| 4.0–6.80 | 1004 | 2400 | 2419 | 0.27 | 6144 |
| 6.8–10.30 | 1506 | 2400 | 5443 | 0.27 | 13,825 |
| 10.3–14.6 | 1781 | 2400 | 7612 | 0.27 | 19,334 |
| 14.6–20.1 | 1007 | 2400 | 2433 | 0.27 | 6179 |
| 20.1–26.8 | 1798 | 2400 | 7758 | 0.27 | 19,705 |
| 26.8–35.3 | 1151 | 2400 | 3179 | 0.27 | 8074 |
| 35.3–46.0 | 1420 | 2400 | 4839 | 0.27 | 12,291 |

**Table 3** Dynamic properties of dam foundation layers using MAWS tests

| Depth (m) | Shear wave velocity (m/s) | Density (kg/m$^3$) | Shear modulus (MN/m$^2$) | Poisson Ratio | Young's modulus (MN/m$^2$) |
|---|---|---|---|---|---|
| 0.0–0.9 | 559 | 1700 | 531 | 0.30 | 1380 |
| 0.9–1.9 | 770 | 1700 | 1007 | 0.30 | 2618 |
| 1.9–3.3 | 750 | 1700 | 956 | 0.30 | 2485 |
| 3.3–5.0 | 737 | 1700 | 923 | 0.30 | 2399 |
| 5.0–7.0 | 606 | 1700 | 624 | 0.30 | 1622 |
| 7.0–9.7 | 402 | 2400 | 274 | 0.30 | 712 |
| 9.7–12.9 | 1075 | 2400 | 2311 | 0.28 | 5916 |
| 12.9–17.0 | 1153 | 2400 | 3190 | 0.28 | 8166 |
| 17.0–22.2 | 1731 | 2400 | 7191 | 0.28 | 18,408 |

## 4 Methodology and Modeling

The following sections briefly describe the methodology used for analyzing Chikkahole dam. Stability analyses carried out by using finite element method using GEOSTUDIO 2012 software product such as SLOPE/W, QUAKE/W [2] and computer program RSPMatch 2005 [1].

### 4.1 *Pseudo-static Analysis*

In absence of knowledge of seismic response against earthquake, the seismic stability of earthen dam can be analyzed by Pseudo static approach in which earthquake

**Table 4** Geotechnical characteristics of various components of dam and foundation

| Component | Unit weight (kN/m$^3$) | Cohesion (kN/m$^2$) | Angle of internal friction (degrees) |
|---|---|---|---|
| *Dam sections* | | | |
| Stone Masonry (Central core) | 25 | 100 | 40 |
| Homogeneous fill in the earthen flank | 18 | 5.0 | 25 |
| Inclined filter (Type B) | 18 | 0.0 | 30 |
| Rock toe | 20 | 0.0 | 30 |
| Stone pitching on u/s slope of dam | 20 | 0.0 | 35 |
| Thick macadam road | 22 | 2.0 | 38 |
| *Foundation of dam* | | | |
| From EGL to 9 m depth (clayey /Silty sand) | 17 | 5.0 | 25 |
| 9 m–22 m depth (Soft disintegrated weathered rock) | 22 | 2.0 | 38 |
| 22 m–30 m depth (Jointed hard rock) | 25 | 2.0 | 45 |

loading is represented by constant horizontal coefficients $K_h$ and $K_v$ [3] are defined as the ratio of the earthquake acceleration in respective directions with the gravitational acceleration. Stability is determined in terms of factor of safety which is the minimum ratio of resisting forces over driving forces on all kinematically possible sliding surfaces.

## 4.2 Sliding Block Method

This method involves evaluation of permanent deformation during an earthquake and comparing it with the permissible deformation [4]. This is usually carried out by the Newmark's sliding block analysis wherein the potential failure mass is treated as a rigid body is assumed to correspond to average acceleration time history of failure mass. Deformation accumulates when the inertial load due to earthquake acceleration exceeds the plastic resistance at the interface between the sliding block and the underlying stable body [3].

## 4.3 Dynamic Analysis

At present, no strong motion records are available in the Chikkahole dam site area. Therefore, synthetic earthquake motions at subsoil layer have been developed. This involved selection of seed acceleration for response spectra matching. Computer program RSPMatch [1] has been used to develop spectrum compatible ground motions.

**Selection of peak ground acceleration**

According to seismic hazard zonation map published by Bureau of Indian Standard [5], the Chikkahole dam region falls in the seismic zone (Zone II) indicating zone factor 0.1. The corresponding peak ground acceleration (PGA) 0.18 g. (Whittier Narrow-1987) and target response spectra at 5% damping for hard rock has been selected to develop synthetic ground motion. Figure 3 shows the time history of selected strong motion.

**Generation of synthetic earthquake motion**

RSPMatch 2005 program is used generation of spectrum compatible synthetic earthquake motion. As a basic step, the methodology requires the use of strong motion records available from historical earthquakes. While selecting a suitable strong motion, several important factors are considered. These include similar magnitude, peak acceleration close to the target value, similar fault distance and similar site conditions [6]. The response spectrum of selected earthquake motions before and after matching the target spectra [5] are shown in Fig. 4 and spectrum compatible synthetic earth-quake motions are shown in Fig. 5. These spectrum compatible synthetic earthquake motions are used for the dynamic (QUAKE/W) analysis.

Within this framework, the Chikkahole dam has been analyzed for following cases:

Case 1 Seismic stability by pseudo-static method for steady seepage condition with full reservoir capacity at Ch:1500ft (gorge portion) of dam

Case 2 Seismic stability by pseudo-static method for seepage condition with sudden drawdown level at Ch:1500ft (gorge portion) of dam

Case 3 Seismic stability analysis by pseudo-static method for steady seepage condition with full reservoir capacity at Ch:1700ft (right bank) of dam

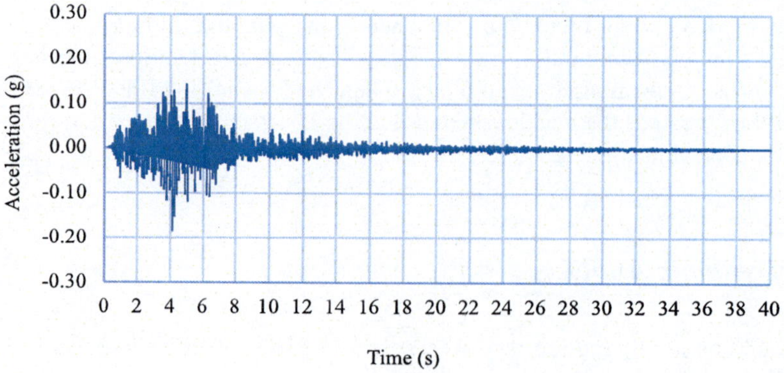


**Fig. 3** Time history of selected strong motion (Whittier Narrows earthquake, $a_{max} = 0.186$ g)

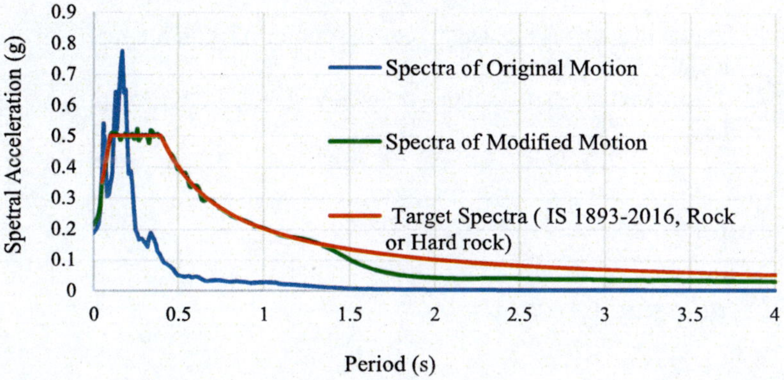


**Fig. 4** Response spectrum before and after matching

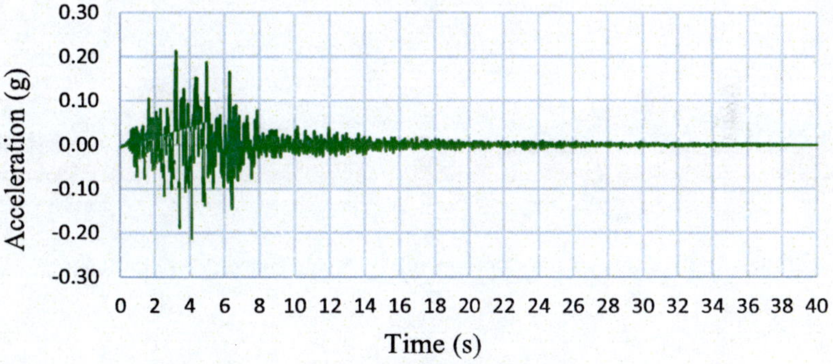


**Fig. 5** Spectra of modified motion (Whittier Narrows earthquake, $a_{max} = 0.214$ g)

Case 4 Seismic stability analysis by pseudo-static method for sudden drawdown level at Ch:1700ft (right bank) of dam

Case 5 Newmark's sliding block method for steady seepage condition with full reservoir capacity at Ch:1500ft (gorge portion) of dam

Case 6 Newmark's sliding block method for seepage condition with sudden drawdown level at Ch:1500ft (gorge portion) of dam

Case 7 Newmark's sliding block method for steady seepage condition with full reservoir capacity at Ch:1700ft (right bank) of dam

Case 8 Newmark's sliding block method for sudden drawdown level at Ch:1700ft (right bank) of dam

Case 9 Dynamic analysis for steady seepage condition with full reservoir capacity at Ch:1500ft (gorge portion) of dam

Case 10 Dynamic analysis for seepage condition with sudden drawdown level at Ch:1500ft (gorge portion) of dam

Case 11 Dynamic analysis for steady seepage condition with full reservoir capacity at Ch:1700ft (right bank) of dam

Case 12 Dynamic analysis for sudden drawdown level at Ch:1700ft (right bank) of dam

# 5 Results

## 5.1 *Pseudo-static Analysis*

Summary of pseudo-static results in terms of factor of safety (FOS) for full reservoir level and for sudden drawdown level at upstream slopes at 1500Ft and at 1700 Ft are as shown in Figs. 6, 7, 8 and 9 and the results are tabulated Table 5.

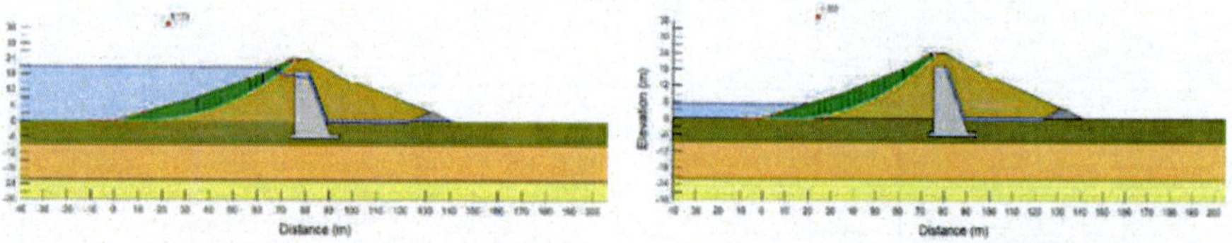


**Fig. 6** Pseudo-static stability of u/s slope for full reservoir and sudden drawdown at chainage 1500ft

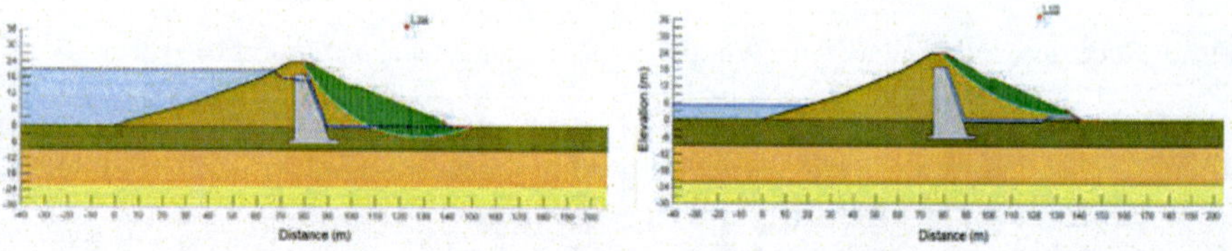


**Fig. 7** Pseudo-static stability of d/s slope for full reservoir and sudden drawdown at chainage 1500ft

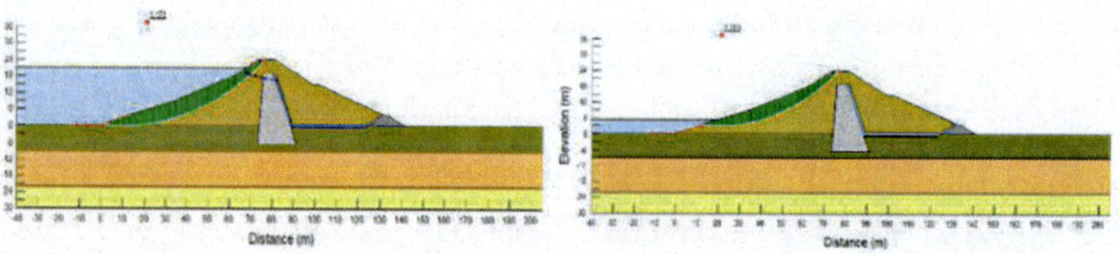


**Fig. 8** Pseudo-static stability of u/s slope for full reservoir and sudden drawdown at chainage 1700ft

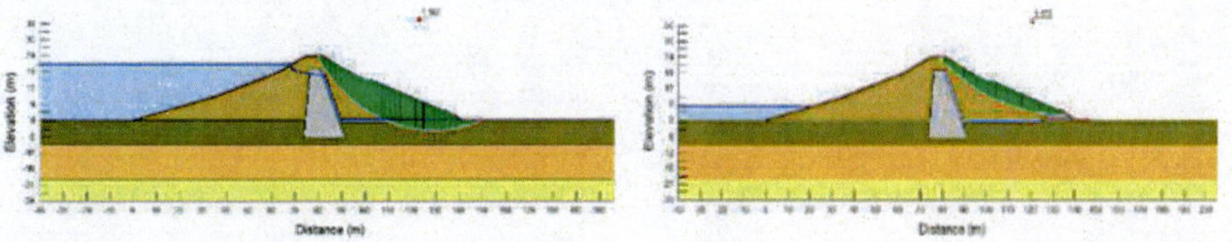


**Fig. 9** Pseudo-static stability of d/s slope for full reservoir and sudden drawdown at chainage 1700ft

**Table 5** Summary of results from pseudo-static loading for different stages

| Case /Stage | FOS (Upstream) | FOS (Downstream) |
|---|---|---|
| Case 1 | 1.17 | 1.16 |
| Case 2 | 1.30 | 1.17 |
| Case 3 | 1.17 | 1.16 |
| Case 4 | 1.30 | 1.17 |

## *5.2 Sliding Block Method*

Deformation calculated along the failure plane by deformation-based methods. By determining the yield acceleration ($a_y$) and the peak ground acceleration ($a_{\max}$) for the dam, the permanent displacements of the dam can be calculated using relationship suggested by Hynes-Griffin and Franklin, as shown in Fig. 10; results are tabulated in Table 6.

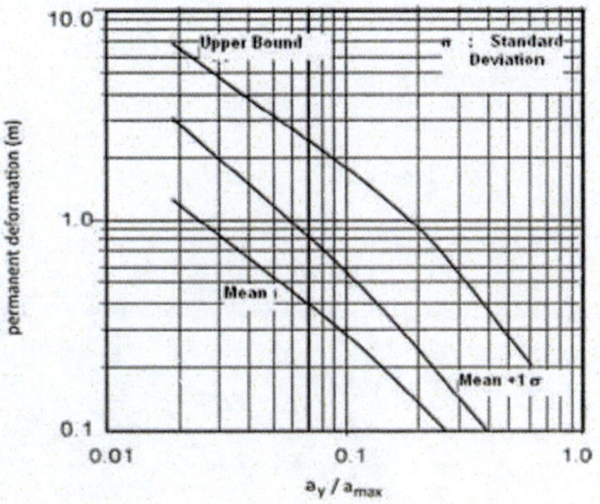


**Fig. 10** Relationship between ratio of yield to peak accelerations and permanent deformation (Hynes-Griffin and Franklin 1984)

**Table 6** Results from Newmark's sliding block method

| Case /Stage | Permanent displacement (Upstream)—mm | Permanent displacement (Downstream)—mm |
|---|---|---|
| Case 5 | 185 | 150 |
| Case 6 | 100 | 140 |
| Case 7 | 190 | 150 |
| Case 8 | 100 | 150 |

## *5.3 Dynamic Analysis*

The seismic stability of dam using time history analysis of upstream and downstream slopes of the dam in terms of FOS along with the critical slip surface displacement for full reservoir capacity and for sudden drawdown condition for the dam section at chainage 1500ft and at chainage 1700ft is tabulated in Table 7. The displacements can be viewed as a deformed mesh as shown in Fig. 11 obtained by the QUAKE/W.

**Table 7** Results of dynamic factor of safety and displacement under dynamic loading for different stages

| Case /Stage | Upstream | | Downstream | |
|---|---|---|---|---|
| | FOS | Permanent displacement (Upstream) mm | FOS | Permanent displacement (Downstream) mm |
| Case 9 | 2.25 | 15.6 | 1.51 | 16.9 |
| Case 10 | 3.07 | 0.40 | 1.42 | 20.6 |
| Case 11 | 1.74 | 83.9 | 1.48 | 15.3 |
| Case 12 | 3.47 | 0.39 | 1.46 | 9.68 |

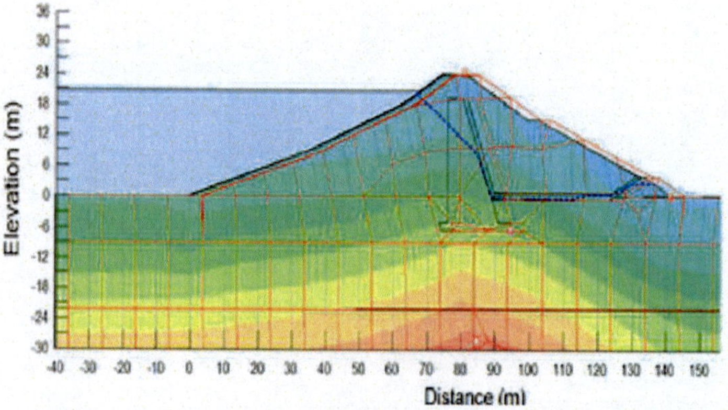


**Fig. 11** Typical deformed mesh of dam form dynamic analysis at chainage 1500ft

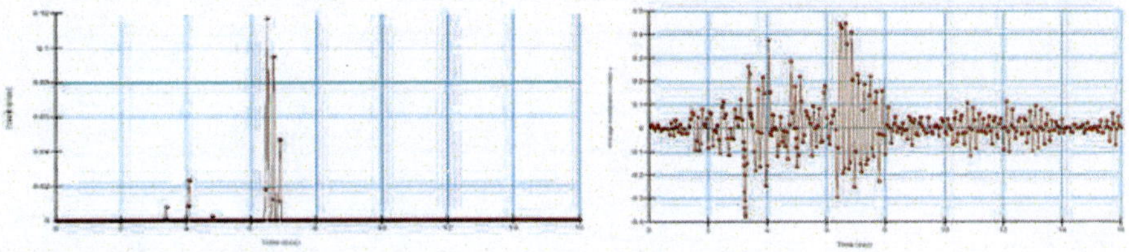

**Fig. 12** Typical velocity and average acceleration versus time plot at toe of the dam obtained from QUAKE/W

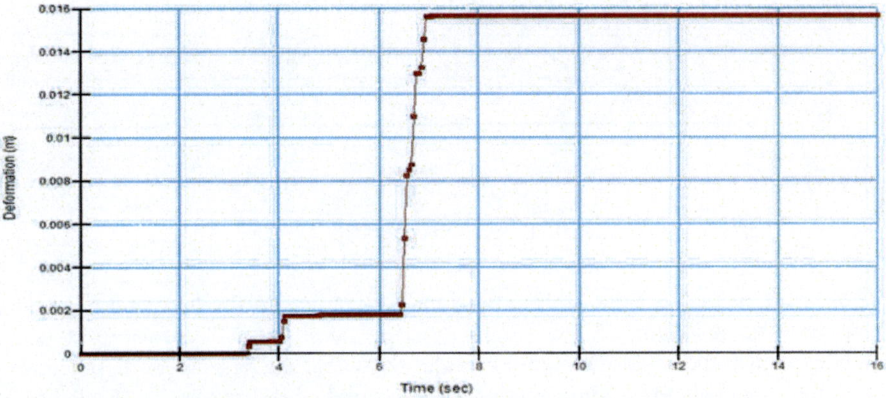


**Fig. 13** Typical deformation plot for different time interval during dynamic loading

Seismic force is a very sudden impulse load; the effects of such events are measured with a seismograph, which can measure ground accelerations and velocities (Fig. 12) at the toe of the dam.

QUAKE/W can by numerical integration convert acceleration or velocity time history records into a displacement versus time records as shown in Fig. 13. Figure 14 shows factor of safety changes during a dynamic loading.

## 6 Conclusions

In this study, the stability condition of the Chikkahole dam against the design earthquake was studied using the pseudo-static method, sliding block model and by time history analysis, at gorge portion (chainage 1500ft) and at right bank (chainage 1700ft) based on the definite site characteristics by using the test results of MASW survey and geotechnical investigations; following conclusions have been drawn;

1. The results of pseudo-static stability analysis of the two dam sections (at chainage 1500ft and at chainage 1700ft) in terms of factor of safety for all the cases are more

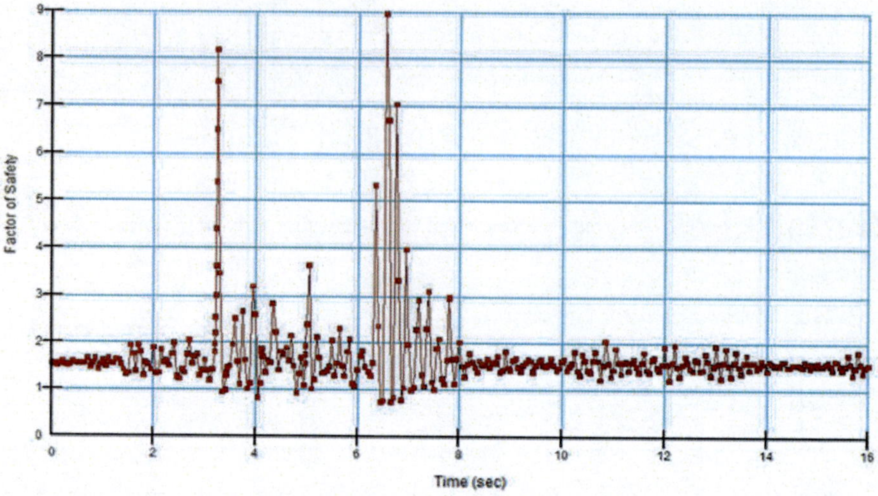


**Fig. 14** Typical factor of safety versus time plot for dynamic loading

than 1.0, as per IS 7894-2002 provision. This indicates, the dam is stable under pseudo-static loading.

2. Newmark's sliding block method results show that the permanent displacement of the dam slope for seismic loading, which is well within the limits (the displacements are lesser than 1 m as per IITK-GSDMA Guideline). Thus indicating the dam slopes are safe against the deformation during the seismic loading condition.
3. The factor of safety from the dynamic analysis is excess of minimum desired values of factor of safety (FoS) of 1.0 as per IS 7894-2002 and critical slip surface deformation is lesser than 1.0 m as suggested by IITK-GSDMA Guideline. Thus dam performance during the extreme event may be considered as acceptable.

## References

1. RSPMatch 2005 Program manual V1; Program Version 11/9/2005
2. GEOSTUDIO 2012 Software product manual
3. Mukherjee S (2013) "Seismic slope stability analysis of earth dam: some modern practices". Int. J. Recent Adv Mech Eng (IJMECH) 2(1)
4. Patel KS, Prof. Sanghvi CS (2012) "Seismic slope stability analysis of Kaswati earth dam". Int J Adv Eng Res Stud, E-ISSN2249–8974
5. Indian standard criteria for earthquake resistant design of structure, IS 1893-2016. Bureau of Indian standards, New Delhi
6. Govindaraju L, Bhattacharya S (2012) "Site-specific earthquake response study for hazard assessment in Kolkata city, India". J Int Soc Prev Mitig Nat Hazards 61
7. Indian standard code of practice for stability analysis of earth dams, IS 7894–1975 (Reaffirmed 2002) Bureau of Indian standards, New Delhi

8. Chandra R, Ghosh B, Prasad SK (2016) "Stability analysis of slopes and seismic analysis of earthen dams using geo studio". Int J Trend Res Dev 3(4), ISSN: 2394–9333
9. Vedavathi RP, Dr. Ramesh V (2015) "Seismic performance study on earthen-dams using geo studio". Int. J. Civ Struct Eng Res 3(1):327–331
10. IITK-GSDMA GUIDELINES for "Seismic design of earth dams and embankments, Indian Institute of Technology Kanpur, August 2005; Revised on May 2007

# Behaviour of RC Frame Subjected to Fire Following Earthquake

**Haveri Manish, M. Swathi, and S. D. Anitha Kumari**

**Abstract** Natural disasters are the second largest enemy of human kind after devastating wars. Natural disasters such as floods, earthquakes, tornadoes, volcanic eruptions, hurricanes, tsunami and landslides have significant effect on a building leading to partial damage or complete collapse. Earthquake effect on a building triggers fire which can completely damage the strength of the building. Post-Earthquake fire can lead to collapse of the building which is partially damaged due to the prior ground-motion. This in turn accounts to the decrease of stability of the structure to withstand the superimposed loads. Fiber Reinforced concrete (FRC) has good resistance to fire and can withstand more loads when compared to nominal concrete. Polypropylene fibers have great resistance to fire and are widely used as fire proofing material. Hence this work is an approach to strengthen the Reinforced concrete (RC) frame with FRC to reduce the damage caused by fire loads imposed on the frame after the occurrence of earthquake. FRC has good resistance to fire and can withstand more loads when compared to nominal concrete. The analysis of RC frame for earthquake loads, fire loads and fire following earthquake are done in ANSYS software and the same is compared with the experimental results of FRC frame. It was found that the cyclic loading deformation of FRC frame reduces by 44.33%, fire loading deformation reduces by 80% and Fire Following Earthquake (FFE) loading deformation by 60%.

**Keywords** Fire following earthquake · Fiber reinforced concrete · Fire loads

## 1 Introduction

Over the centuries, some towns and cities have repeatedly been struck and sometimes devastated by major earthquakes. The difference between damage and devastation depends not only on the magnitude of the earthquake, but also on local geology and on building techniques. Earthquakes do not cause fires directly. It is the damage to

H. Manish (✉) · M. Swathi · S. D. A. Kumari
M. S. Ramaiah University of Applied Sciences, Bangalore, India
e-mail: haveri.manish@gmail.com

 
T. G. Sitharam et al. (eds.), *Seismic Design and Performance*, Lecture Notes in Civil Engineering 120, https://doi.org/10.1007/978-981-33-4005-3_28

structures containing flammables like wooden walls and ignition sources. The effects of post-earthquake fire on buildings are of two kinds, one is the damage due to the burning of non-structural material, which possesses material value and the other is the damage caused due to the additional loads (fire) on the building. These extra loads are normally loads that the structures are not designed for, and when combined with the earthquake damage can lead to the collapse of buildings. The structural strength of concrete is enhanced with the addition of fibers which increases the tensile property of concrete making it stronger to take more loads when subjected to unexpected severe loading conditions. Different fibers available in the market perform differently to fire with regard to the thermal conductivity and bonding with concrete. Fibers such as steel fibers, carbon fibers, polypropylene fibers are generally used in concrete to enhance its properties. Experiments have proved that polypropylene fibers are more resistant to fire when compared to steel fibers and vinyl fibers. This study focusses on the effect of fibers as major strengthening materials in buildings with RC frame which are subjected to earthquake followed by fire.

Yoshitake et al. [1] presented the behavior of fiber reinforced concrete under elevated temperatures. The study compares the performance of plain concrete and different Fiber Reinforced Concrete viz., Steel FRC, Polypropylene FRC and Vinyl FRC to the effect of fire in a tunnel lining. It is reported that polypropylene fiber reinforced concrete exhibited high resistance to fire compared to steel fiber reinforced concrete and plain concrete. Zhang et al. [2] explains the results of experimental setup of an RC frame that has exposed to different temperatures and static load. It was concluded that fire load and higher temperature of the specimen had large influence on the mechanical properties of the frame reducing its strength and stiffness. Ye et al. [3] explains the non-linear analysis of a RC frame subjected to fire adopting finite element method. The structural deformation and post fire ultimate loadings were investigated. It was suggested to increase the beam and column rotation in order to increase the ductility of the structure so as to resist the changes of the deformation and internal forces. Mostafaei and Kabeyasawa [4] conducted shake table experiments on a six-story RC frame subjected to Kobe earthquake. The same frame was subjected to fire following earthquake loading and its performance was evaluated using numerical methods. It was concluded that fire on the damaged structure had considerable effects on the material properties due to degradation and rapid heat penetration. This reduced the load carrying capacity of the frame. Kumar [5] reported experimental studies of a single story sub-assemblage of a G + 3 RC frame subjected to Post-Earthquake Fire (PEF). It was observed that despite spalling of concrete, the roof slab remained structurally stable to carry the superimposed loads whereas the roof beam suffered maximum degradation of concrete which was confirmed through visual inspection also. Behnam et al. [6] reported results of PEF based on FEMA356 sequential analysis where the structure is subjected to different magnitude of earthquake followed by 4 h fire analysis. It was noticed that fire resistance of structure was initially 240 min which reduced to 90mins after exposed to earthquake of magnitude 0.35 g. Similarly, fire resistance of structure was initially 210 min which reduced to 60 min after exposed to earthquake of magnitude 0.25 g. Raouffard and Nishiyama [7] describes the results of fire test carried out to evaluate the performance of moment-resisting RC frame at elevated temperatures. From the experiment it was observed that the thermal gradient within the frame was non-linear where the end sections of the beam attain lower temperature compared to the mid sections due to which there were flexural cracks in mid-section.

Behrouz et al. [6] describes the numerical and experimental studies on the structural resistance of two RC beam-column joints with similar geometry configuration, reinforcement and concrete strength. The second is strengthened by means of carbon fibre-reinforced polymer (CFRP) at the vicinity of the joint in order to relocate the plastic hinge away from the column extremity towards the beam. Both specimens are subjected to a cyclic load. The post-earthquake fire (PEF) resistance of the specimens subjected to various damage levels such as immediate occupancy (IO), life safety (LS) and collapse prevention (CP) is then evaluated based on finite element analysis. The resistances of frame to different damage levels were noticed and were observed that the resistance of frame is 25% greater at the life-safety and 35% more in collapse prevention levels of damage. Poorna and Prasad [8] explain the behavior of RCC slabs for different grades and different concrete cover. The results showed that deflection of slab decreases as the grade of concrete increases and decreases as the concrete cover increases. Kadir [9] explains about the structural damage that is caused by fires occurring after the earthquake on a Roorkee frame test. The analysis of these two sequential events i.e., fire and earthquake loads was done using Finite element methods. The studies indicate that the structural behavior during a FFE is very damaging and hence more insight is required into the performance of structures and the ways to improve the mechanical behavior of such structures. This study focusses on a G + 2 structure subjected to earthquake followed by fire loading condition with and without fiber reinforcement.

## 2 Validation and Methodology

Kadir [9] has reported experimental results using Roorkee frame. These results are used for the validation of the numerical model developed in this study. The experimental set up consists of a frame subjected to cyclic loading at the slab level at the height of 4.3 m using hydraulic jack system. The frame is later subjected to temperature loads. Figure 1 shows the experimental setup of the frame. Cyclic loading was achieved by applying a lateral load to the frame using a hydraulic jack at the slab level in load control mode. The temperatures in the different structural elements are recorded in the frame using different Thermocouples at different locations after exposing it to fire in furnace. The sectional details of the frame are shown in Table 1. The displacement history values used for the study are shown in Table 2. The maximum base shear obtained from the experiment is 315 kN.

To simulate the mechanical behavior and validation, a G + 1 frame is modeled using solid 365 element and is shown in Fig. 2. The total height of the single bay frame is 5.8 m and width 3 m and the grade of concrete used is M-30. The description of loads and dimensions are same as used in the experiment. The frame was analyzed

**Fig. 1** Roorkee frame experimental set up [9]

**Table 1** Description of the experimental setup [9]

| Building elements (mm) | Material | Load details |
|---|---|---|
| Column—300 × 300 | Grade of concreteș—M30 | Floor finish—1 kN/m$^2$ |
| Beam—230 × 230 | Density of concrete—25 kN/m$^2$ | Live load—2 kN/m$^2$ |
| Thickness of slab—120 | | |

**Table 2** Displacement history for cyclic loading

| Time (s) | Displacement (mm) |
|---|---|
| 0 | −19 |
| 100 | 19 |
| 200 | −30 |
| 300 | 35 |
| 400 | −39 |
| 500 | 49.8 |
| 600 | −61 |
| 700 | 68.8 |
| 800 | −79.6 |
| 900 | 95 |

using numerical software for both 3D and 2D models to compare the results. Cyclic loading was given which was obtained from the test. The frame was subjected to equivalent cyclic lateral displacements in displacement control mode. Figure 3 shows the capacity curves for the displacement history obtained from experimental test results and 3D simulation. The results show that the maximum base shear obtained

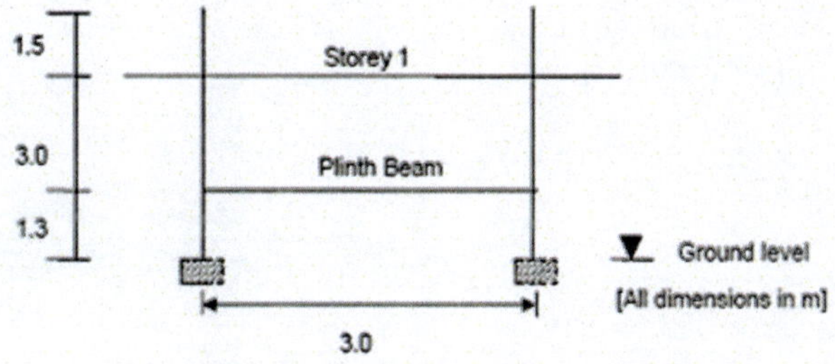


**Fig. 2** Details of the RC frame

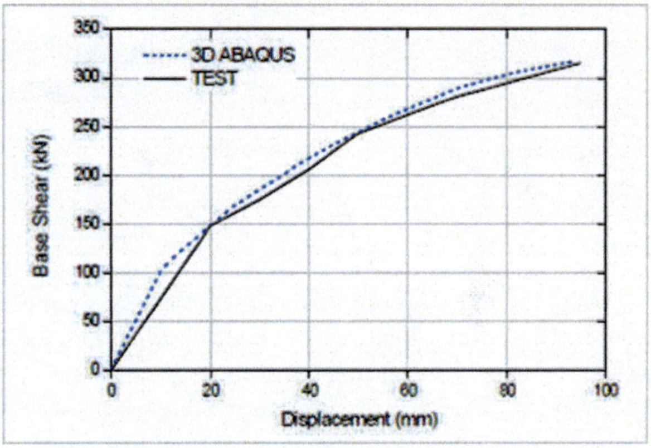


**Fig. 3** Comparison of test results and numerical results

for the frame was 317 kN. Following the application of earthquake loads, fire loads are applied to the frame. The temperatures recorded from the thermocouples for different structural elements are input in software to see the deformations. The model in ANSYS is shown in Fig. 4.

M-30 grade concrete is used for the frame and its behavior is simulated by multi-linear isotropic hardening properties. Stress-strain values for the concrete are assigned for M-30 concrete. Fixed supports can resist vertical, horizontal forces and moment as they restrain both rotation and translation. The structure only needs one fixed support in order to be stable and is used in this study. Connection between the elements should be defined to ensure the connectivity of the members and their load transfer.

The accuracy that can be obtained from any FEA model is directly related to the finite element mesh that is used. Types of meshes considered here are Quadrilateral meshing, tetrahedral meshing and Quad-dominant meshing. It was observed that quad-dominant meshing gave the most accurate results compared to the other two methods but took excess time for the simulation. Quadrilateral meshing showed approximate results and consumed the least time of all the three methods. Tetrahedral meshing showed considerably accurate results but less accurate than quad-dominant method. However, tetrahedral meshing was considered since it consumed lesser time and provided satisfactory results.

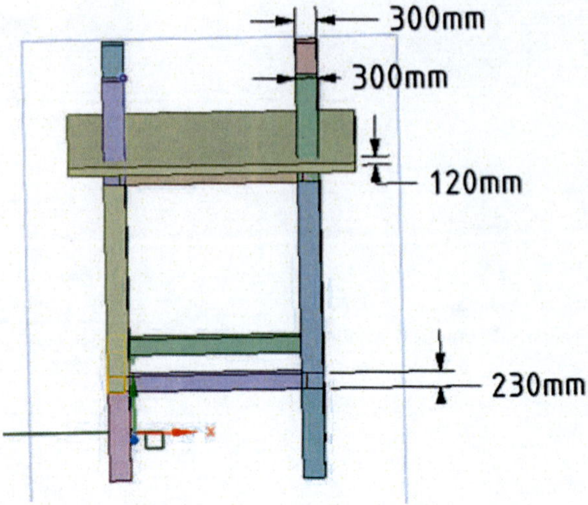


**Fig. 4** 3D ANSYS model (All Dimensions in mm)

Once the model is defined and the boundary conditions are given, the structure is subjected to cyclic loading. In experiment, cyclic loading was achieved by applying a lateral load to the frame using a hydraulic jack at the slab level in load control mode. In the numerical simulation, the frame was subjected to equivalent cyclic lateral displacements in displacement control mode.

The results of base shear and displacement are shown in Table 3.

Figure 5 shows the comparison of capacity curves obtained from ANSYS and experiment. The graph is plotted for force reactions (Base shear) and the displacement input. The base shear obtained for the RC frame is 364 kN from the Numerical analysis. The graph indicates that the numerical evaluation is almost convergent with the test results.

The comparison of the numerical results (ANSYS and ABAQUS) with that of experimental results for cyclic loading is shown in Table 3.

A G + 2 Frame with the same section properties is considered for further study. The frame was subjected to same boundary conditions and loading conditions which was considered in validation. The total deformation observed is 103.35 mm is shown in Fig. 6 and the capacity curve for cyclic loading of G + 2 frame is shown in Fig. 7.

**Table 3** Comparison of results: Experimental, ABAQUS and ANSYS simulations

| Analysis type | Base shear (kN) | Displacement (mm) | Percentage error of base shear | Percentage error of displacement |
|---|---|---|---|---|
| Experimental results | 315 | 80 | | |
| ABAQUS 3D | 317 | 80 | 0.6 | |
| ANSYS Workbench | 330 | 79.566 | 4.5 | 0.54 |

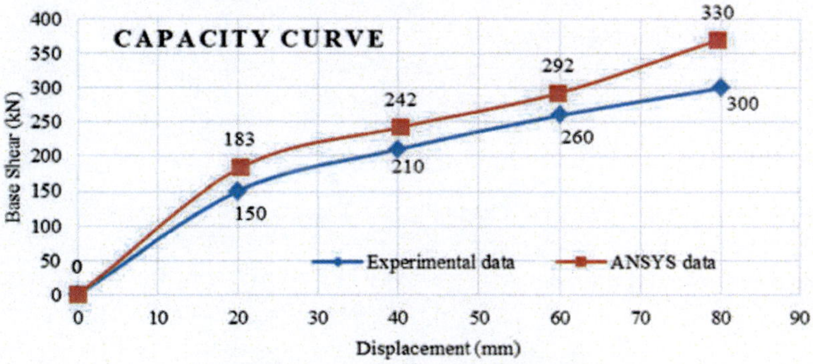


Fig. 5 Capacity curves from experiment and ANSYS simulations

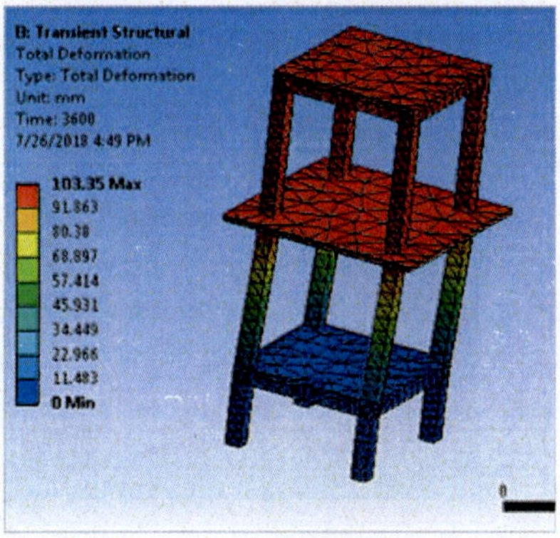


Fig. 6 Total deformation of G + 2 frame for cyclic loading

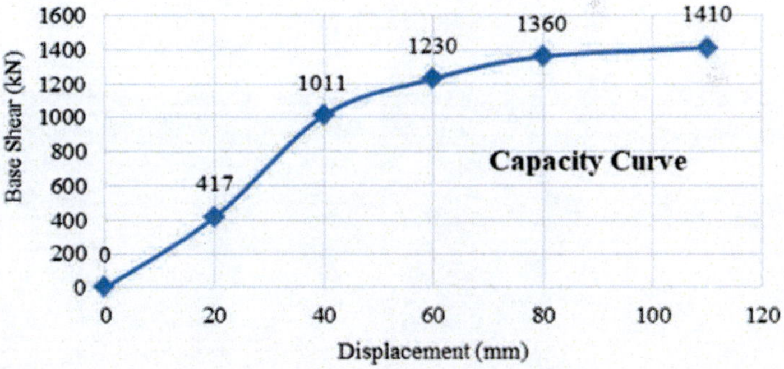


Fig. 7 Capacity curve for cyclic loading for G + 2 frame

**Table 4** Temperature load of the slab

| Sl. No | Time (s) | Temperature (°C) |
|---|---|---|
| 1 | 0 | 0 |
| 2 | 600 | 310 |
| 3 | 1200 | 600 |
| 4 | 1800 | 800 |
| 5 | 2400 | 950 |
| 6 | 3000 | 1000 |
| 7 | 3600 | 900 |

The temperatures in various structural elements are recorded in the frame using different Thermocouples at different locations. The temperature variation in the slab as obtained from the experiment [9] is input in ANSYS. The temperatures recorded in the slab are shown in Table 4.

Transient thermal is used for the fire analysis in ANSYS. The results from transient structural are linked to the transient thermal for the input of loads for post-earthquake fire. The deformation of the G + 2 slab due to fire in the RC slab is shown in Fig. 8. A comparison between the experimental and numerical simulation indicates around 30% variation in the fire deformation. This may be due to the disparity in the distribution of the temperature across the experimental and numerical sections. The effect of temperature and slow cooling in real situations lead to more deformations compared to FEM modeling.

The total deformation of the slab due to the temperature variation and is already subjected to earthquake load is shown in Fig. 9. Due to the effect of earthquake, the structure already underwent considerable damage. The same structure when subjected to further extreme loading of fire deforms considerably and the maximum deformation observed in this case was 182.18 mm.

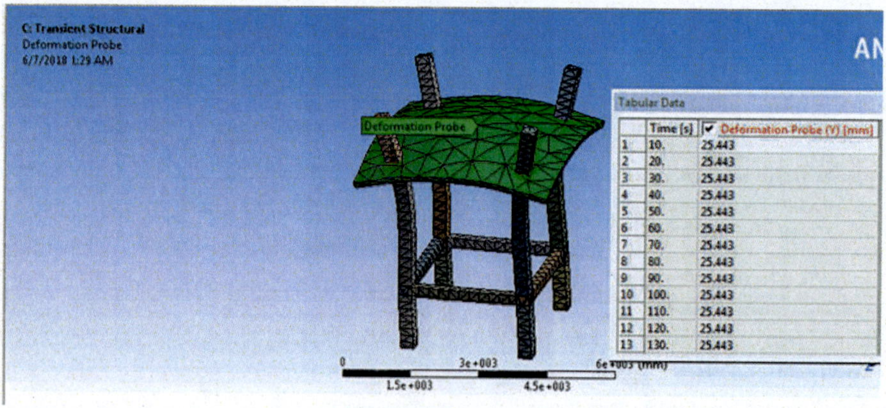


**Fig. 8** Deformation due to temperature on the slab

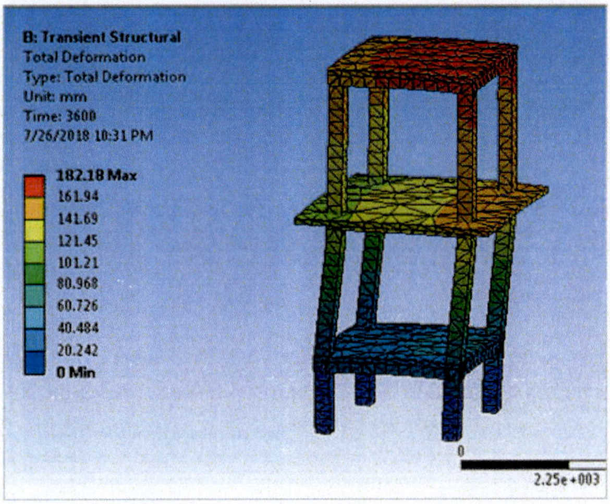


**Fig. 9** Total deformation of RC frame subjected to FFE loading

The same model was exposed to ISO 834 standard fire curve using the nominal fire curve equation given by:

$$T = 345\ \log_{10}(8t + 1) + 2 \tag{1}$$

where $T$ is the temperature in °C and t is the time in minutes. Figure 10 shows the deformation on the RC frame when loaded with temperature according to ISO 834.

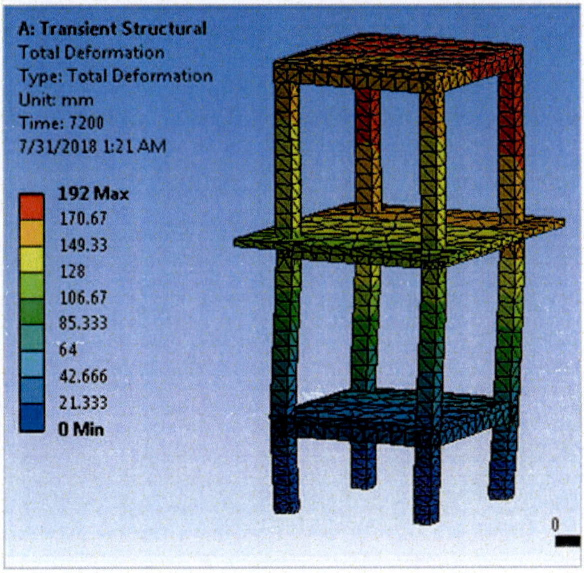


**Fig. 10** Total deformation of RC frame subjected to FFE loading (As per ISO curve)

**Table 5** Temperature loading for the structural components

| Time (s) | Beam (°C) | Column (°C) | Slab (°C) |
|---|---|---|---|
| 0 | 0 | 0 | 0 |
| 600 | 100 | 130 | 310 |
| 1200 | 270 | 350 | 600 |
| 1800 | 400 | 500 | 800 |
| 2400 | 480 | 650 | 950 |
| 3000 | 550 | 780 | 1000 |
| 3600 | 550 | 830 | 900 |

The maximum deformation observed in this case is 192 mm and closely matches with the behavior predicted by experimental temperature values which is around 182 mm.

Table 5 shows the temperature loading for the various structural components in the frame.

## 3 Analysis of FRC Frame

Fibers when added to concrete make the concrete more tough and improve the ductility and its resistance to thermal shock, fatigue and impact loads to a greater extent. Widely used fibers in concreting are steel fibers, polypropylene fibers, glass fibers, organic and carbon fibers. Polypropylene fibers are found to have excellent resistance for fire and help reduce the spalling of concrete in cases of severe fire. They have high density and melting point is as high as 165 °C.

The analysis of the G + 2 frame was done for the performance in cyclic loading, fire loading and fire following earthquake loading. The results of total deformation for the respective loading were found to be 57.529, 21.253 and 72.05 mm. Figure 11 shows the total deformation of the G + 2 frame for FFE loading.

Table 6 shows the comparison of deformation values of the G + 2 frame when subjected to different loading conditions.

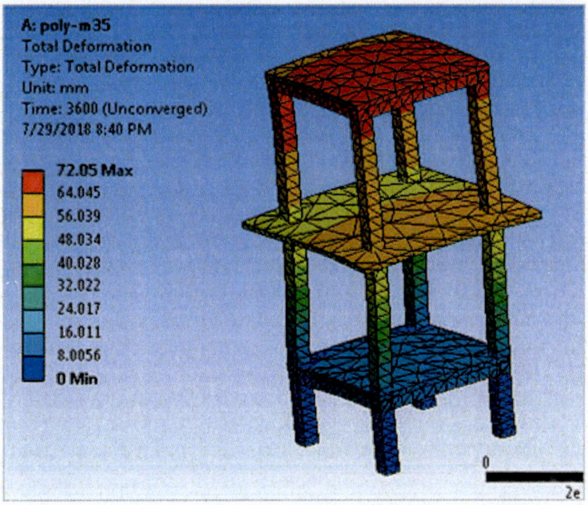


**Fig. 11** Total deformation of the FRC frame

**Table 6** Comparison of deformation for cyclic, fire and FFE loading in FRC frame

| Loading type | RC frame (mm) | FRC frame (mm) | % decrease |
|---|---|---|---|
| Cyclic loading | 103.35 | 57.529 | 44.33 |
| Fire loading | 106.29 | 21.253 | 80 |
| Fire Following Earthquake(FFE) | 182.18 | 72.05 | 60 |

## 4 Conclusion

The building when exposed to fire after suffering from earthquake loads have a greater effect on the stability of the structure since the structural elements already undergo degradation due to dynamic loading. The results from the analysis of the frame clearly show that the vulnerability of the frame to fire after the exposure to earthquake loading is greater. The deformation of frame for FFE loading was found to be 42% more than only fire loading [10]. The polypropylene fibers in the FRC frame help concrete withstand the fire loads to a greater extent. The decrease in total deformation of the FRC frame was found to be 44.3% in cyclic loading, 80% in fire loading and 60% in FFE loading. The results conclude that the addition of polypropylene fibers can make the concrete hard against the fire following earthquake loading.

## References

1. Yoshitake I, Baba K, Ito T, Nakagawa K (2005) Behavior of fiber reinforced concrete under fire temperature, Yamaguchi University

2. Zhang X, Qing-Qing S, Zhong-Yi Li, Song-Hua T,Ying-She L (2014) Experimental study on fire resistance of reinforced concrete frame structure. International Conference on Mechanics and Civil Engineering
3. Ye Z, Huang R, Jiang Y, Lv L (2015) Nonlinear anal-ysis of reinforced concrete frame subjected to fire. International Conference on Mechatronics, Electronic, Industrial and Control Engineering, pp 1148–1152
4. Mostafaei H, Kabeyasawa T (2010) Performance of a six-story reinforcement concrete structure in post-earthquake fire. In: Proceedings of the 9th U.S. national and 10th Canadian conference on earthquake engineering, pp 01–10, Canada
5. Kumar V (2012) Behaviour of full scale reinforced concrete frame under stimulated post-erathquake fire. World conference on earthquake engineering, Lisboa
6. Behnam Behrouz, Ronagh H, Lim PJ (2015) Numerical evaluation of the post-earthquake fire resistance of CFRP strengthened reinforced concrete joints based on experimental observations. Eur J Environ Civil Eng 1(1–19). https://doi.org/10.1080/19648189.2015.1018448
7. Raouffard MM, Nishiyama M (2015) Fire resistance of reinforced concrete frames subjected to service load: part 1. Experimental study. J Adv Concr Technol 13:554–563
8. Poorna S, Prasad R (2016) Effect of fire on RC slab. Int J Innov Res Sci Technol 3(4):179–183
9. Kadir MA (2013) Fire resistance of earthquake damaged reinforced concrete frames. The University of Edinburgh
10. Samir S (2011) Effect of polypropylene fibers on concrete fire resistance. J Civ Eng Manage 17:259–264

# Analysis of Urban Tunnels Under Different Loading Conditions

**Swapnil Mishra, K. S. Rao, N. K. Gupta, and R. R. Dhamne**

**Abstract** Underground structures are divided into fully buried structure and partially buried structures which can be any structure of diverse shapes such as, shelters, basement structures, underground mall facilities, underground parking spaces, silos, storage facilities, retention basins, shafts, tunnels, pipes, underground railway, metro stations, to mention a few. Underground tunnels are one of the most frequent modes for modern transportation. These tunnels are subjected for varying loading conditions from static to quasi-static and dynamic loading. These loading conditions are differentiated on the basis of strain rate which varies from $10^{-9}$ to $10^{4/}$sec. It is very important to determine static and dynamic response of lining under different loads for better performance and safety of tunnel. The response of tunnel can be determined numerically and then compared with experimental results. Crack patterns and deformation at tunnel crown are determined for surface static and dynamic load. Crack zone under loading condition is related to the material strength of the surrounding rock mass. An attempt has made in the study to analyze the response of lining under static and impact loading through experimentation.

**Keywords** Response of lining · Loading conditions · Crack pattern · Numerical analysis

## 1 Introduction

Large deformation in underground structures at shallow depth has always been a threat to engineering safety in the process of tunnel construction, especially when the tunnel is crossing the highly stress induced area, e.g., joints/fracture zone and

S. Mishra (✉)
Civil Engineering Department, PDPU, Gandhinagar, Gujarat, India
e-mail: swapnil.mishra@pdpu.ac.in

K. S. Rao · N. K. Gupta · R. R. Dhamne
Civil Engineering Department, IIT Delhi, New Delhi, India

R. R. Dhamne
Applied Mechanics Department, IIT Delhi, New Delhi, India

 
T. G. Sitharam et al. (eds.), *Seismic Design and Performance*, Lecture Notes in Civil Engineering 120, https://doi.org/10.1007/978-981-33-4005-3_29

weak/soft rocks. The deformation in rock mass is dependent on the basic engineering properties such as water softening, uniaxial compressive strength (UCS), the rheological characteristics. Therefore, the primary objective of design engineer is to counter these deformations by minimizing the stress concentration in rock mass. The problem of stress concentration is countered by creating a stress field as uniformly distributed as possible in the excavation of underground structures. Hence, the optimum support systems can be provided without compromising the safety of the structure. Different parameters that influence the safety of a tunnel are rock type (geological and engineering properties), lining properties, depth of overburden, in situ stresses, loading conditions, etc. The tunnel in soft rock mass at shallow depths is highly sensitive to disturbance. Under the influence of disturbance and continuous deformation, the mechanical properties (Elastic modulus, UCS, tensile strength, cohesion, and friction angle) of the surrounding rock mass are deteriorated. The tunnels in soft ground are typically circular in shape because this shape has inherently greater strength and ability to readjust to subsequent load changes.

The fracture behavior and the impact resistance of the geologic materials under ballistic conditions depend on various factors such as strength, density, dynamic response. Due to impact loading, structures experience high strain rate, and there is an apparent increase of the dynamic strength when geo-materials such as rock and concrete are subjected to high strain rate. Therefore, the dynamic strength augmentation of geo-materials has drawn enormous concern in structural design in impact loading condition.

Although numerous experimental and numerical studies are performed on the effect of impact loading on above-ground structures but experimental study for impact loading condition on underground structures are rare. The main reasons behind this are (i) full-scale experiments will be costly, and (ii) model tests will be unrealistic if the effect of overburden is not considered. Hence, numerical simulation is becoming essential in engineering analysis and design of complex conditions. The present study is an attempt to simulate and observe the effect of impact load from projectile penetration on shallow tunnels in rock and sensitivity of the rock and tunnel using ABAQUS/CAE software. Idris et al. [1] conducted a study on the numerical modeling of masonry joints degradation in built tunnels. In this study, an attempt is made to study the influence of the mechanical properties of the masonry joints on the tunnel behavior by the help of experimental design method distinct elements numerical modeling. Elshamy et al. [2] conducted a study to determine the deformation behavior of different shapes of twin tunnels in soft clay. A numerical model is constructed as a plane strain problem taking into consideration the soil-structure interaction. The study was done to examine the effect of thickness-radius ratio on the tunnel deformation behavior, forces acting on the tunnel and the stress distribution. The settlement of the surface was noticed for various shapes and dimension of different twin tunnels. Three different tunnel shapes (circular, elliptical, and rectangle with circular arch) were selected for this study.

Chehade and Shahrour [3] studied the interaction between the twin tunnels with the help of numerical software. The study was done to know how the position of the twin tunnels and their construction procedure may affect the settlement behavior of the soil structures. Mishra et al. [4]) conducted a study to determine the deformation behavior of shallow tunnels under static and dynamic loading. For this study, ABACUS is used as modeling software which is based on FEM numerical tool. In this study, natural soft rock conditions are simulated in laboratory by using Plaster of Paris and mix material as tunnel model material. Then, the results which are obtained from numerical modeling are then compared with experimental results. Tiwari et al. [5] conducted a study on analyze the behavior of twin tunnel under dynamic conditions. The twin tunnel consists of tunnel reinforced concrete lining which is subjected to internal blast loading. Numerical modeling is done with the help of ABACUS. Pressure acting in the reinforced concrete lining and the surrounding soil mass, which is caused by the explosive induced shock wave, was determined for both tunnels. And from the results, it was concluded that the damage and deformation of reinforced concrete lining and soil mass depend upon charge weight and the clearance between the tunnels. Shalabi [6] conducted a study on the interaction behavior of shallow twin tunnels of circular shape by using FE analysis. The study was conducted to determine the effect on the stability behavior of tunnel when the spacing between the tunnels varies. Mishra [7] conducted a study to determine the effects of static and cyclic loading conditions on lined and unlined tunnels samples. From the results, it was concluded that in case of unlined tunnels, the number of cycles required to damage the tunnel is less from lined samples; this is due to the resistance offered by the lining material in case of lined samples.

To date, research efforts have been focused on experimental methods to quantify the response of rock materials to impact loading, [8–13]. In recent years, some researchers have emphasized on theoretical models for the evolution of rock damage under impact loading [14, 15]; Hu and Li [16].

However, an issue in these analytical models is that there are too many parameters involved and it is difficult to ascertain them. In addition, only a few of these investigations deals with practical applications. Some researchers have employed the numerical modeling tools in order to determine the behavior of rock materials under impact loading, [15]; Beus et al. [12, 17–20], Mishra et al. [4]. Numerous authors have looked into stability of tunnel in soft soil such as Broms and Bennermark [21], Atkinson and Potts [22], Poulos and Davis [23], Kimura and Mair (1981), Leca and Dormieux [24], Anagnostou and Kovári [25], Jancsecz and Steiner [26], Chambon and Cort [27], Broere [28], and Mollon et al. [29]. Many researchers observed that the induced stresses at civil and mining sites around the world are such that the ratio $k = \sigma h/\sigma v$ is relatively high at shallow depth (up to 500 m) [30, 31] and reduces at greater depth, i.e., more than 500 m, where $\sigma_h = \sigma_v$ (hydrostatic condition).

Presently, the tunnels are designed for gravitational loads and impact/blast loads. Several researchers have investigated such type of loading conditions numerically as well as analytically. Ahmed and Iskander [32] investigated the effect of ground movements induced by tunneling, on structures (above ground and underground) through physical modeling. Most of such the experimentations are performed on the soils [33]; however, the underground structures are constructed in weak rock mass or soft rocks. In such fragile conditions, the underground structures are subjected to different type of loading (earthquake loading, blast loading, impact loading, loading

**Table 1** Different loading conditions (Rao et al. [34])

| Loading | Strain rate | Type |
|---|---|---|
| Static | 0 to 0.1 | Construction or demolition of heavy/multi story structures (Overburden pressure) continuous vehicular movements/High traffic volume |
| Quasi—Static/Dynamic | 0.1 to 10,0000 | Cyclic loading/Missile attacks/Bomb penetration (Under terrorist activities)<br>Excavation and construction by blasting Seismic load/Heavy machine foundation |

due to increase in the litho-static pressure) for which the study of strength–deformation behavior of rock mass, and its effect on both lined and unlined underground structures (Table 1) is also essential. Therefore, it is essential to quantify the damage occurring in such loading and ground conditions. The mechanical properties of rock changes under varying burial depths and different strain rates from creep to shock ($10^{-5}$–$10^{5}$). Due to this rate dependency behavior of rock, noticeable variations are marked as increase in mechanical properties including tensile strength, compression strength, and failure strain and fracture energy by increasing the strain rates.

In the present work, an attempt is made to simulate the in situ condition through the physical modeling, to understand the tunnel deformation behavior under different loading conditions in soft and weathered rocks. The mechanical properties of rock changes under varying burial depths and different strain rates from creep to shock ($10^{-5}$–$10^{5}$).

Due to this rate dependency behavior of rock, noticeable variations are marked as increase in mechanical properties including tensile strength, compression strength, and failure strain and fracture energy by increasing the strain rates. Further, the analysis is also performed using finite element method and compared with the experimental results. In the present study, numerical simulations of soft and weathered rock subjected to static and dynamic load through drop load are carried out. Sensitivity of the rock and tunnel is investigated. The tunnels are considered to be located at varying depths with C/D ratio from 0.25 to 0.50, and the effect of cover is studied. A synthetic rock is prepared in the laboratory by mixing Plaster of Paris (PoP), Badarpur sand, and Kaolinite clay in order to simulate the soft rock. The material used as soft rock is selected according to its stress strain behavior and its ductile nature. The composition of material for synthetic rock is 50% PoP, 35% sand, and 15% clay by weight, and the water-powder ratio considered is 0.6. Models having dimensions of 30 cm × 30 cm × 35 cm were fixed for experimentation. For tunnel lining, polyvinyl chloride (PVC) pipe of circular opening having 5 cm diameter with thickness of 1.5 mm.

## 2 Model Material Characterization and Physical Modeling

Since rock mass found at shallow depth which is exposed to change climatic conditions and hence is more prone to weathering. Mechanical properties of such rocks found to be continuously deteoriated with respect to time. Thus, even small type of disturbance in such rocks will engross into large deformations and damage to the structures. Most of the urban metro tunnels, subways, and parking areas are constructed at shallow depth surrounded by weak and weathered rock mass. It is very difficult to anticipate the effect of impact load on natural rock mass from field, in laboratory. Due to fragile nature of such rock mass, bringing and testing them in the laboratory are very challenging. Therefore, to overcome this difficulty, laboratory samples are made from synthetic mixture of PoP, clay, sand, and mica (Fig. 1).

In order to simulate the real field conditions, three geo-materials are modeled covering the range of very low strength rock mass (from soil to soil forming rocks like mudstones, sandstones, highly weathered quartzite, or basalt).

In the present research, after several permutations and combinations of the mixture of above materials (Figs. 2 and 3) with varying water-powder ratio, finally combination of fine sand, mica, passing through 425 micron sieve, and clay and PoP is chosen. All the engineering and morphological properties of the laboratory modeled geo-materials are determined (Table 2). Cylindrical samples of NX size are tested for characterization purpose. The exact value of shear wave velocity is given in Table 2. Tunnel rock model (Fig. 4) is casted in the laboratory in Perspex mold, with selected geo-material as surrounding rock mass. For simulating lined tunnels in field, PVC hollow tubes are selected as a lining material. With varying burial depths, different models have been casted. Size of the model is kept as 30 cm × 30 cm × 35 cm with tunnel embedded at depth of 2.5 cm, 3.5 cm, 5.0 cm from the top (Fig. 5). Dimension of tunnel is taken as 35 cm length and 5 cm$\phi$.

**Fig. 1** Different constituent material for modeling geo-material

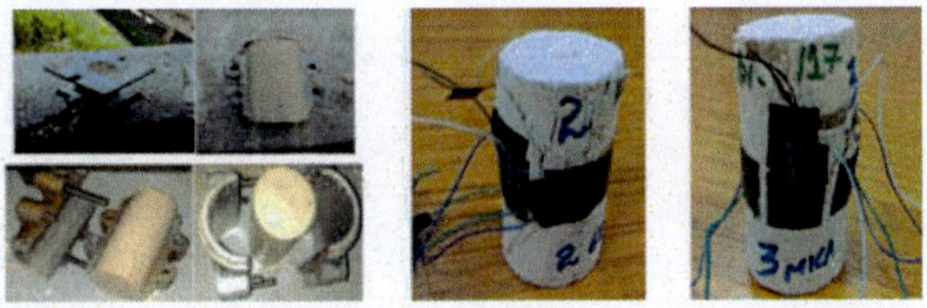

**Fig. 2** Preparation of samples and testing for characterization

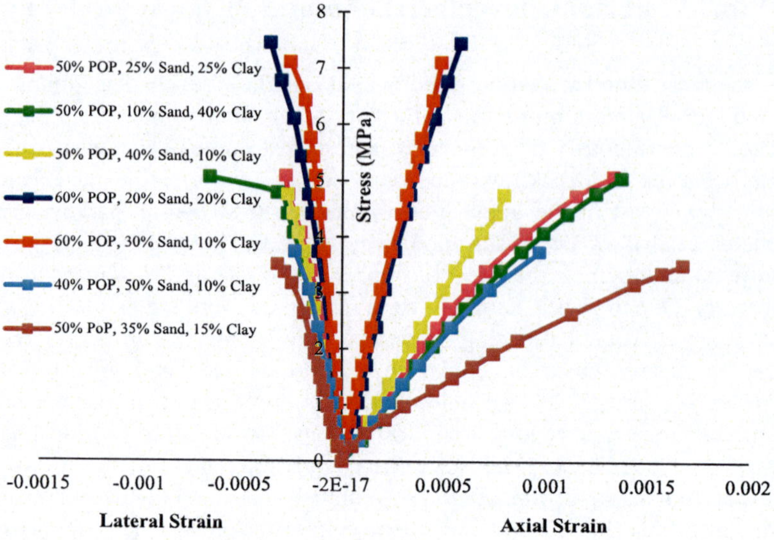


**Fig. 3** Stress-strain behavior of some of the laboratory modeled geo-material

**Table 2** Engineering properties of selected geo-material

| Property | GM1 |
|---|---|
| Density (g/cm$^3$) | 1.41 |
| Sonic wave velocity (m/s) | 1669.62 |
| Tensile strength (MPa) | 0.861 |
| UCS (MPa) | 4.48 |
| Modulus $E_{t50}$ (MPa) | 3153.8 |
| Poisson's ratio ν | 0.16 |
| Cohesion, c (MPa) | 1.81 |
| Friction angle φ (°) | 42.00 |
| Deere-Miller's chart [35] | EH |

## 3 Experimental Testing

The static testing is performed under displacement controlled mode and the loading rate is kept as 0.01 mm/sec. The confinement is applied on two faces of the model with the help of platens of dimension 300 mm × 200 mm × 20 mm. The force applied by these platens just before the testing is maintained as 1 kN, to simulate the confinement condition (Fig. 6a). The typical load versus deformation curves are plotted for the lined and unlined physical model, for different locations. Experiments have been performed using impact testing facilities in which cylindrical–hemispherical nosed drop hammer of 5 cm tup diameter was attached with the load assembly, and disk weights may be added according to requirement (Fig. 6b).

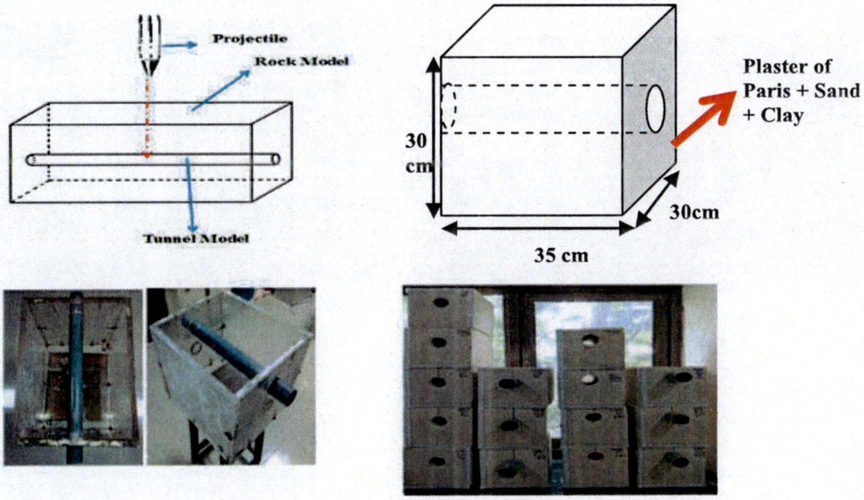


**Fig. 4** Schematic representation of tunnel rock model casted in laboratory

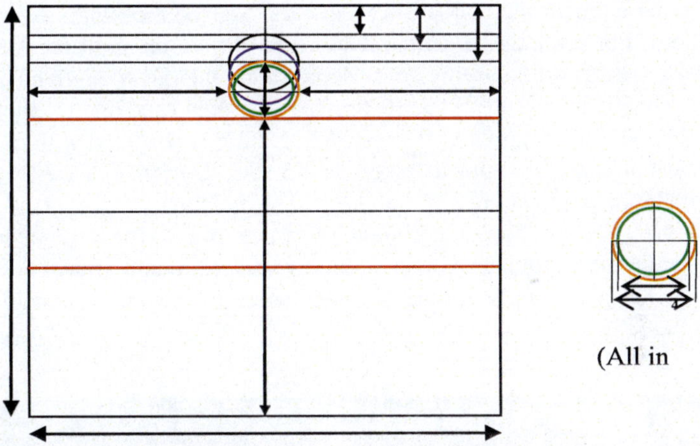


**Fig. 5** Rock tunnel model with varying burial depth along (x-x); (b) Tunnel and lining

The load–deformation behavior, vertical deformation at the loading area, fractured zone length, and the effect of lining are examined for each test in order to analysis the fracture threshold. Further, the numerical validation is performed for the model. The load cell and LVDTs are calibrated every time before the start of the experiments. Tunnels at higher depth are safe against any projectile hitting or impact loading. Drop energy get dissipated slowly, and spalling occurs from inside. It is observed that as the C/D ratio increases; low deformations would occur in tunnels. Crack length decreases gradually with the increase in the depth of overlying strata. Deformations at higher depth will only occur at very high impact load or drop energy.

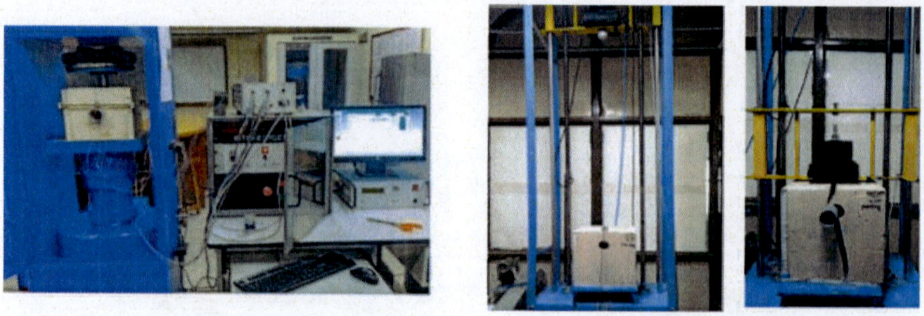

(a) Static Loading (b) Impact (Dynamic) Loading

**Fig. 6** Rock tunnel model when subjected to **a** Static loading **b** Dynamic loading

## 4 Validation of Experimental Testing

The rock and tunnel assembly, steel patch (static), and drop hammer were modeled in three dimension using ABAQUS explicit code. ABAQUS/CAE is continuum approach finite element method (FEM) software. So, the rock mass is assumed to be continuous. For static and dynamic loading condition, three different models of rock tunnel assembly with varying cover depths and material strength having cross-sectional area 350 mm × 300 mm × 300 mm, with unlined and lined tunnel scenario are created having *C*/*D* ratio 1.0, 0.7, and 0.5. Lining is modeled with diameter as 50 mm, and thickness of lining is taken as 1.5 mm. Likewise, a steel hammer is created with hemispherical nose of diameter 50 mm. Schematic view is represented below in Fig. 7. The top horizontal boundary was kept free for displacement in all directions while the bottom horizontal boundary was restrained against both vertical and horizontal displacements. All the side boundaries are allowed to undergo vertical displacements.

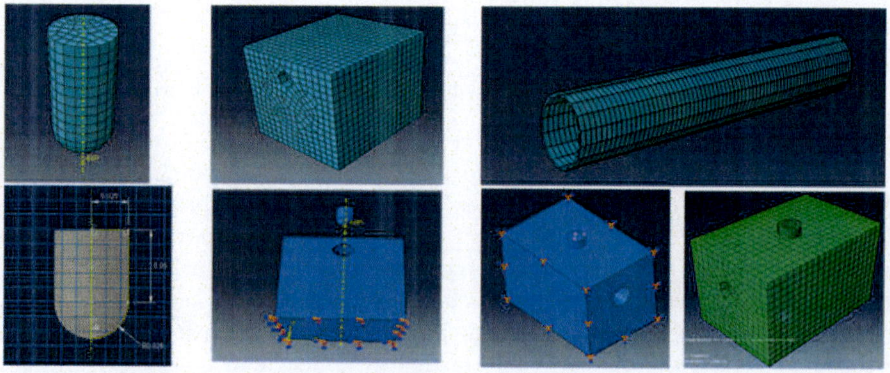

**Fig. 7** Schematic view of rock tunnel model, hammer, tunnel lining with meshing

The validation of numerical model is carried out by using the experimental data from Rao et al. [36] and Mishra et al. [37]. It is worth noting that the pattern as well as the magnitude of vertical deformation computed/observed from the numerical model is in good agreement with that of physical modeling, as shown in Fig. 8 and Table 3.

The displacement of the crown of tunnel along the tunnel axis and transverse to the tunnel axis is observed and quantified. It should be noted that the punching behavior observed below the patch plate in the physical model is attributed to the densification of the material due to collapse of voids during slow rate of loading. This punching behavior is also captured by numerical model.

In the present study, an attempt is made to understand the tunnel damage under dynamic loading conditions, to simulate the natural and artificial stress states such as overburden load, building loads, earthquake loads, and blast loads. The behavior of top surface tunnel lining surrounded by different rock mass under varying impact loads is compared physically as well as numerically, and the results obtained from both studies are critically analyzed. The displacement of the crown just below the loaded area is highest. Hence, deformation of top surface of lined and unlined tunnels under varying cover depth surrounded by three different synthetic rock mass has been compared. Numerical analysis predicts values on slightly higher side than the experimental predictions due to different assumptions on which analyses are based.

Therefore, a deformation value obtained from numerical analysis for unlined and lined tunnels is considered as representative values for the study. Nature of failure and crack propagation can be easily predicted from the deformation profile. For lined tunnels with varying surrounding rock masses and drop energies, the difference between the ranges of deformations obtained numerically is 2–7% more than that of the deformations obtained experimentally. This difference is larger in the case of unlined tunnels. Particularly, for unlined tunnels with $C/D$ 1.0, in GM1 material

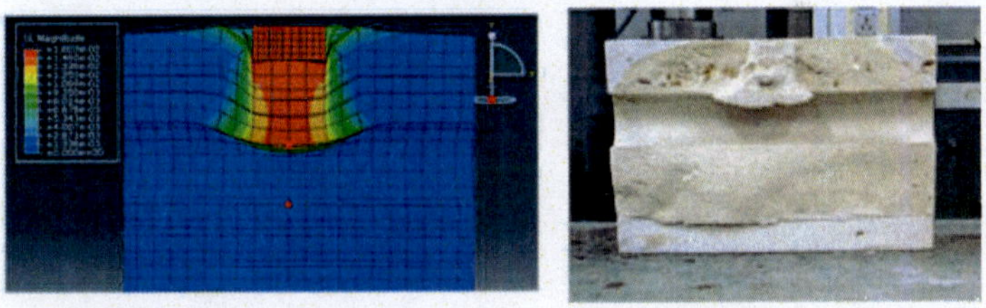

(a) Numerical Modelling (b) Experimentation

**Fig. 8** Comparison of yielded zone in a failed rock tunnel model

**Table 3** Comparison between numerical and experimental result

| | Experimental | Numerical |
|---|---|---|
| Length of cracked zone along the tunnel axis | 97 mm | 100 mm |
| Length of cracked zone transverse to the tunnel axis | 41 mm | 43 mm |

difference varies from 10 to 15% (Fig. 9). For unlined tunnels at deeper cover, numerical results hold good agreement.

After extensive numerical simulations and validation of results with experimental data, effect of different loading conditions has observed in the tunnel crown just beneath the loading. Load-displacement curves and energy-displacement curves are plotted and compared for all the loading conditions experimentally and numerically. Variations of displacement of crown are analyzed when the rock tunnel model is subjected to 20 kg load (both static and impact) with *C/D* 1.0 as shown in Figs. 10 and 11. From Fig. 10, it was observed that peak load for experimental impact loading in unlined case is 16.8 kN with displacement 1.3 mm in the crown, while peak load

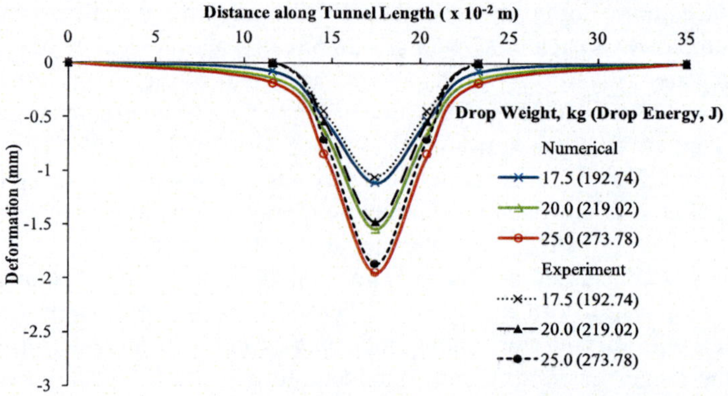


**Fig. 9** Numerical validations for deformation in GM1 for lined tunnel for impact load

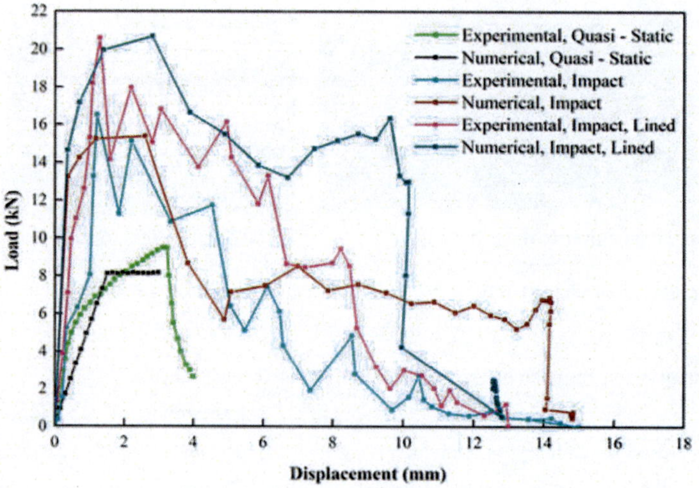


**Fig. 10** Variation of load versus displacement for different loading conditions

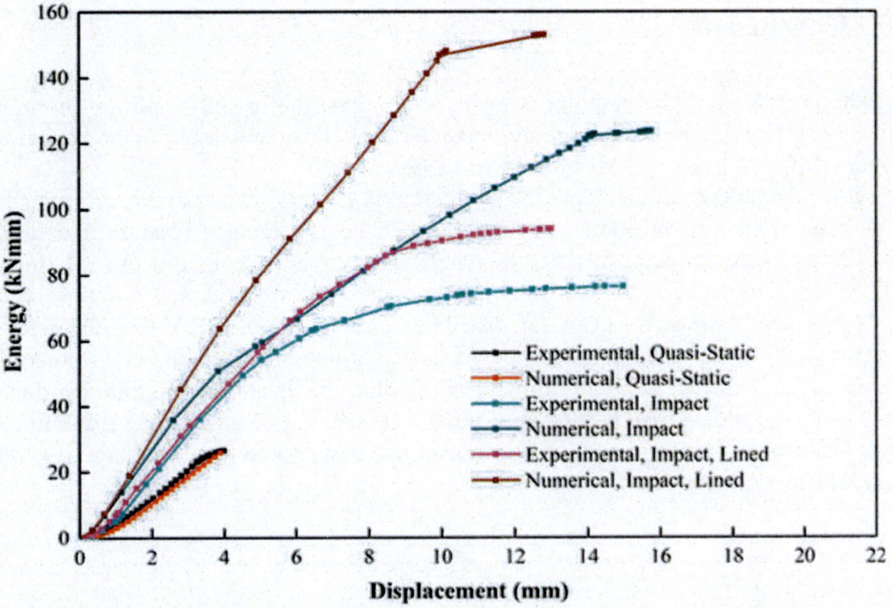


**Fig. 11** Variation of energy versus displacement for different loading conditions

for static loading is measured as 9.6 kN with displacement 3.5 mm in the tunnel crown.

It predicts that tunnel under static loading shows more displacement under small load. For lined case, peak load is more because of the resistance offered by the lining to the rock mass. Similar behavior was observed in numerical simulations also. Peak load for lined case is more, which is 21.46 kN with displacement measured as 3.4 mm, while for unlined case, it is 15.7 kN with displacement 1.8 mm.

When compared with static loading case, peak load for numerical static analysis is reported as 8 kN with displacement 1.85 mm. It was noticed that experimental values and numerical values hold good agreement in predicting displacements with varying loading conditions. Energy-displacement curve is plotted by integrating load-displacement curve. It is clear from Fig. 11 that more displacement is observed in the crown of unlined tunnel as compare to lined tunnel for the same drop energy. In terms of loading condition, it was noticed that crown encounters more displacement for impact loading.

## 5 Conclusion

In the present study, an attempt is made to simulate the in situ condition through the physical and numerical modeling, to understand the tunnel deformation behavior under different loading conditions in soft rocks.

It is concluded that the depth of overburden is a very critical parameter in quantification of fracture threshold and degree of damage. It is observed that the punching under the loading area occurred, which shows rock type of punching shear failure. The deformations of lined and unlined tunnels are investigated, and it is observed that the tunnel follows the Gaussian distribution along the length. Further, the deformation and degree of damage are determined for lined and unlined conditions. It is observed that the surrounding material is failing in shear, and the failed surface has taken elliptical shape at the top of tunnel. Hence, the fractured zone for unlined physical model is compared with numerical model, and both the results are in good agreement with each other.

## References

1. Idris J, Verdel T, Al-Heib M (2009) Numerical modeling of masonry joints degradation in built tunnels. Tunn Undergr Space Technol 24:617–626
2. Elshamy A, Attai G, Fawzy H, Abdel K (2013) Behavior of different shape of twin tunnels in soft clay soil. Int J Eng Innovative Technol (IJEIT), vol 2
3. Chehade F, Shahrour I (2008) Numerical analysis of the interaction between twin-tunnels: Influence of the relative position and construction procedure. Tunn Undergr Space Technol 23:210–214
4. Mishra S, Challa A, Singh A, Kumar A, Rao KS, Gupta NK (2017) Parametric study of lined and unlined tunnels at shallow depths under coupled static and cyclic loading conditions. Tunn Himalayan Geol 340
5. Tiwari R, Chakraborty T, Matsagar V (2016) Dynamic analysis of a twin tunnel in soil subjected to internal blast loading. Indian Geotech J 46(4):369–380
6. Shalabi F (2017) Interaction of twin circular shallow tunnels in soil-parametric study. Open J Civ Eng 7:100–115
7. Mishra S (2019) Physical and numerical modeling of tunnels under blast and impact loads. Ph.D. Thesis. Department of Civil Engineering, Indian Institute of Technology Delhi
8. Camacho GT, Ortiz M (1996) Computational modeling of impact damage in brittle materials. Int J Solids Struct 33(20–22):2899–2938
9. Xia K, Ahrens TJ (2001) Impact induced damage beneath craters. Geophys Res Lett 28(18):3525–3527
10. Momber AW (2003) Damage to rocks and cementitious materials from solid impact. Int J Rock Mech Rock Eng 37(1):57–82
11. Grange S, Forquin P, Mencacci S, Hild F (2008) On the dynamic fragmentation of two limestones using edge-on impact tests. Int J Impact 35(9):977–991
12. Gao F, Hou H, Yang X (2010) Numerical analysis of dynamic mechanical properties for rock sample under strong impact loading. Int J Inf Eng Electron Bus 2:10–16
13. Cao P, Wang Y, Xia K (2011) Testing study on damage fracture responses of rock under dynamic impact loading. Adv Mater Res 255–260:1815–1819
14. Ahrens TJ, Rubin AM (1993) Impact induced tensional failure in rock. J Geophys Res 98(E1):1185–1203

15. Taylor LM, Chen EP, Kuszmaul JS (1986) Microcracking-induced damage accumulation in brittle rock under dynamic loading. Comput Methods Appl Mech Eng 55(3):301–320
16. Hu L, Xi Li (2006) Damage and fragmentation of rock under experiencing impact load. J Central South Univ Technol 13(4):432–437
17. Beus MJ, Iverson SR, Dreschler A, Scott VA (1999) Static and dynamic loads in ore and waste rock passes in underground mines. In: Rock Mechanics for Industry, Proceedings of the 37th U.S. Rock Mechanics Symposium, Vail, Colorado
18. Nazeri H, Mustoe GGW, Rozgonyi TG, Wienecke CJ (2002) Implementation of a discrete element methodology for the modeling of gravity flow of ore in ore passes. In: Proceedings of the 5th North American rock mechanics symposium and 17th tunneling association of Canada, Toronto, Canada
19. Wang Y, Tonon F (2010) Discrete element modeling of rock fragmentation upon impact in rock fall analysis. Int J Rock Mech Rock Eng 44(1):23–35
20. Esmaieli K, Hadjigeorgiou J (2011) Selecting ore pass-finger raise configurations in underground mines. Rock Mech Rock Eng 44(3):291–303
21. Broms BB, Bennermark H (1967) Stability of clay at vertical openings. J Soil Mech Found Div
22. Atkinson JH, Potts DM (1977) Stability of a shallow circular tunnel in cohesion less soil. Geotechnique 27(2):203–215
23. Poulos HG, Davis EH (1980) Pile foundation analysis and design (No. Monograph)
24. Leca E, Dormieux L (1990) Upper and lower bound solutions for the face stability of shallow circular tunnels in frictional material. Geotechnique 40(4):581–606
25. Anagnostou G, Kovári K (1994) The face stability of slurry-shield-driven tunnels. Tunn Undergr Space Technol 9(2):165–174
26. Jancsecz S, Steiner W (1994) Face support for a large mix-shield in heterogeneous ground conditions. In Tunnelling'94. Springer, Boston, MA, pp 531–550
27. Chambon P, Corte JF (1994) Shallow tunnels in cohesionless soil: stability of tunnel face. J Geotech Eng 120(7):1148–1165
28. Broere W (2003) Influence of excess pore pressures on the stability of the tunnel face. Reclaiming Undergr Space 2:759–765
29. Mollon G, Dias Soubra A H (2009) Probabilistic analysis of circular tunnels in homogeneous soil using response surface methodology. J Geotech Geoenviron Eng 135(9):1314–1325
30. Herget G (1988) Stresses in rock. Balkema, Rotterdam, p 179P
31. Brown ET, Hoek E (1978) Trends in relationships between measured in-situ stresses and depth. Int J Rock Mech Min Sci Geomech 15(4):211–215
32. Ahmed M, Iskander M (2011) Analysis of tunneling induced ground movements using transparent soil models. ASCE J Geotech Geoenviron Eng 137(5):525–535
33. Meguid MA, Saada O, Nunes MA, Mattar J (2008) Physical modeling of tunnels in soft ground: a review. Tunn Undergr Space Technol 23(2):185–198
34. Rao KS, Mishra S, Gupta NK (2015) Fracture propagation of rock under dynamic regime. 50th International conference, Engineering Geology New Millenium (EGNM), Special publication of Journal of Engineering Geology. ISSN: 0970–5317, Delhi
35. Deere DU, Miller RP (1968) Engineering classification and index properties for intact rock. Illinois Univ at Urbana Dept of Civil Engineering
36. Rao KS, Mishra S, Gupta NK (2016) Effect of different loading conditions on tunnel lining in soft rocks. ISRM International Symposium EUROCK 2016, Cappadocia, Turkey. Taylor and Francis Group, London, ISBN 978-1-138-03265-1
37. Mishra S, Rao KS, Gupta NK (2018) Damage to shallow tunnels in different geomaterials under static and dynamic loading. TWST_2017_198. Thin Walled Struct 126:138–149. https://doi.org/10.1016/j.tws.2017.11.051

# Bidirectional Static Load Test (Bdslt) Using Instrumented Pile: An Innovate Testing Method in the Construction Industry

**Anil Cherian**

**Abstract** The application of a static load test in the optimisation and verification of pile design is mandatory in any foundation industry. Bi-Directional Static Load Test (BDSLT) is carried out using a hydraulically driven, purpose built, calibrated, sacrificial loading cell installed within the foundation unit. As the load is applied to the cell, it begins working in two directions; upward against upper side shear and downward against base resistance and lower side shear. Bidirectional Static Load Test (BDSLT) is a modern full-scale proofing load test method in the realm of performance-based pile design. Due to the many advantages over the conventional top-down load test, the Bidirectional Static Loading Test (BDSLT) using the hydraulic jacks is becoming an increasingly popular way to determine the ultimate capacity of deep foundations. Bidirectional loading tests are now becoming common practice around the Middle East and other countries, more convenient than traditional top-down loading tests. Strainstall has pioneered a practice of bidirectional test in conjunction with strain gauge instrumentation, which has vastly contributed to the knowledge and state of the art of pile response to load. This test is widely used in both preliminary and working compression and tension tests. BDSLT was carried out for a 2000 mm diameter pile in Dubai to a test load of 15,700 KN along with twenty levels of strain gauges to understand the geotechnical behaviour of the pile, also discuss the influence of different parameters on strain gauge reading and its possible solutions.

**Keywords** Bidirectional · Foundation · Pile · Strain gauge

## 1 Introduction

Bidirectional Static Loading Test (BDSLT) using sacrificial hydraulic cell is globally used in the foundation pile testing industry. The first commercial development came in the early 1980s in Brazil and about a decade later in the USA [1, 6, 10]. Due to the many advantages over the conventional top-down load test, the Bidirectional

A. Cherian (✉)
Strainstall Middle East LLC (James Fisher and Sons plc, UK), Dubai, United Arab Emirates
e-mail: dranilct@gmail.com

 
T. G. Sitharam et al. (eds.), *Seismic Design and Performance*, Lecture Notes in Civil Engineering 120, https://doi.org/10.1007/978-981-33-4005-3_30

Static Loading Test (BDSLT) using the hydraulic jacks is becoming an increasingly widespread way to determine the ultimate capacity of deep foundations. The ultimate capacity of a deep foundation is often demarcated with reference to settlement criteria at the pile head. This method is internationally accepted and referred in the standards such as ASTM D8169 [4], ICE manual [8] and federation of piling specialists [5]. Bidirectional loading tests are now becoming common practice around the world and more convenient than old-fashioned top-down loading tests. This test method is frequently used in both preliminary and production compression, tension, raked and continuous flight auger piles [3]. BDSLT using strain gauge instrumentation is a contemporary tool in optimising the design and providing validation of suitability and constructability in the foundation industry. Nevertheless, when these strains are used to calculate the load or skin friction in the pile, the results can occasionally be puzzling and challenging to interpret.

## 2 Aim and Objectives

This paper articulates the result of the application of a bidirectional static load test that carried out in a 379 m tall, 77 storey mixed-use tower development, Dubai, UAE to understand the geotechnical behaviour of the pile. Also describe the common problems associated with the strain gauge data interpretation during the test.

The local geology is characterised by the presence of the Barzaman formation which is encountered at depths greater than 23 m. The Barzaman formations include reddish brown conglomerates, brecciated dolomitic calcisitites and breccia with clasts of coarse gravel and cobble size of limestone [9]. The Barzaman formation is overlain by the reddish brown sandstones, which are extremely weak to weak with localised medium beds of calcilutite breccia. The sandstones are fine to medium sand size with a cementing material that imparts a very inconsistent strength to the rock. The reddish brown sandstones are overlain by a brown to light brown calcarenite. Localised medium beds of imperfectly laminated or massive calcarenite with fine to medium clasts are also encountered [2].

## 3 Methodology

BDSLT was performed using a multiple expandable jack assembly embedded within the test pile. As pressure is applied, the jack assembly expands and loads are applied to the foundation in an upward and downward direction. In the present project, the bidirectional static load test was carried out by Strainstall Middle East, Dubai from 12th to 13th October 2014. The pile details as follows:

| Pile type | Pile dia (mm) | Cut off level (m DMD) | Toe Level (m DMD) | Pile length (m) | Working load (kN) | Test load (kN) |
|---|---|---|---|---|---|---|
| TP—1 | 2000 | −13.09 | −78.00 | 64.91 | 62,800 | 157,000 |

Test was performed using a hydraulic jack assembly comprising of eight 900 tonne capacity bidirectional jacks placed at −56.50 m DMD (Dubai Municipality Datum). The upward and downward movements were measured at the level of the jack assembly using telltale rods in conjunction with displacement transducers. Twenty levels of Geokon, USA vibrating wire-type strain gauges comprising four units at each level were also installed on the test pile to measure strains at nominated locations. The data obtained from the site was analysed using an equivalent top loading method to identify the elastic settlement [7]. The strain gauge data was used to analyse skin friction distribution along the pile shaft as well as estimation of axial load distribution.

## 4 Important Outcomes

The calculated settlement and shaft friction at different load are provided below Table 1.

Based on the equivalent top load settlement analysis, at 100% of the working load, a total settlement of 17.60 mm was calculated. At 250%, the total settlement of 38.60 mm was calculated.

For the pile section above the jack, a unit skin friction ranges from 12 kN/m$^2$ to 1193 kN/m$^2$ was calculated at 250% of the working load. For the pile section below the jack, a unit skin friction from 6 kN/m$^2$ to 1077 kN/m$^2$ was calculated at 250% of the working load. The skin friction values are increasing linearly and do not show any evidence of geotechnical failure. The maximum settlement obtained was 38.60 mm, indicates that the pile can be still loaded to mobilize the ultimate shaft friction along the pile length. The result shows that strain gauge reading is influenced by the residual load, shape of the pile, position of the gauges, concrete modulus, etc. Ignoring these aspects would introduce errors in the shaft resistance determined from the strain gauge readings. The preliminary test pile, which was carried out prior to the installation of production piles, verified the pile capacity, shaft friction and

**Table 1** Calculated settlement from BDSLT

| Total settlement (mm) | | |
|---|---|---|
| Load percentage | Load (KN) | Settlement (mm) |
| 100 | 62,800 | 17.60 |
| 150 | 94,200 | 27.70 |
| 200 | 125,600 | 35.00 |
| 250 | 157,000 | 38.60 |

load-settlement behaviour, confirmed the effectiveness of the design and provided assurance of the satisfactory performance of the foundations.

## 5 Conclusions

From the results, it is identified that vibrating wire strain gages can provide very reliable and useful data for analysing drilled shaft performance. The residual load can significantly influence the axial load distribution as interpreted by strain gauge data. Conventional methods used to estimate both area and modulus of concrete can lead to significant errors in loads computed from strain data. The modulus of concrete can vary with a wide range and is also a function of the imposed load. The shape of a drilled shaft can influence both the strain and the calculated shear loads and hence caliper logging data to be checked during analysis. It is important to note that the lack of such information may cause a significant variation in the geotechnical behaviour of the pile during its working lifecycle. Based on the results, it is identified that the BDSLT method with strain gauge instrumentation is an effective contemporary tool in optimising the design and providing valuable information to the foundation industry.

## References

1. Amir JM (1983) Interpretation of load tests on piles in rock. Proc of the 7th Asian regional conference on soil mechanics and foundation engineering, Haifa, pp 235–238
2. Cherian A (2018) Value engineering of foundation design using bidirectional static load test (BDSLT): a case study from Dubai, United Arab Emirates. Int J Earth Sci Eng 11(4):270–273
3. Cherian A (2019) Application of bidirectional static load test (BDSLT) in the middle east foundation industry. DFI 2019 Middle East Conference, Dubai, pp 23–25
4. D 8169 (2018) Standard test methods for deep foundations under bi-directional static axial compressive load, ASTM standard, pp 1–13
5. FPS: Federation of Piling Specialists (2006), pp 16–17
6. Gibson GL, Devenny DW (1973) Concrete to bedrock testing by jacking from the bottom of a borehole. Can Geotech J 10(2):304–306
7. Hoyoung S, Rozbeh BM, William DL (2016) Assessment of methods for construction of an equivalent top loading curve from O-cell test data. Soils Found 56(5):889–903
8. ICE: Manual of Geotechnical Engineering (2012), pp 1458–1460
9. Macklin S, Gaba A (2009) Engineering in the Barzaman formation, coastal Dubai, UAE. Civ Eng 162(6):18–24
10. Osterberg JO (1998) The Osterberg load test method for drilled shaft and driven piles—the first ten years. Deep Foundation Institute, 7th international conference and exhibition on piling and deep foundations, Vienna, Austria

# Seismic Behaviour of RC Structure Under Disconnected Piled Raft System

**H. Ashok kumar and Visuvasam Joseph Antony**

**Abstract** Structures are widely considered to be designed on assuming fixed base condition which is usually limited to hard stratum. The recent evolution lead to initiation on soil structure interaction (SSI) study which is tremendously growing interest in development of huge structures to build on soft soils. Piled raft system is mostly adopted as a foundation system for such highrise building. In this paper, A 12 storey RC structure building is modelled with connected piled raft system and disconnected piled raft system with no structural connection between them where a provision of cushioned layer with granular soil is introduced. By equivalent static method both connected piled raft system and disconnected piled raft system are studied and compared by varied configuration spacing between piles, and cushion thickness are examined by finite element method using Plaxis 3D software. Results showed that the displacement in the structure, inter storey drift, pile displacement on connected and disconnected pile raft system. DPR has performed better when cushion thickness is increased.

**Keywords** Soil-Structure interaction · Cushioned · Disconnected piled raft system · Plaxis 3D

## 1 Introduction

High rise building resting on soft soil usually designed for piled raft system. A safe bearing capacity of soil ensures safety for structure. Pile along with raft are introduced to reduce the settlement effect of structure resting on such soils as indicated by Davis and Poulos [1]. In practice most of pile are connected to raft which ensured some part of load is transferred by pile. A very vast study is carried on connected piled raft system where else only limited study is done on disconnected piled raft system.

H. Ashok kumar (✉) · V. J. Antony
Vellore Institute of Technology, Vellore, India
e-mail: ashokkumarh7@gmail.com

V. J. Antony
e-mail: visuvasam.j@vit.ac.in

 
T. G. Sitharam et al. (eds.), *Seismic Design and Performance*, Lecture Notes in Civil Engineering 120, https://doi.org/10.1007/978-981-33-4005-3_31

Earlier study on disconnected piled raft (DPR) system was done by Wong et al. [2] by considering the pile as a reinforcement to homogenous sub soil which act as stiffener and part of load is carried by pile and seen reduction in settlement. A lower factor of safety is considered for pile as no structural connection between them. Cao et al. [3] an experimental study is carried on sand which resulted in reduction in differential settlement of pile and bending moment on raft. Negative skin friction which occurred on top pile which helped in load transfer mechanism. Jung In Choi et al. [4] carried load test and numerical test and suggested DPR system is an effective method of ground reinforcement.

Vincenzo Fioravante and Daniela Giretti [5] considered centrifuge model tests on dry dense sand, it does not ensure displacement compatibility where negative skin friction is responsible for pile settlement if granular layer which interposed is not stiff enough then efficiency of DPR will be lower than connected pile raft. Eslami and Calehi Malekshah [6] had done a numerical modelling by considering soil on modified coulomb theory results showed if lower cushion thickness is used the deeper the neutral plane in pile. By increasing cushion, the axial stress lower in DPR system then connected pile. Ata et al. [7] suggested that sharing of load between cushion and pile is affected by its thickness as the cushion increases the axial load at pile head reduces. Hor et al. [8] analysed using Plaxis 3D and concluded that cushion made of high strength granular layer well compacted for maximize the load transfer and pile should be embedded in hard layer for improved efficiency of pile.

In this paper, a numerical analysis is performed to investigate the behaviour of connected piled raft (CPR) and disconnected piled raft (DPR) system of RC framed structure resting on soil profile consisting of sand. A 12 storey RC framed structure is modelled in Plaxis 3D by varying cushion thickness and pile spacing. The structure displacement, pile displacement, foundation settlement, bending moment in pile are studied. Both connected and disconnected piled raft system are compared.

## 2 Soil Profile

A homogenous soil layer consists of uniform fine sand for a depth of 100 m with relative density 50% is modelled as Mohr Coulomb model in Plaxis 3D as shown in Fig. 1. Table 1 lists the properties used in numerical modelling for soil and cushion.

## 3 Characteristics of Model

### *3.1 Superstructure*

A 12 storey RC framed structure of 4 bays in both X and Y direction with a bay length of 4.5 m each and storey height of 3.5 m is modelled. The size of column and

**Fig. 1** Plaxis 3D model

**Table 1** Properties of soil and cushion

| Characteristics | Soil | Cushion |
|---|---|---|
| Unit Weight, $\gamma$ (kN/m$^3$) | 17.16 | 18.34 |
| Elastic modulus, E (MPa) | 30 | 320 |
| Cohesion, c (kN/m$^2$) | 0 | 0 |
| Critical friction angle, $\phi$ (°) | 33° | 33° |
| Angle of dilation, $\Psi$(°) | 3° | 3° |
| Poisson's ratio, $\nu$ | 0.3 | 0.35 |

beam is described in Table 2 which is adopted for structure. The thickness of slab used is 150 mm. Grade of concrete M25 and steel grade of Fe415 is provided for the structural members of the building. The brick load acting on beam is 14.49 Kn/m and live load of 3 Kn/m$^2$ and floor finish of 1 Kn/m$^2$ is acting on slab. Structural analysis is carried out using SAP2000 software considering the base as fixed base consist of equivalent static method and response spectrum method as per IS 1893. Table 3 describes the seismic parameter adopted for analysis.

**Table 2** Dimension of column and beam

| Storey | Column (mm) | Beam (mm) |
|---|---|---|
| 1–3 | 550 × 550 | 300 × 450 |
| 4–6 | 500 × 500 | 300 × 450 |
| 7–9 | 450 × 450 | 300 × 450 |
| 10–12 | 400 × 400 | 300 × 450 |

**Table 3** Seismic parameter

| Seismic zone | III |
| --- | --- |
| Zone factor | 0.16 |
| Time period | 0.89 s |
| Response reduction factor | 5 |
| Importance factor | 1.5 |

### *3.2 Foundation*

A total load of 125 MN acting on foundation for a base shear of 2608 Kn to which a raft foundation of 21 × 21 m with thickness of 1 m is designed made of concrete grade M25 and steel grade Fe415. Piles are introduced as settlement reducer as suggested by Burland et al. [9]. Pile with length of 26 m and diameter of 0.8 m is designed with group of nine piles. For study, two cases as piles are connected to raft without any provision of cushion and disconnected to raft with provision of cushion are modelled in Plaxis 3D for the soil properties of Table 1. The spacing between pile are varied as 3d, 5d, 7d, 9d. The cushion beneath the raft is varied as 0.5 m, 1 m, 2 m. Cushion consist of Gravel layer filled between raft and pile.

### *3.3 Numerical Modelling*

Plaxis 3D is a geotechnical software which uses finite element method. As soil is a complex material which exhibits non-linear behaviour FEM provides various results like displacement, bending moment of structure and soil failure point, settlement, stress and strain and even dynamic analysis can be carried. Mohr coulomb material model is considered for sandy soil. For the soil model, soil contour made of 147 m × 105 m × 100 m was considered on which structure is resting to which a static analysis is carried in FEM software through medium meshing.

## 4 Discussion of Results

The results are obtained from Plaxis 3D are discussed in terms of storey displacement, Inter storey drift, pile displacement,

## 4.1 Storey Displacement

Displacement of structure for raft alone without cushion and with cushion is compared in Fig. 2 it shows raft with cushion 2 m shown a minimal displacement when compared to other. In case of pile is introduced beneath the raft with variation in spacing between pile it seen when cushion thickness in disconnected piled raft is minimum the displacement of structure exceeds connected pile by 6% as shown in Figs. 3, 4, 5 and 6. When cushion thickness is kept at 2 m, it can be seen it shows

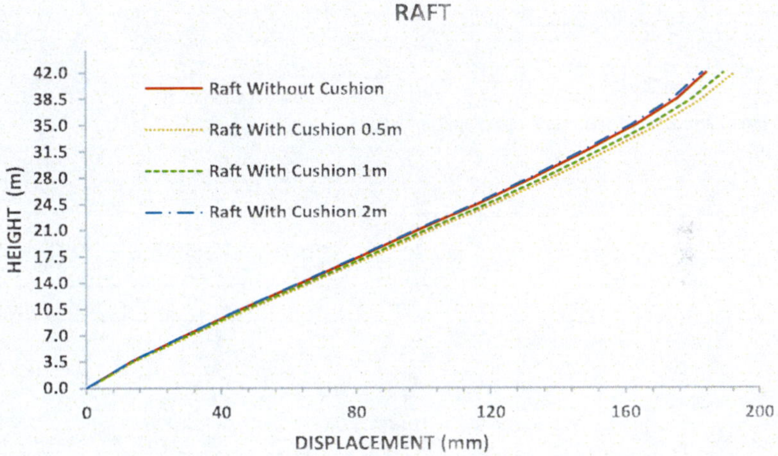


**Fig. 2** Displacement of structure in raft alone

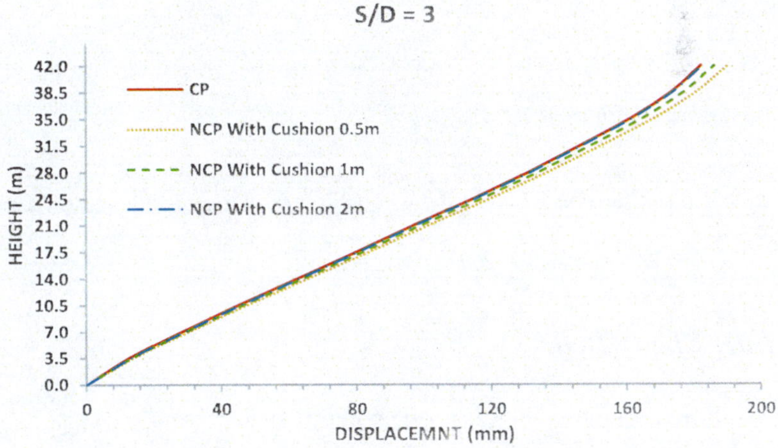


**Fig. 3** Displacement of structure in 3d spacing

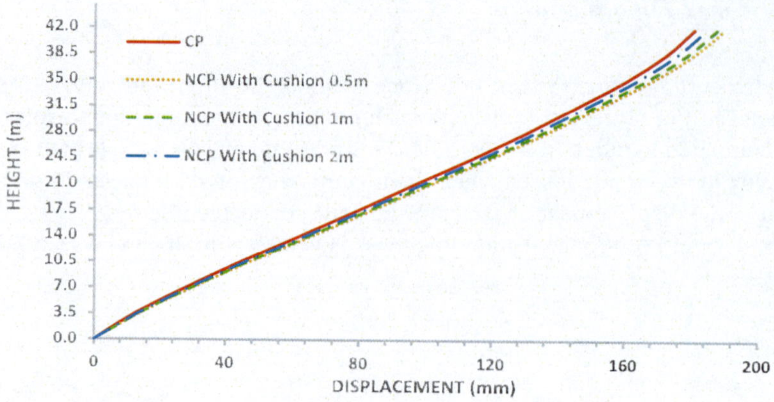


**Fig. 4** Displacement of structure in 5d spacing

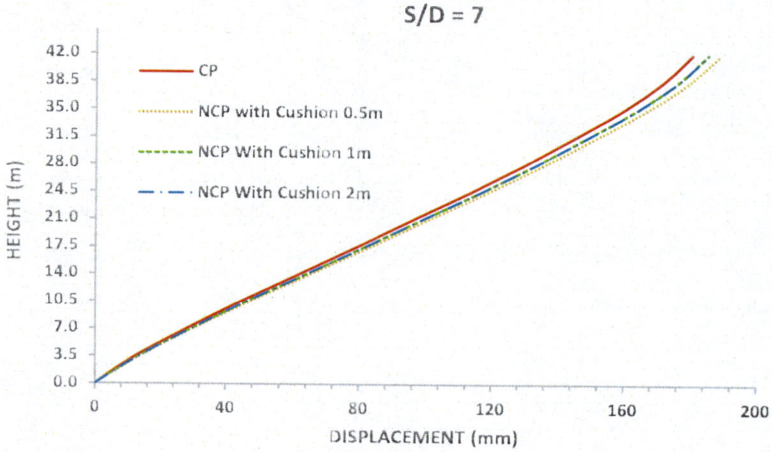


**Fig. 5** Displacement of structure in 7d spacing

displacement less than the CPR. It shows cushion thickness influence on structure displacement.

## 4.2 *Inter Storey Drift*

Inter storey drift of structure analysed in SAP2000 is shown in Fig. 7 which shows the structure does not exceeds the limit prescribed in IS code 0.004. However, in Plaxis 3D by considering soil structure interaction it becomes necessary as drift between each storey increases and exceeds the limits as shown in Fig. 8.

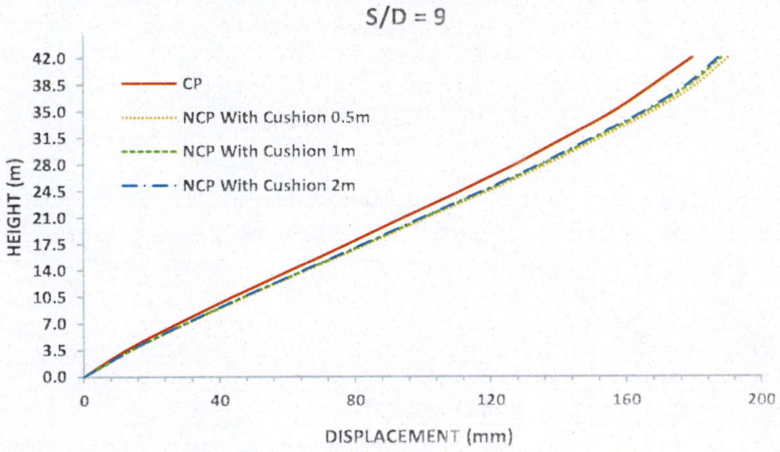


**Fig. 6** Displacement of structure in 9d spacing

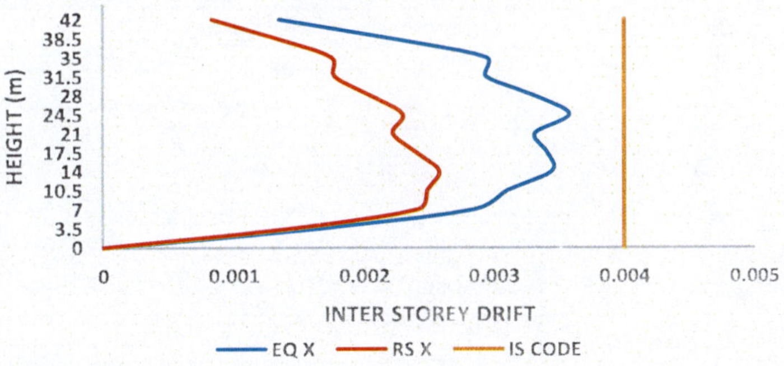


**Fig. 7** Inter storey drift from SAP2000

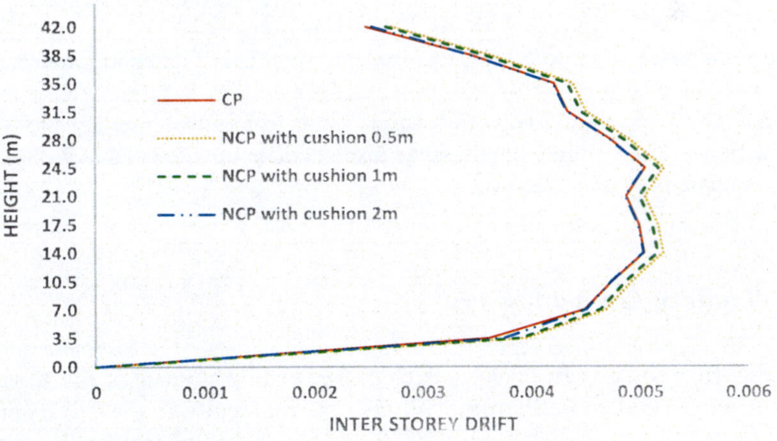


**Fig. 8** Inter storey drift from Plaxis 3D

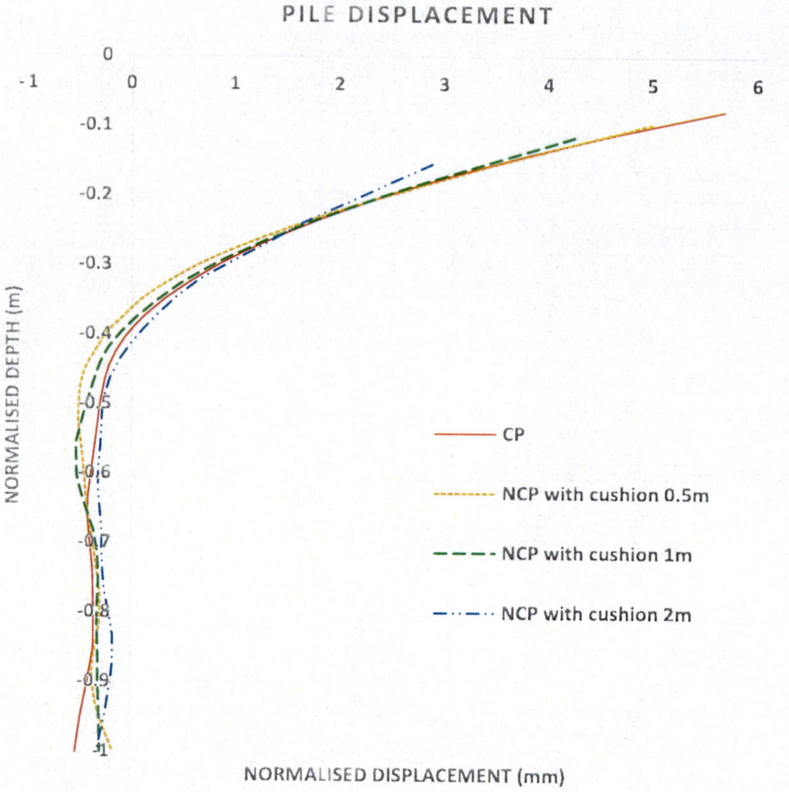


**Fig. 9** Displacement in pile

## 4.3 *Pile Displacement*

Figure 9 indicates the pile displacement at top of pile is higher in connected pile when disconnected pile with cushion of thickness 0.5 m is introduced it reduced marginally by 14%. Similarly, with cushion 1 m is introduced it reduced to 34% on introduction of 2 m cushion thickness the displacement on pile head reduced to 98% when compared to connected pile.

## 4.4 *Bending Moment in Pile*

From Fig. 10, when pile are connected to raft the bending moment on pile top is high and it requires rigid connection to withstand the moment when it is disconnected piled raft system the bending moment is zero at top of the pile which pile act as soil

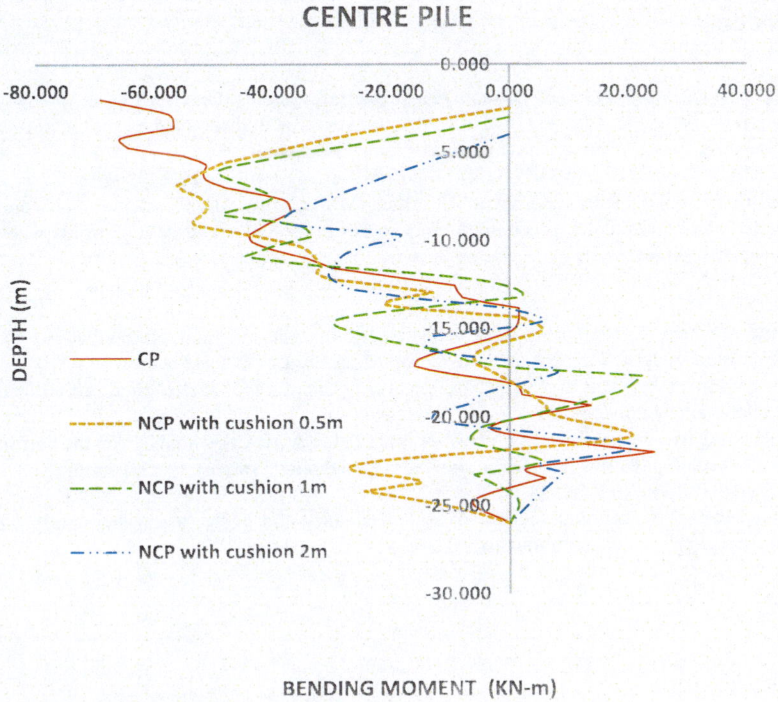


**Fig. 10** Bending moment in pile

reinforcement instead of structural member. As the cushion thickness is increased, it reduces the bending moment in pile.

# 5 Conclusion

This study showed the comparison between connected pile raft system and disconnected pile raft system with cushion on different spacing configuration of pile and the following conclusion are

1. As cushion reduced DPR performance reduced. At cushion 2 m, the displacement and inter storey drift of DPR are smaller compared to CPR.
2. Pile head displacement in DPR is lower compared to CPR.
3. In DPR bending moment, pile top is neglected hence act as soil reinforcement but in CPR system bending moment is higher hence it requires good structural connection.

## References


1. Davis EH, Poulos HG (1972) The analysis of pile raft systems. Geomech J G2:21–27
2. Wong IH, Chang MF, Cao XD (2000) Raft foundations with disconnected settlement-reducing piles. MEng thesis, Nanyang Technological Univ, Singapore
3. Cao XD, Wong IH, Chang MF (2004) J Geotech Geoenviron Eng 130:129–138
4. Choi JI, Min KH, Kim SH, Kwon OS, Kim MM (2009) ASCE international foundation congress and equipment expo contemporary topics in ground modification, problem soils, and geo-support, pp 305–312
5. Fioravante V, Giretti D (2010) Contact versus noncontact piled raft foundations. Can Geotech J 47:1271–1287
6. Eslami A, Salehi Malekshah S (2011) Analysis of non-connected piled raft foundations (NCPRF) with cushion by finite element method. Comp Meth Civ Eng 2(2):153–168
7. Ata A, Badrawi E, Nabil M (2015) Numerical analysis of unconnected piled raft with cushion. Ain Shams Eng J pp 421–428
8. Hor B, Song MJ, Jung MH, Song YH, Park YH (2015) A 3D FEM analysis on the performance of disconnected piled raft foundation. Japanese geotechnical society special publication. https://doi.org/10.3208/jgssp.KOR-21
9. Burland et al. International society for soil mechanics and geotechnical engineering. https://www.issmge.org/uploads/publications/1/36/1977_03_0011.pdf

# Application of Image Analysis to Investigate the Scale Effect of Open Driven Piles Under Different Infill Conditions

**G. Sreelakshmi and M. N. Asha**

**Abstract** Hollow piles with open ends are used in many civil engineering applications by virtue of its less mass density and easy drivability. Because of the soil plug formed at the ends of the open-ended hollow pile, the bearing capacity of such piles is evaluated similar to solid piles. Field tests on pile can be used to determine its load carrying capacity but it fails to predict the crushing and densification of soils that happens at the pile tip. In the present work, PIV technique is used to assess plugging effect of pile in cohesion-less soil prepared at different densities. A steel tank of dimensions 450 mm × 200 mm in plan, and 450 mm deep is used for performing experiments. Wood's scaling law is used to scale the model sections and three different scaling factors of 10, 5.7, and 4.5 are chosen for the study. Piles are driven using impact loads, and this process is captured through digital camera. Later the images are analysed to understand shear strain and volumetric strain developed around pile soil interface. From the experimental study, it is observed that the extent of soil plugging is dependent upon the pile diameter and it influences the pile capacity as well.

**Keywords** Image analysis · Open driven piles · SPT · Scaling factors · Strain contours

## 1 Introduction

Piles are structural members of timber, concrete, and/or steel that transmit loads vertically along the pile shaft or directly to a lower stratum through the pile tip. The vertical distribution of the load is through friction, and direct transfer is through end-bearing action. This distinction is developed for only convenience since all piles carry load as a combination of side resistance and point bearing when it penetrates through

G. Sreelakshmi (✉)
Department of Civil Engineering (VTU RRC), CMR Institute of Technology, Bengaluru, India
e-mail: sreelakshmi.g@cmrit.ac.in

M. N. Asha
Department of Civil Engineering, CMR Institute of Technology, Bengaluru, India

 
T. G. Sitharam et al. (eds.), *Seismic Design and Performance*, Lecture Notes in Civil Engineering 120, https://doi.org/10.1007/978-981-33-4005-3_32

softer strata to hard strata. Piles are generally classified based on its function, material properties, and applications such as timber piles, steel piles, cast in situ concrete piles and pre cast piles. Among these, steel pipe piles are generally adopted since they offer lesser driving resistances, greater rigidity, lesser fabrication cost and higher resistances to compression, tension, bending, and shear. These pipe piles are either welded or seamless steel pipes, which are driven either open-end or closed-end. Closed-end pipe piles are usually filled with concrete after driving. But in the case of open-end piles, dense soil plug is developed at some depth below the pile top. Due to this, driving resistance of an open-ended pipe is same as that of a closed-end pile since the soil plug inside open-ended pile behaves like a pile with driving plate or shoe. Soil plug formed in an open-ended pile is dependent upon the volume of soil displaced by the pile; amount and type of overburden material and friction between inner pile perimeter and soil plug [1]. When frictional resistance on pile perimeter becomes larger than weight of plug, the soil plug becomes a part of the pile cross section. It is seen that when pile cuts soil, the soil plug so formed does not fill pile completely since it is compressed because of vibration and side friction on the interior walls [2].

## 2 Literature Survey

Numerous experimental studies have been carried out to investigate the plugging effect on hollow driven pile by researchers [3–5]. They concluded that open-ended piles behave as closed pile under static loading and requires less installation effort as the number of blows needed is lesser for an open-ended pile than for a closed one [5]. The scaling effects of driven piles have been investigated by few researchers in sandy soil. The formula for ultimate point resistance was modified with reduction factors based on pile diameter [6]. It is estimated that for large diameter piles, there is a decrease in pile resistance due to lesser plugging effect. Tejchman and Tejchman [7] has studied the scaling effect of axially loaded open-ended piles in pulling and pushing conditions. Results indicated that when pile diameter is increased, there is a decrease in bearing capacity of pile which is more pronounced in pushing than pulling conditions in denser media. It is also concluded that with increase in pile diameter, there is a decrease in volume change of infill around the shaft region. Hight et al. [8] has investigated the effect of pile diameter on sand plug density and end-bearing capacity of open-ended driven piles. A finite element analysis is also carried out in same study to examine the sand plug behaviour and concluded that tip resistance reduced with increase in pile diameter. This is mainly due to result of non-development of stable arches (soil plug) in large diameters or because of development of large soil displacements around the pile driving zone. The twentieth century marked the beginning of use of different image-based tools in the field of geotechnical engineering. The technique was applied to visualise the displacement and deformations developed in subsurface structures. Many image processing techniques like X-ray, stereo-photogrammetric techniques, digital image correlation techniques (DIC), and computer-based image processing techniques were used to measure planar deformation fields in element and model test studies. One of the techniques which are currently adopted in civil engineering field is particle image velocimetry (PIV). Geo-PIV software is a MAT LAB-based tool developed by [9] working on the principles of PIV. Plane-strain model fits well since strain components developed along horizontal and vertical components can be monitored during pile penetrations. Plane-strain model studies were carried out in a tank of 1000 mm length and 745 mm depth, and Geo-PIV is used for analysing images. It is concluded that image analysis technique offered sufficient precision to measure pre-failure strains and offered flexibility to capture non-homogenous soil deformations [10]. Researchers [11–15] found that image-based deformation measurement is suited best for geotechnical applications and demonstrated that this enhanced measurement enables interpretation of all deformation modes, including rigid-body displacements, rotations, and strains. Manh Ngoc Tran [12] investigated the possibility of installing suction caissons in layered deposits and used PIV technique to capture soil deformations within the caisson. Visualisation of soil behaviour helped researchers to identify macroscopic failure mechanisms, provided stronger experimental support for geotechnical analysis, and develop complex constitutive or discrete element models [13, 14]. The pile soil interaction behaviour on hollow and solid closed ended single pile and pile groups is investigated through experimental methods and analysed through image analysis it is observed that surface heave developed around hollow pile under axisymmetric loading is more than solid pile due to lesser mass density of hollow pile than solid piles [15]. Motamedinia et al. [16] has carried DIC measurements on half-cut double-helix anchors in a sand tank and used PIV analysis to study displacement fields developed around the piles. The failure surfaces are analysed through displacement strain fields and findings indicated that helix spacing has a significant effect on pile soil interface behaviour.

From the literature, it is observed that soil plug plays a critical role as far as bearing capacity of open-ended piles is considered. Though the plug formed varies with respect to different sizes of piles, while computing capacity capacity, open-ended piles are considered equivalent to solid piles. This paper aims at investigating the feasibility of using image analysis in analysing the strain contours and volumetric strains developed in infill soils while driving open-ended piles of different sizes. The variation in the formation of strain contours with respect to different infill density is also studied.

## 3 Methodology

The methodology adopted in the present experimental study can be sequenced as follows:

a. Characterisation of infill medium
b. Modelling and design of steel tank in accordance with [17].

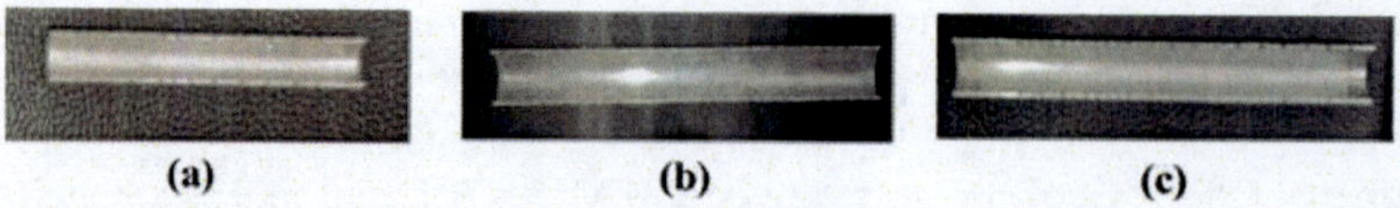

**Fig. 1** **a** 12 mm pile and 42.6 gm hammer mass for n = 10 **b** 19 mm dia pile and 360 gm hammer mass for n = 5.7 **c** 32 mm dia pile and 760 gm hammer mass for n = 4.6

c. Image capture and acquisition with high resolution digital camera and analyse soil deformation profiles with Geo-PIV tool
d. Effect of scaling factors on strain contours under different infill densities
e. Comparison of engineering and volumetric strains developed in plugged soil under different soil densities.

## 4 Details of Experimental Work

### *4.1 Standard Penetration Test Rammer Model and Modelled Piles as Per Scaling Ratio 10, 5.7, 4.6*

A prototype RCC pile of 250 mm outer diameter is modelled as aluminium hollow half pile sections using Wood's scaling laws [18]. For scaling, the n factor used in the present study are 10, 5.7, and 4.5, and accordingly, dimensions of three different half-section aluminium hollow piles have been arrived. A length to diameter ratio of 10 is used and the outer and inner diameter of the pile sections are 12 mm, 11.5 mm; 19 mm, 18.5 mm; 32 mm, 31.5 mm. Markings are made at 10 mm interval to observe plugging and penetration. By applying scaling factors, hammer of masses 42.6 g, 360 g, and 760 g falling from a height of 100 mm has been used to apply impact loads for driving the hollow piles. Figure 1a–c shows standard penetration hammer models and piles.

### *4.2 Infill Characterisation*

Sand is used as infill and its shear parameters and index properties are determined as per Codal provisions. Table 1 shows properties of infill materials. The infill density is fixed through pluviation method with a constant height of fall viz., 5 cm, 13.3 cm, and 28.3 cm for achieving low, medium, and high densities. Pluviation method, or raining technique, is a method commonly used to prepare sand/cohesion-less materials. Soil is rained down from a designated height to maintain consistent impact energy through a funnel into tank. Test setup for calibration of density is represented in Fig. 2.

The interfacial friction is dependent upon the infill, the pile material, and its surface roughness. To estimate the same, direct shear tests is performed with the pile

**Table 1** Properties of infill materials

| Description | Value |
|---|---|
| Coefficient of curvature | 0.723 |
| Coefficient of uniformity | 5.64 |
| Sand classification | Well graded |
| Specific gravity | 2.6 |
| Soil-plate friction angle with and without plate 5 cm, in degrees, (1.3 g/cc) | 43.60, 18.50 |
| Soil-plate friction angle with and without plate 13.3 cm in degrees, (1.4 g/cc) | 42.10, 17.20 |
| Soil-plate friction angle with and without plate 28.3 cm in degrees, (1.5 g/cc) | 40.02, 16.7 |

**Fig. 2** Test setup for calibration of density

material at the interface. Five aluminium specimens of size 6 cm × 6 cm × 1.4 cm are prepared. The specimens are placed in the lower half of the direct shear box, and the upper half of the shear box was filled with sandy soils at predetermined density. The direct shear test setup is shown in Fig. 3. When a shearing force is applied to the lower box through the geared jack, the movement of the lower part of the box is transmitted.

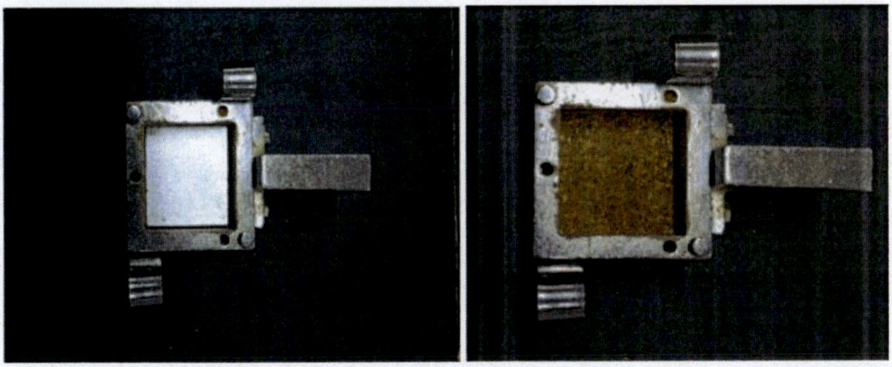

**Fig. 3** Direct shear test with and without aluminium plate

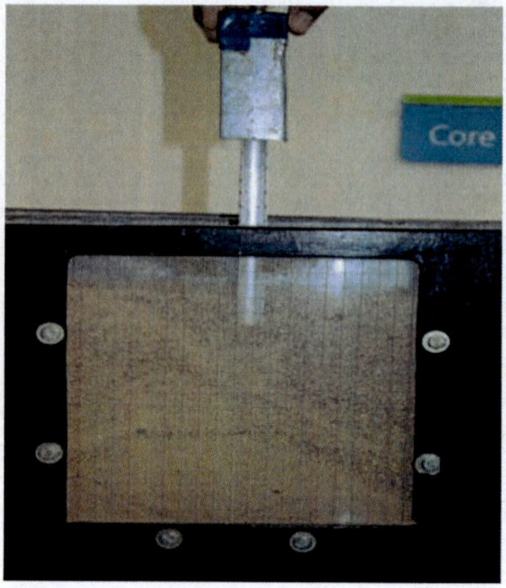


**Fig. 4** Application of impact load on the half-section pile

## 4.3 Test Set up

A plane-strain calibration chamber of dimensions 600 mm × 200 mm in plan and 450 mm depth [17] is fabricated with vertical partitions. Grid marking is done on front perpex sheet to track particle movement through observational window. In experimental studies, hollow open-ended piles are used as they enable the capture of strain actions at the interface region as well as the measurement of the plug along the pile shaft by means of image analysis. Piles are driven through sand infill by means of impact load applied using a hammer whose modelled energy corresponds to that used in standard penetration test. The model test set up used in the present study is presented in Fig. 4.

## 4.4 Geo-PIV

The Geo-PIV software implements the principles of particle image velocimetry (PIV) in a style suited to the analysis of geotechnical tests. Particle image velocimetry (PIV) is a non-intrusive method which provides instantaneous velocity vector measurements in a cross-section of an observation field. Here, principles of PIV are used to obtain displacement data from sequences of digital images captured during geotechnical model and element tests. Geo-PIV is a MAT LAB module, which runs at the MAT LAB command line. The development and performance of the software are described in detail by [14]. The initial image is divided up into a mesh of PIV test patches. PIV operates by tracking the texture (i.e., the spatial variation of brightness) within an image of soil through a series of images. The initial image is divided up into a mesh of PIV test patches. To find the displaced location of patch with initial coordinates (u1, v1) in a subsequent image, following operations are carried out. The correlation between the patch extracted from image 1 (time = t1) and a larger patch from the same part of image 2 (time = t2) is evaluated. The location at which the highest correlation is found indicates the displaced position of the patch (u2, v2). The location of the correlation peak is established to sub-pixel precision by fitting a bi cubic interpolation around the highest integer peak. This operation is repeated for the entire mesh of patches within the image, then repeated for each image within the series, to produce complete trajectories of each test patch here PIV works by tracking texture of soil particles through a sequence of images, and image analysis approach using digital image correlation is to obtain engineering and volumetric strain components around pile section.

## 5 Results and Discussions

### 5.1 *Effect of Scaling Factors on Shear Strain Components*

Figure 5a–c shows the shear strain developed in the soil domain around hollow half pile section of 12 mm, 19 mm, and 32 mm diameters at a density of 1.3 g/cc at the end of 30 blows as per SPT test procedure.

In the case of open-ended piles, soil plug is formed inside the pile shaft as a result of pile driving. A frictional resistance is developed by soil plug along the inner pile shaft which prevents further soil intrusion. A close examination of Fig. 5a–c reveals that the high shear strain contours(red contours) developed within the pile decreases with pile size. This emphasises that with increase in pile diameter, there is reduction in plugging effect causing greater displacement of soil around pile soil interface. Also, the zone of influence is more for larger pile diameter as expected and the PIV technique could capture the same. Analysis of the shear contours reveal that the formation of soil plug is significant for 12 mm and 19 mm pile (approximately in the order of 100 pixel in depth direction). Whereas for 32 mm, the length of soil plug length is very less and is in the order of 60 pixel scale with respect to depth.

### 5.2 *Effect of Infill Density on Shear Strain*

Figure 6a, b shows the variation of shear strain with respect to infill densities 1.3 g/cc and 1.5 g/cc for hollow pile of diameter 19 mm. From figure, it has been observed that as density increases from 1.3 g/cc to 1.5 g/cc, there is decrease in area of influence of engineering shear strain contours at the pile soil interface. This is evident from reduction in greatest shear strain from 160 pixels to 120 pixels (red colour) and lower

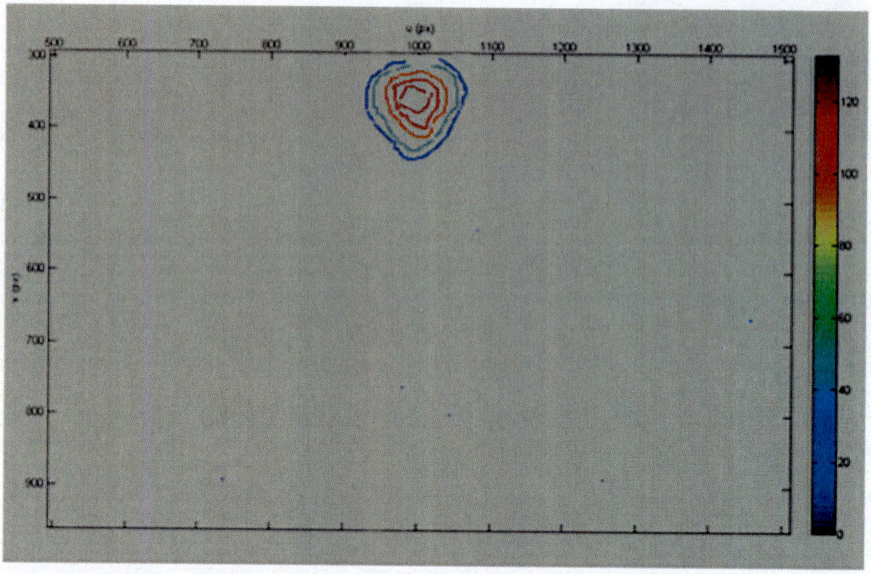

**(a)**.Variation of shear strain for 12 mm diameter pile with infill density 1.3 g/cc

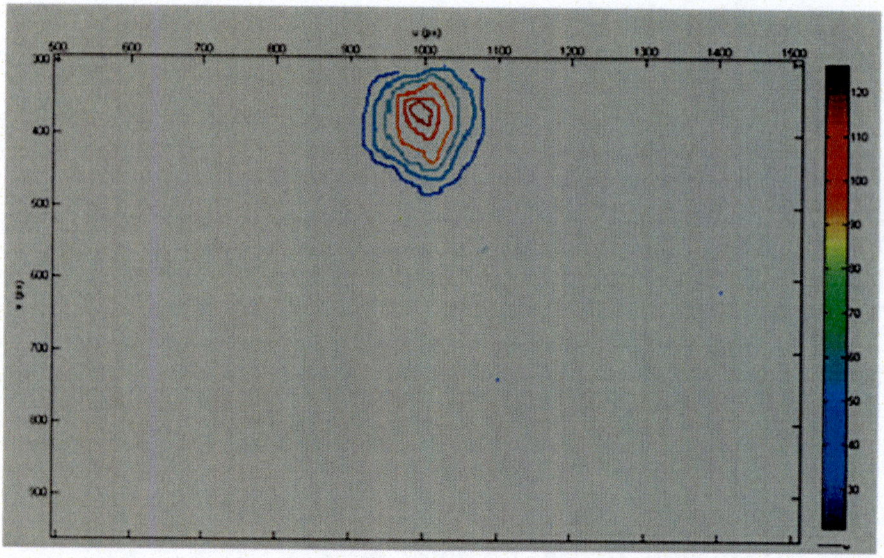

**(b)**.Variation of shear strain for 19 mm diameter pile with infill density 1.3 g/cc

**Fig. 5** **a** Variation of shear strain for 12 mm diameter pile with infill density 1.3 g/cc, **b** Variation of shear strain for 19 mm diameter pile with infill density 1.3 g/cc, **c** Variation of shear strain for 32 mm diameter pile with infill density 1.3 g/cc

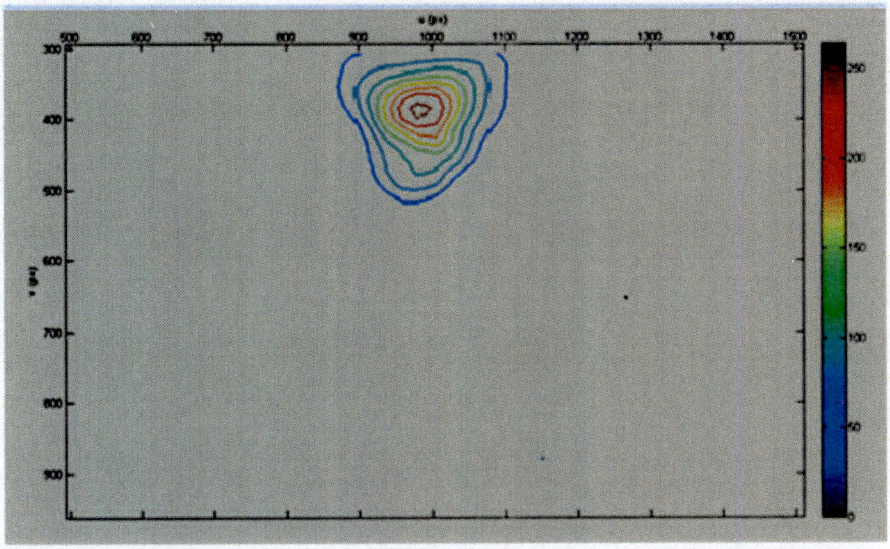


**(c)**.Variation of shear strain for 32 mm diameter pile with infill density 1.3 g/cc

**Fig. 5** (continued)

shear strain from 40 pixels to 30 pixels (dark blue). This is because, loose packing of sand grains happen at low densities and due to this grain to grain contact will be less. Hence, the shear flow of sand grain will be high at low densities. This behaviour will be in contrary to the behaviour of sand grains at higher densities. Hence, resistance to penetration increases due to greater frictional resistance offered by dense packing of soil grains at higher densities.

### 5.3 *Effect of Pile Diameters and Infill Density on Volumetric Strains*

Figure 7 a–c shows the volumetric strain developed in soil domain for two different diameters of hollow pile viz. 12 mm (outer diameter) and 32 mm (outer diameter). However, for pile diameter of 12 mm, the volumetric strains developed for two different densities are also compared.

A comparison of the figures reveal that the ratio of zone of influence of volumetric strain to pile diameters are in the ratio of 10–20 (pixels/mm), and it is observed that this zone of influence is proportional to pile diameter and infill density. A close examination of strain plots reveals that the image analysis could capture the formation of shear bands and the associated dilation of the infill. The dilation is more for larger diameter piles because the disturbance caused due to pile driving is more at the base

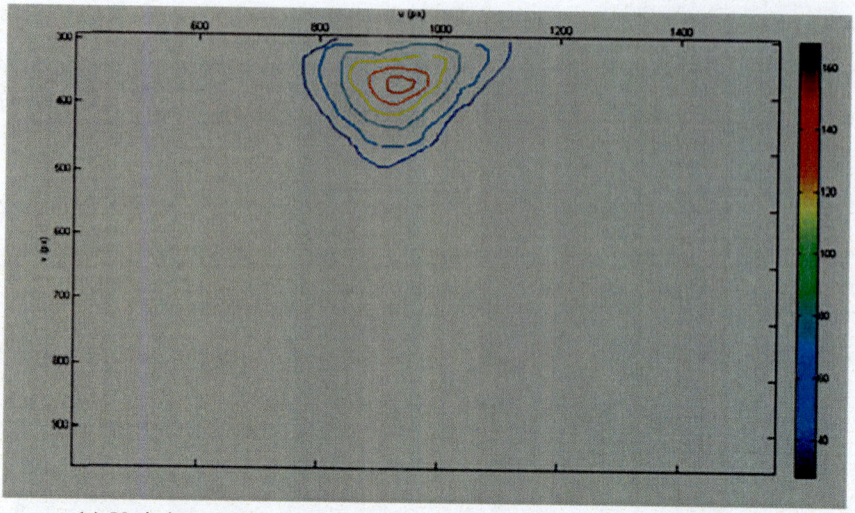


**(a)**.Variation of shear strain for 19 mm diameter pile with infill density 1.3 g/cc

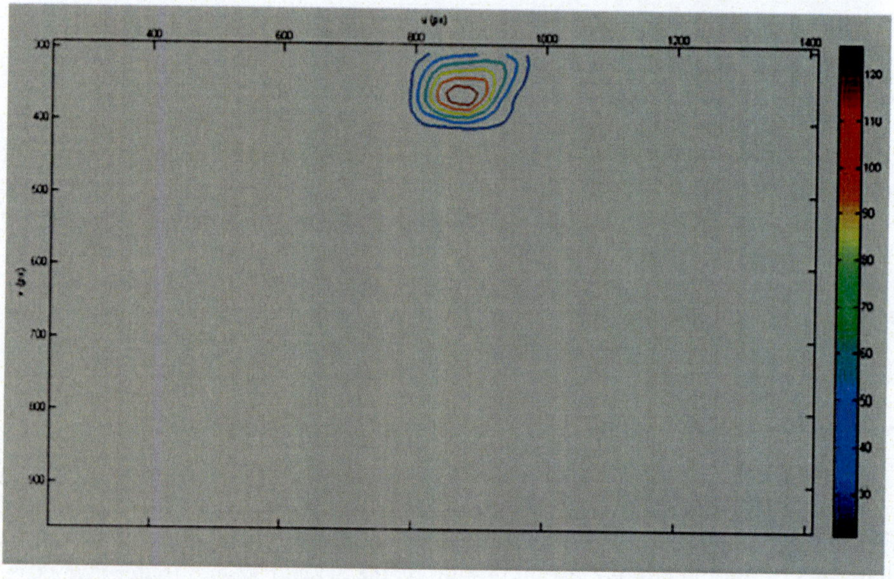


**(b)**.Variation of shear strain for 19 mm diameter pile with infill density 1.5 g/cc

**Fig. 6** **a** Variation of shear strain for 19 mm diameter pile with infill density 1.3 g/cc, **b** Variation of shear strain for 19 mm diameter pile with infill density 1.5 g/cc

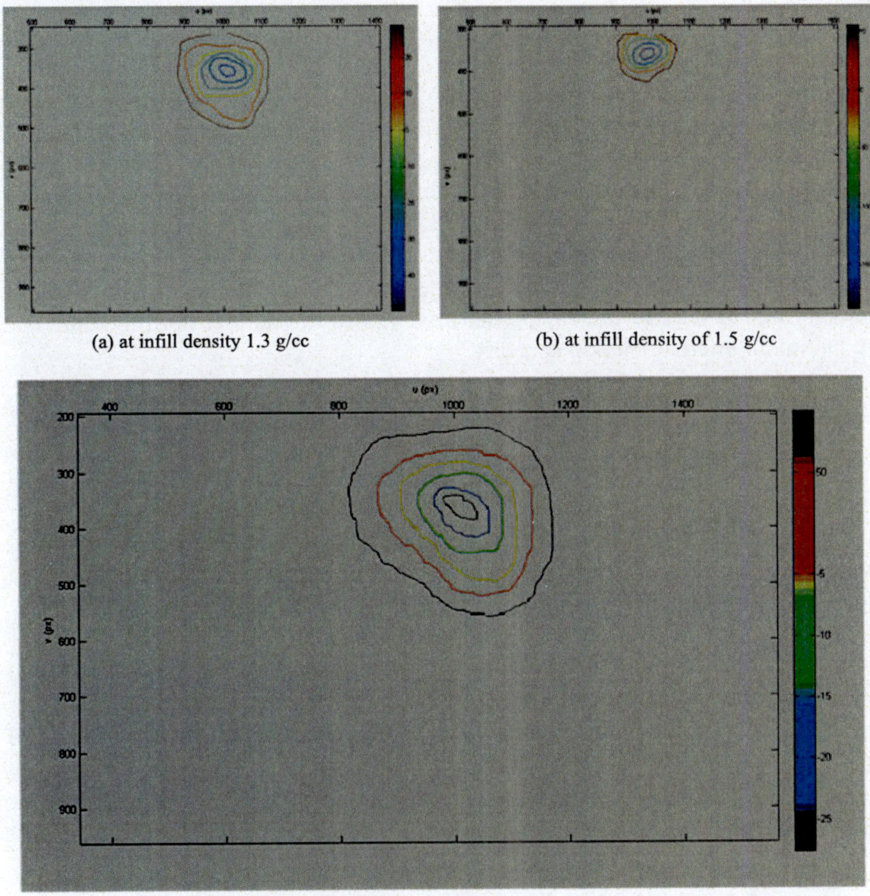


(a) at infill density 1.3 g/cc

(b) at infill density of 1.5 g/cc

(c) Volumetric strain around 32 mm diameter pile with infill density 1.3 g/cc

**Fig. 7** Volumetric strain around 12 mm diameter pile **a** at infill density 1.3 g/cc, **b** at infill density of 1.5 g/cc, **c** Volumetric strain around 32 mm diameter pile with infill density 1.3 g/cc

for such piles. This causes more shear flow in the zone of influence and hence a higher negative values for volume change. For a same pile diameter, with increase in density, the densification induced to pile driving is lesser and hence a lesser value of negative volume change.

## 6 Conclusions

The present study reveals the suitability of image analysis in capturing the deformation of infill while driving an open-ended pile. When pile diameter is increased, there is reduction in plugging effect causing greater displacement of soil around pile soil interface. This is also evident through shear strain contours which show increase in area of influence with increasing diameter. It is also observed that as density increases, there is decrease in area of influence of shear strain contours at the pile soil interface. This is because resistance to penetration also increases due to greater frictional resistance to driving is offered due to dense packing of soil grains.

It is observed that dilatancy of soil is closely related to the pile diameter and infill density. With increase in pile diameter, there is a reduction in soil plugging causing greater movement of soil around pile shaft. This phenomenon can affect the load carrying capacity of large diameter driven piles in granular infill medium due to reduction in internal friction capacity.

## 7 Scope for Future Study

The study indicates that the formation of soil plug is dependent upon the pile diameter as well as the density of the infill. However, the current code of practice does not account for this plug length and considers a plugged open-ended pile as a solid pile. This is a major concern in the design of small displacement piles which is mainly used for different on-shore and off-shore applications. To strengthen experimental observations, numerical simulations needs to be carried out, wherein soil plugging can be correlated to pile diameter and pile capacity. These small and real scale level simulations will be scope for future research works.

## References

1. Bowles JE (1997) Foundation analysis and design. The Mcgraw Hill publications, Illinois
2. Paikowsky SG, Whitman RV, Baligh MM (1989) A new look at the phenomenon of offshore pile plugging. Mar Geotech 8:213–230. https://doi.org/10.1139/cgj-2015-0162
3. Brucy F, Meunier J, Nauroy JF (1991) Behaviours of pile plug in sandy soils during and after driving. International proceedings of 23rd annual offshore technology conference, Houston, 1:145–154. https://doi.org/10.4043/6514-ms
4. Randolph MF, Leong EC, Houlsby GT (1991) One-dimensional analysis of soil plugs in pipe piles. Geotechnique 41(4):587–598. https://doi.org/10.1680/geot.1991.41.4.587
5. Raines RD, Ugaz OG, O'Neill MW (1992) Driving characteristics of open-toe piles in dense sand. ASCE J Geotech Eng 118(1): 72–88. https://doi.org/10.1061/(asce)0733-9410(1992)118:1(72)
6. Meyerhof GG (1983) Scale effects of ultimate pile capacity. J Geotech Engg 109:797–806. https://doi.org/10.1061/(asce)0733-9410(1983)109:6(797)
7. Tejchman A, Tejchman J (1994) Scale effect in pile model tests due to different pile and grain diameters, XIIII, CSMFE. New Delhi, India/ XIII CIMSTF, 1994, New Delhi
8. Hight DW, Lawrence DM, Farquhar GB, Mulligan WE, GuE SS, Potts DM (1996) Evidence for scale effects in the end bearing capacity of open-ended piles in sand. Offshore Technology Conference Annual OTC m Houston, TX, USA. https://doi.org/10.4043/7975-ms

9. Lehane BM, Gavin KG (2001) Base resistance of jacked pipe piles in sand. J Geotech Geoenviron Eng, ASCE 127(6):473–479. https://doi.org/10.1061/(asce)1090-0241(2001)127:6(473)
10. Szechy CH (1959) Tests with tubular piles. Acta Technica, Vol. 24:181–219. https://doi.org/10.1139/cgj-2014-0041
11. Stanier SA, Blaber J, Take WA, White DJ (2015) Improved image-based deformation measurement for geotechnical applications. Can Geotech J 53:727–739. https://doi.org/10.1139/cgj-2015-0253
12. Manh Ngoc Tran BB (2005) Installation of suction caissons in dense sand and the influence of silt and cemented layers. Ph.D. thesis
13. White DJ, Bolton MD (2004) Displacement and strain paths during plane-strain model pile installation in sand. Geotechnique 54(6):375–397. https://doi.org/10.1680/geot.2004.54.6.375
14. Stanier SA, Blaber J, Take WA, White DJ (2015) Improved image-based deformation measurement for geotechnical applications. Can Geotech J 2016 53(5):727–739. https://doi.org/10.1139/cgj-2015-0253
15. Sreelakshmi G, Asha MN, Viswanath D (2017) Investigations on pile-soil interaction using image analysis, Proceedings of second pan-American conference on unsaturated soils 2017, November 12–15, 2017 | Dallas, Texas, pp 466–476 https://doi.org/10.1061/9780784481684.047
16. Motamedinia H, Nader H, Habibagahi G (2018) A study on failure surface of helical anchors in sand by PIV/DIC technique. Int J Civ Eng, pp 1–15. https://doi.org/10.1007/s40999-018-0380-2
17. Parkin AK, Lunne T (1982) Boundary effects in the laboratory calibration of a cone penetrometer for sand. Proceedings of 2nd European symposium on penetration testing, Netherlands
18. Wood DM, Crew A, Taylor C (2002) Shaking table testing of geotechnical models. Int J Phys Model Geotech 1:1–13. https://doi.org/10.1680/ijpmg.2002.020101

# Behaviour of Disconnected Piled Raft Foundation on Sandy Soil: An Experimental Study

**Mainak Chowdhury, Plaban Deb, and Sujit Kumar Pal**

**Abstract** Disconnected piled raft foundation (DPRF) is an efficient and economical foundation option in todays' era where the piles are separated from the raft with the help of intermediate cushion layer. The cushion layer helps to distribute the vertical pressure uniformly on the raft bottom and acts as soil reinforcements. The load-settlement characteristics and load sharing mechanism of this type of foundation are complex in nature. Therefore, the major objective of this paper is to investigate the load-settlement and load-sharing behaviour of the disconnected piled raft foundation and to obtain the optimum thickness of cushion layer. In order to study the behaviour of disconnected piled raft foundation, a series of laboratory model tests are conducted on sandy soil with different thicknesses of the cushion layer. The influence of number of piles, pile spacing and the thickness of cushion layer on the load-settlement and load-sharing mechanism of the disconnected piled raft are illustrated in this paper. The test results show that with the increase of cushion layer thickness up to a certain value, the load carrying capacity of the disconnected piled raft increases and the disconnected piled raft foundation considerably reduces the settlement and improves the performance of the foundation.

**Keywords** Disconnected piled raft · Sandy soil · Cushion layer · Load sharing mechanism

M. Chowdhury (✉) · P. Deb · S. K. Pal
Department of Civil Engineering, NIT Agartala, Tripura 799046, India
e-mail: seemainak6@gmail.com

P. Deb
e-mail: plaban930@gmail.com

S. K. Pal
e-mail: skpal1963@gmail.com

 
T. G. Sitharam et al. (eds.), *Seismic Design and Performance*, Lecture Notes in Civil Engineering 120, https://doi.org/10.1007/978-981-33-4005-3_33

# 1 Introduction

Foundations are used for providing a support for structures which transfers their load to soil or rock strata that has adequate bearing capacity and appropriate settlement characteristics. They are mainly of two types, i.e., shallow foundations and deep foundations. In general, raft foundations (also referred to as mat foundations) are shallow foundations formed by a reinforced concrete slab of uniform thickness covering a wide area, often the entire area of a building. This 'raft' uniformly disperses the load imposed by a number of columns or walls over the entire area of the foundation and can be considered to 'float' on the ground like a raft. However, where a typical raft foundation does not provide the required support, it can be enhanced by the addition of piles which is also known as a piled raft foundation. Burland et al. [1] was the first one who came up with the concept of using piles under rafts to reduce the overall settlement of the structure. There is an increased need for optimised foundation design of the building structure worldwide and is becoming an important issue in engineering practice. Until recently, there were some separately used systems like shallow foundations such as rafts and deep foundations such as piles. However, lately the foundation engineers tend to combine these two separate systems. In urban areas, the principle question related to the design of high-rise buildings on settlement sensitive soils is the cost-optimised reduction of settlements to prevent possible damage and to minimise deformations of adjacent structures and the high-rise building itself. One method is to combine conventionally bored piles with a raft foundation resulting in a combined piled raft foundation. A piled raft foundation is a combination of a shallow foundation and a deep foundation with the best characteristics of each of its components. The piled raft foundation is a composite construction consisting of three bearing elements, piles, raft, and subsoil. Unlike the traditional design of foundation where the load is carried either by the raft or by the piles, in the design of a piled raft foundation the load share between the piles, and the raft is taken into account. In this foundation, the piles are used to reduce the settlements and the raft is used to provide the adequate bearing capacity.

Several studies have undertaken to investigate pile-supported rafts, in which the load coming from the structure are shared by the pile and the raft, have shown that such a hybrid foundation system provides an efficient way of supporting highly-loaded rafts by sharing the load with the raft, the number of piles needed under the raft foundation is reduced, and the spacing between them is increased. This saves pile costs.

Nakai et al. [2] inferred that even when piles are not connected with the raft, there is a significant contribution to the dynamic soil-structure interaction. Fioravante et al. [3] performed their experiments with an interposed layer in between the piles and raft and concluded that in case of non-contact piles, the cushion layer diffuses the pressure and produce shallow settlements. Denzfouli et al. [4] conducted experiments on unconnected piled raft foundation on sandy soil and inferred that with an optimum cushion thickness and pile spacing, lower settlement is observed. Azizkandi et al. [5] studied connected and unconnected piled raft foundation using numerical methods and concluded that for optimum design, axial stiffness of pile, settlement and stress along pile length are the most important factors of design. Cao et al. [6] assessed the settlement reduction in case of disconnected piles by varying the raft stiffness, pile configuration and pile length and also evaluated the bending moment and axial forces within the raft and the piles with the help of strain gauge readings. Tradigo et al. [7, 8] used numerical approach and inferred that by using disconnected piled raft foundation the overall settlement reduces. As per the research carried out by various researchers, it was concluded that a lot of research work has been done on connected piled raft foundation and significant contributions have been made by studying different parameters.

On the other hand, it was found that limited research work has been carried out on unconnected piled raft foundation. Hence, in this paper, the influence of number of piles, pile spacing and the thickness of cushion layer on the load-settlement behaviour of the disconnected piled raft are illustrated.

## 2 Experimental Test Program and Set-up

The aim and objective of the experimental study is to determine and analyse the load-settlement behaviour of the disconnected piled raft foundation system, the influence of number of piles, pile spacing and the thickness of cushion layer on the load-settlement of the disconnected piled raft. A total of 21 tests have been conducted in the test set-up with different pile configurations, spacings and cushion thickness. The program of the different configurations of piles and rafts are illustrated in Table 1. The materials that were used for the experiment were sandy soil, gravel, steel pipes and mild steel plates. Sandy soil was used as the base of the foundation, and gravel

**Table 1** Model test program

| Type of foundation | Pile configuration | Spacing/Diameter Ratio | Thickness of gravel layer |
|---|---|---|---|
| Raft Foundation | – | – | 5 cm, 10 cm, 15 cm |
| Disconnected Piled Raft Foundation (DC-PRF1) | 2 × 2 | 3 | 5 cm, 10 cm, 15 cm |
| Disconnected Piled Raft Foundation (DC-PRF2) | 2 × 2 | 4 | 5 cm, 10 cm, 15 cm |
| Disconnected Piled Raft Foundation (DC-PRF3) | 2 × 2 | 5 | 5 cm, 10 cm, 15 cm |
| Disconnected Piled Raft Foundation (DC-PRF4) | 3 × 3 | 3 | 5 cm, 10 cm, 15 cm |
| Disconnected Piled Raft Foundation (DC-PRF5) | 3 × 3 | 4 | 5 cm, 10 cm, 15 cm |
| Disconnected Piled Raft Foundation (DC-PRF6) | 3 × 3 | 5 | 5 cm, 10 cm, 15 cm |

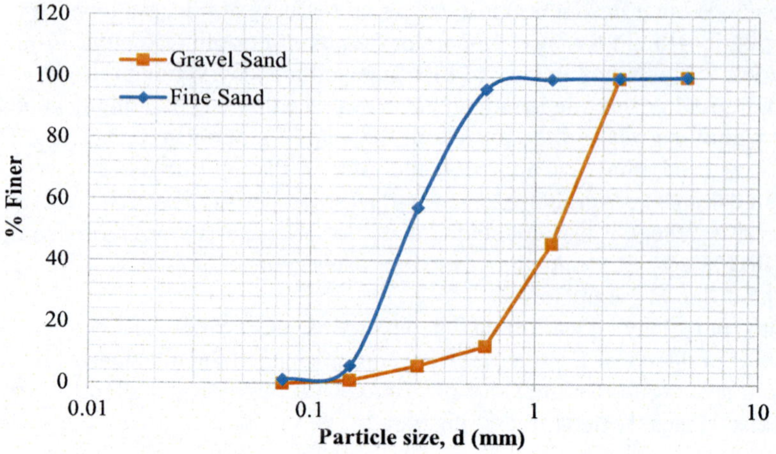


**Fig. 1** Particle size distribution curve for sand and gravel

layer were provided as cushion in layers of 5 cm, 10 cm and 15 cm for different pile configurations. The specific gravity of sand was found out to be 2.66, the maximum dry unit weight being 16.97 kN/m$^3$. Dry sieving method was used to find out the particle size distribution curve which is shown in Fig. 1 and the coefficient of uniformity and the coefficient of curvature was found out to be 2.91 and 1.25, respectively. The angle of internal friction at a unit weight of 16.97 kN/m$^3$ was found out to be 25.77°. The model raft is made up of mild steel having a square dimension of 250 mm × 250 mm with 10 mm thickness. The raft is shown in Fig. 2. The model pile that was used in the project work is smooth, hollow and made of mild steel and has a circular cross section. The outer diameter of piles was 20 mm with a length of 300 mm. To maintain the different spacing along the pile group, different pile head connectors are made and shown in Fig. 3a. The image of the model pile group with pile head connectors is shown in Fig. 3b.

## 3 Experimental Set-up

The testing tank of size 900 mm × 900 mm × 600 mm made of plexi-glass and steel sheet. The bed is prepared freshly after each test. The test set-up is shown in Fig. 4. The vertical load is given by a hydraulic jack arrangement and measurement of load is made by calibrated load cell to measure a maximum load of 30 kN, and settlement is measured by two Linear Variable Displacement Transducers (LVDTs) measuring settlement of 25 mm which is corresponding to 10% of raft width. The load cell and LVDTs are electronically connected with the data logger. Then, the vertical load is applied though the hydraulic jack at a rate of 1.25 mm/min.

**Fig. 2** Model raft

## 4 Test Procedure

The tests were carried out systematically and the following steps were followed:

I. Firstly, sand was poured in the testing tank by a method called rainfall technique to achieve the required relative density in all the tests. The sand was poured from a height of 270 mm to achieve a unit weight of 16.97 kN/m$^3$. The sand was poured in intervals of 100 mm. Sand was poured till a height of 200 mm from the bottom of the tank and then levelled.

II. Each corner piles of 2 $\times$ 2 pile and 3 $\times$ 3 pile group are instrumented with four strain gauges along the pile length and electronically connected with the data logger which gives axial load value along the pile. The pile spacing is maintained with the help of model pile connectors as shown in Figs. 2 and 3. The experimental set up is shown in Fig. 5. The test box is filled with sand up to the tip of the piles and the piles are carefully placed at the centre of the loading arrangement.

III. After placing the piles carefully, sand is poured using the same technique till a height of 500 mm from the bottom of the tank.

IV. Then, a gravel layer (which acts as a cushion) of 5 cm is introduced and is poured in the testing tank using the same rainfall technique from a height of 270 mm to achieve a unit weight of 16.97 kN/m$^3$. A free-board of 50 mm was kept to ensure that after loading the sand and gravel does not topple over the tank.

**Fig. 3** **a** Pile head connectors, **b** Pile connectors with pile

**(a)** Pile head connectors

**(b)** Pile connectors with pile

V. After installation of model piles and filling up of sand and gravel layer up to the required depth, the raft is placed over the gravel layer and placed at the centre of the loading arrangement.

VI. Then, vertical load is applied with the help of a hydraulic jack arrangement and

VII. Measurement of load is made by load cell calibrated to measure a maximum load of 30 kN, and settlement is measured by two Linear Variable Displacement

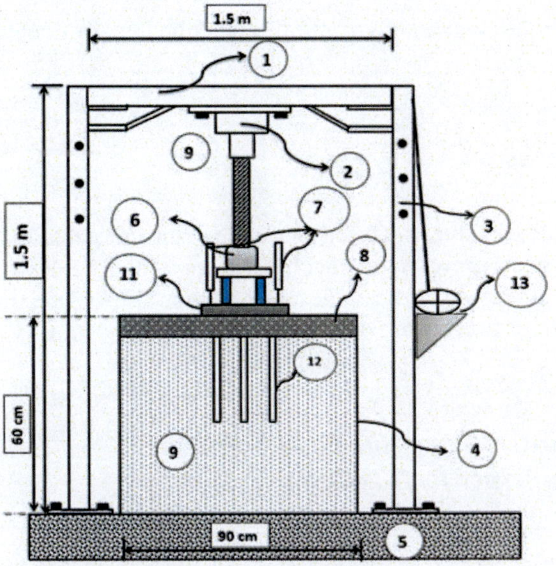


1. **Horizontal loading frame**
2. **Hydraulic jack**
3. **Vertical loading frame**
4. **Soil Tank**
5. **Concrete base**
6. **Load cell**
7. **LVDT**
8. **Gravel layer**
9. **Sand layer**
10. **Loading column**
11. **Raft**
12. **Pile**
13. **Pumping Unit**

**Fig. 4** Schematic representation of model test se-up

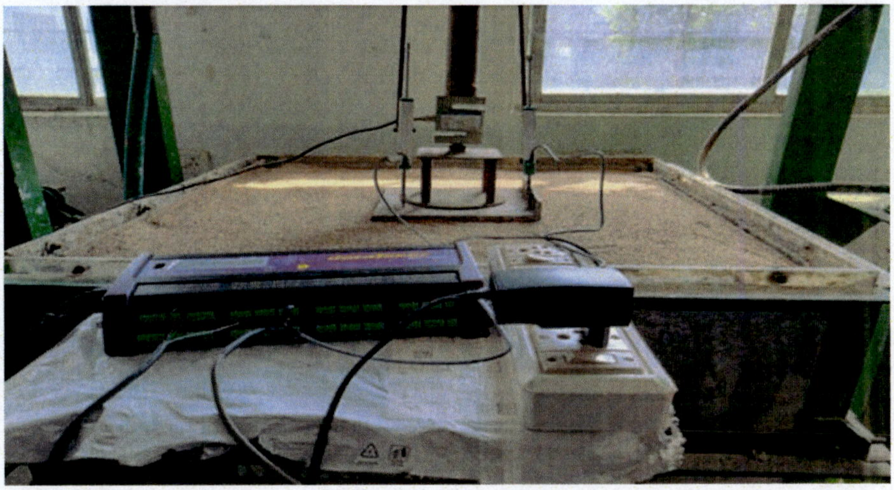

**Fig. 5** Experimental set-up

Transducers (LVDTs) measuring settlement of 25 mm which is corresponding to 10% of raft width. The load cell and LVDTs are electronically connected with the data logger. Then, the vertical load is applied though the hydraulic jack at a rate of 1.25 mm/min. The same procedure is repeated with different pile configurations and different thickness of gravel layer, and the load-settlement characteristics are noted.

## 5 Test Results and Discussion

The results of the load-settlement tests were analysed and some of the interpretations are made. Effects of number of piles, gravel layer thickness, and spacing of piles are analysed, and the load-settlement curve have been discussed herein. Effects of axial force developed along the length of the pile are discussed herein.

### 5.1 *Effect of Depth of Gravel Layer on Load-Settlement Characteristics When Only Raft Is Used*

The effect of depth of gravel layer has been studied in details and the load applied via the loading mechanism which is connected electronically with the data logger, and the corresponding settlement of the raft is recorded with the help of LVDT which is again connected to the data logger electronically. The data is recorded in a computer, and the load-settlement curve is plotted. It is seen that with the increase in depth of gravel layer, the load carrying capacity of the raft increases. The load versus settlement graph is shown in Fig. 6. The main purpose of using gravel layer beneath the raft is to redistribute the vertical stresses between the piles and the surrounding soil. The load capacity of raft is considered at 25 mm settlement of the raft which has been recorded with the help of LVDT and data logger. It has been observed from

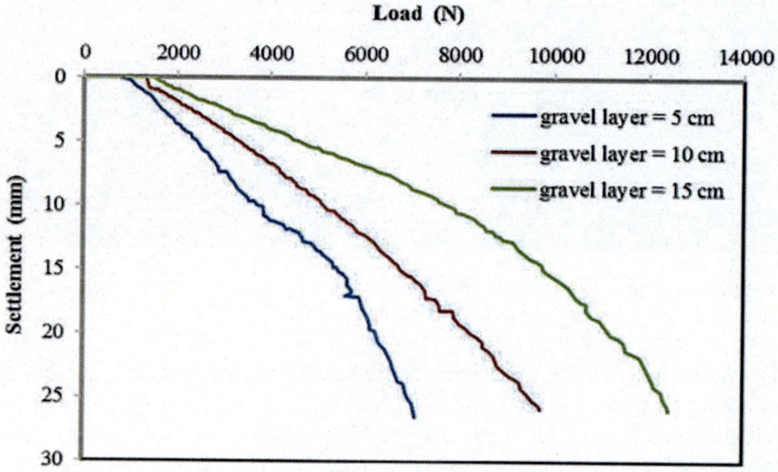


**Fig. 6** Load versus settlement graph for raft with gravel layer 5 cm, 10 cm, 15 cm

Fig. 6 that the load carrying capacity has been found to increase by 92.31% when depth of gravel layer is increased from 5 cm to 15 cm. This is due to the fact that the increase in the gravel layer thickness may improve the load sharing capacity of gravel layer.

## 5.2 *Effect of Pile Configuration, Spacing of Piles and Gravel Layer Thickness on Load-Settlement Characteristics*

From the various load-settlement responses, it is assessed that the vertical load-settlement response of disconnected piled raft is non-linear. It has been observed that with the increase in number of piles, the load bearing capacity of the combined piled raft foundation increases. The corresponding settlement is also lower as the number of piles is increased. It was observed that with an increase in spacing of piles, the load carrying capacity of the disconnected piled raft foundation increases significantly, and the settlement is comparatively lesser. It was also observed that with an increase in the depth of the gravel layer, the load carrying capacity of the disconnected piled raft foundation increases substantially. The load versus settlement graphs for different configurations are shown in Figs. 7, 8, 9, 10, 11 and 12. For 2 × 2 pile configuration and depth of gravel layer of 5 cm, 10 cm and 15 cm, the load carrying capacity has been found at maximum settlement of 25 mm which has been recorded with the help of LVDT and data logger. From Figs. 7, 8 and 9, it has been observed that the load carrying capacity is increased by 103.57%, 67.5% and

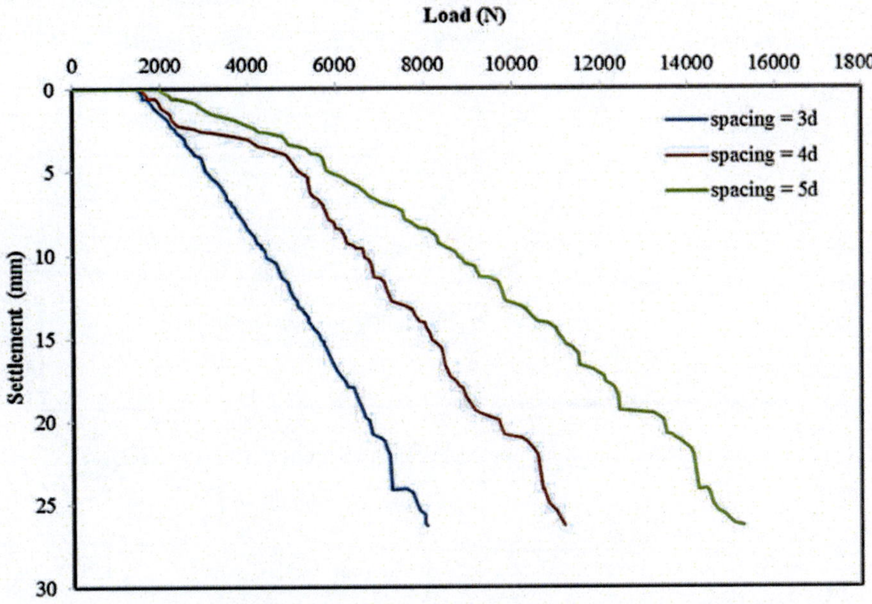


**Fig. 7** Load versus settlement graph for 2 × 2 pile configuration with 5 cm gravel layer

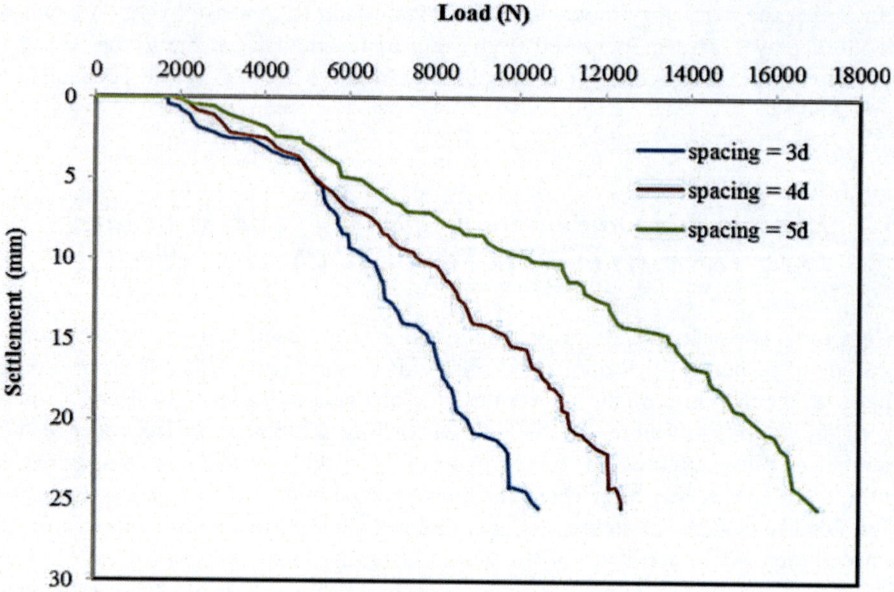


**Fig. 8** Load versus settlement graph for 2 × 2 pile configuration with 10 cm gravel layer

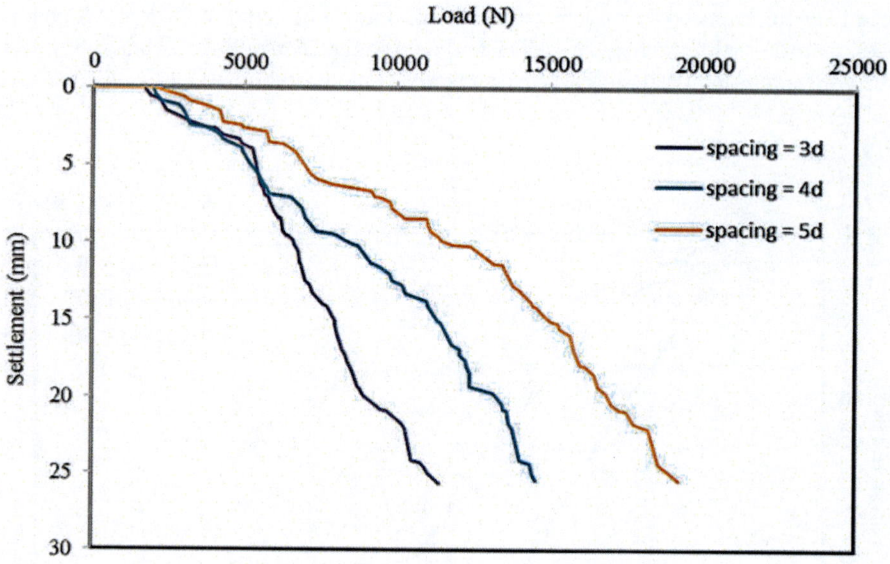


**Fig. 9** Load versus settlement graph for 2 × 2 pile configuration with 15 cm gravel layer

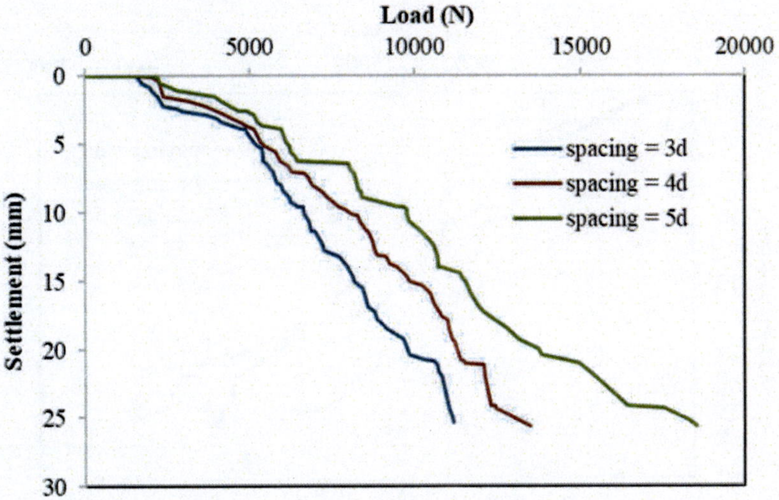


**Fig. 10** Load versus settlement graph for 3 × 3 pile configuration with 5 cm gravel layer

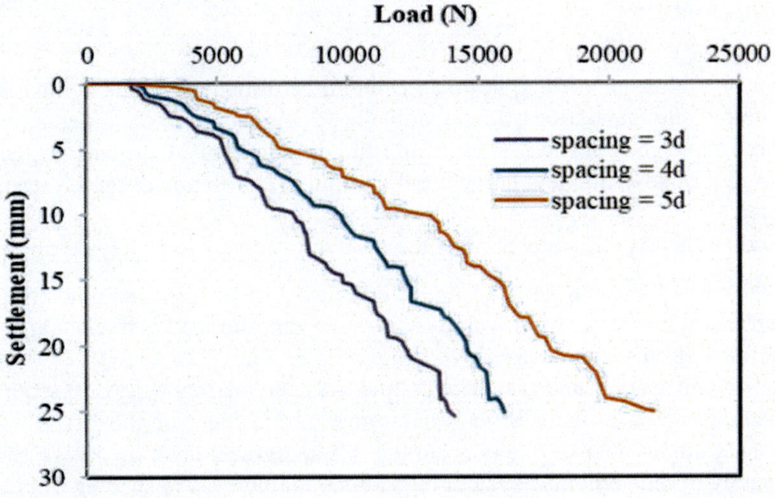


**Fig. 11** Load versus settlement graph for 3 × 3 pile configuration with 10 cm gravel layer

65.9%, respectively, when the spacing between piles has been increased from 3d to 5d. However, for 3 × 3 pile configuration, the load carrying capacity has been increased by 73.81, 62.96 and 32.19% at a depth of 5 cm, 10 cm and 15 cm, when the spacing between piles has been increased from 3d to 5d. The load transferred across the gravel layer will be transmit to the deeper bearing layer through the piles by means of skin friction.

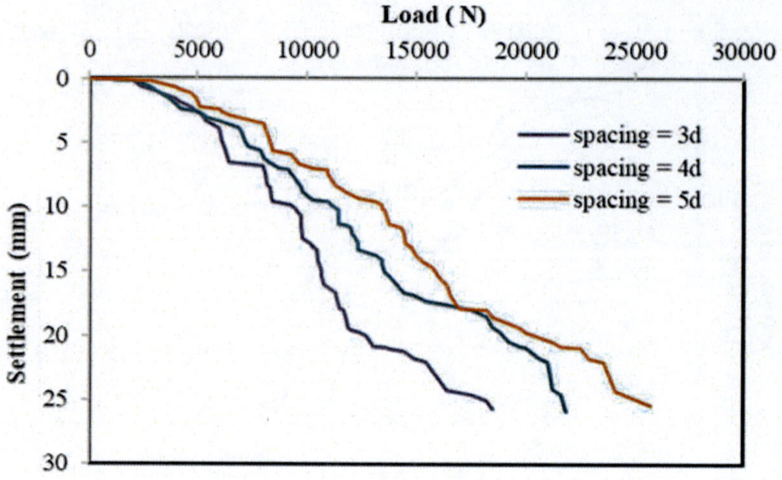


**Fig. 12** Load versus settlement graph for 3 × 3 pile configuration with 15 cm gravel layer

# 6 Conclusion

Piled raft systems are verified to be an economical foundation type comparing the conventional pile foundations, where, only the piles are used for the reducing of both for bearing and reducing settlements and the contribution of the raft is generally disregarded. Load-settlement curve are drawn for each test and observed results are discussed.

Based on the experimental results, the following conclusions may be drawn:

1. In the case of only raft and disconnected raft with four piles and nine piles, with increase in the depth of gravel layer of 5–15 cm, the load carrying capacity of the foundation system increases.
2. At a certain spacing and gravel layer thickness, the load capacity of disconnected piled raft increases with increase in number of piles beneath the raft.
3. At a certain number of pile beneath the raft and gravel layer thickness, the load capacity of the combined system increases with increase in spacing of piles.

# References

1. Burland JB (1995) Piles as settlement reducers. In: 18th Italian Cong Soil Mechanics, Pavia
2. Nakai Sh, Kato H, Ishida R, Mano H, Nagata M (2004) Load bearing mechanism of piled raft foundation during earthquake. In: Proc third UJNR workshop on soil–structure interaction, March 29–30. Menlo Park, California (USA)
3. Fioravante V, Giretti D (2010) Contact versus noncontact piled raft foundations. Can Geotech J 47(11):1271–1287

4. Dezfouli MG, Dehghani M, Asakereh A, Kalantari B (2018) Behavior of geogrid reinforced and unreinforced non-connected pile raft foundation. International Journal of Civil Engineering, pp 1–14
5. Azizkandi AS, Rasouli H, Baziar MH (2018) Load sharing and carrying mechanism of piles in non-connected pile rafts using a numerical approach. Int J Civ Eng, pp 1–16
6. Cao XD, Wong IH, Chang MF (2004) Behavior of model rafts resting on pile-reinforced sand. J Geotech Geoenviron Eng 130(2):129–138
7. Tradigo F, Pisanò F, Di Prisco C, Mussi A (2015) Non-linear soil–structure interaction in disconnected piled raft foundations. Comput Geotech 63:121–134
8. Tradigo F, Pisanò F, di Prisco C (2016) On the use of embedded pile elements for the numerical analysis of disconnected piled rafts. Comput Geotech 72:89–99

# Lateral Performance of Helical Pile in Cohesionless Soil

**Raunak Sinha, S. P. Dave, and S. R. Singh**

**Abstract** Helical pile is one of the deep foundation system used to support the axial compression, axial tension, and lateral loading. Helical piles use as foundation for wind farms, transmission tower, pipelines, for underpinning, monopole, etc., as axial and lateral support. Advantages of the helical pile in construction are its ease of installation, reusability, and noise control. In this study, lateral performance of helical piles on small-scale model tests in cohesionless soil was analyzed. The model tests were done by using variable parameter as helix pitch and helix diameter by considering the relative density constant at 30%. The experimental tests were performed on model piles in a cylindrical tank having 600 mm diameter and 650 mm height. The soil was classified as per the IS classification as poorly graded sand (SP). Lateral load capacity of helical piles were determined by varying helix pitch 10 mm to 30 mm at an interval of 10 mm and varying the helix diameter 40 mm to 60 mm at an interval of 10 mm, respectively. The results are validated with empirical formula of Mittal et al. (2010) and Perko (2009). It is concluded that the ultimate lateral load increases with increase in helix diameter and decreases with increase in helix pitch.

**Keywords** Helical piles · Lateral load test · Relative density · Helix pitch · Helix diameter

R. Sinha · S. R. Singh (✉)
Applied Mechanics Department, L.D. College of Engineering, Ahmedabad 380015, Gujarat, India
e-mail: shalini.2204@gmail.com

R. Sinha
e-mail: raunaksinha24@gmail.com

S. P. Dave (✉)
Government Engineering College, Gandhinagar 382028, Gujarat, India
e-mail: sdave_1966@yahoo.co.in



T. G. Sitharam et al. (eds.), *Seismic Design and Performance*, Lecture Notes in Civil Engineering 120, https://doi.org/10.1007/978-981-33-4005-3_34

## 1 Introduction

Helical piles are being used by civil engineer for nearly two century year ago. Helical piles were first introduced by Irish civil engineer Alexander Mitchell in 1836. Helical piles are also known as helix piles or screw pile. These piles are generally used in deep foundations to support axial loading and lateral loading system. It has been noticed that in recent years, the use of helical pile is highly increased in construction of offshore wind farms. There are many field applications of helical pile such as in transmission towers, underpinning of residential houses, pipelines, etc. Based on material classification, there are three kind of material from which piles are made: timber, steel, and concrete. But in case of helical pile, it is difficult to construct the helices with concrete. So, helical piles are made up of steel material and helices are bolted or welded with the pile shaft. It consists of steel shaft with one or more helices. The central steel shaft can be hollow or solid, and it can be circular, square, rectangular, or triangular in shape. There are few advantages of helical piles over other driven piles are easy to install, accessible to narrow places, reusable, and noise control. Installations of helical piles are done by applying torque using hydraulic motors, it can be hand operated or machine operated.

However, there are many research papers available for the lateral capacity of helix pile. Abdrabbo et al. performed the lateral load test of helical pile having different helices diameter, helices position, number of helices, helices spacing. They concluded that the lateral loading capacity of helical pile is dependent upon helices diameter, helices position, helices number, and helices spacing too [1]. Sakr performed eleven full scale pile load tests including axial, compression, uplift, and lateral load tests in oil sand of single and double helix pile. The lateral load versus deflection behavior of a helical pile is controlled mostly by the size of the shaft. The helixes had a minor effect on the lateral behavior of helical piles [2]. Sakr has also performed the lateral pile load test programs and the field monitoring of helical and driven piles with shaft diameters varied between 219 and 508 mm, installed in firm to very soft clay to very hard clay till was presented in this study. The test results indicated that the effect of embedment depth on the lateral resistance of helical piles is relatively small [3]. Narasimha et al. concluded that load carrying capacities of helical pile in clay are controlled by helical plates. Prasad and Narasimha Rao performed laboratory test with 2 and 4 helices installed in saturated clay. They concluded that the lateral capacity increases from 1.2 to 1.5 times that of plain shaft pile and it increases with embedment depth [4]. Mittal et al. performed an experimental investigation on laterally loaded screw pile in dry sand with number of helices, different embedment length of pile, lateral load applied at different depth, and static lateral load applied at screw pile. They concluded that lateral load increases as the number of helices increases and embedment depth increases [5].

# 2 Experimental Investigation

## 2.1 Test Setup

These small-scale model tests were conducted at LD College of Engineering, Ahmedabad with model piles in cylindrical tank, having inside diameter of 600 mm, outside diameter of 620 mm and height of 650 mm as shown in Fig. 1. The loading frame has been fabricated for applying lateral load on model pile as shown in Fig. 2. And Fig. 3 show that laboratory test setup to find the ultimate lateral capacity of helix pile.

**Fig. 1** Model tank

**Fig. 2** Loading frame

**Fig. 3** Laboratory Test setup for lateral load

**Table 1** Properties of sand

| Soil properties | Test | Determination |
|---|---|---|
| Index properties | Specific gravity | 2.65 |
| | Sieve analysis | Cc = 1.15<br>Cu = 2.52<br>SP |
| | Relative density | $\gamma_{max} = 17.6$ kN/m$^3$<br>$\gamma_{min} = 14.5$ kN/m$^3$ |
| Engineering properties | Direct shear | Φ—34.3° |

## 2.2 *Sand Properties*

The sand was taken from Vasna Barrage, Ahmedabad. The soil was classified from IS classification as poorly graded sand SP. The index and engineering properties of sand used for the study are shown in Table 1.

## 2.3 *Model Piles*

The model piles were made mild steel helical piles with diameter shaft of 20 mm. The helices were fabricated from steel plates having thickness of 2 mm. The pile lengths are considered 400 mm corresponding to L/d ratio of 15. The tests are performed on pile having size of different pitch at 10 mm, 20 mm, 30 mm, and keeping the helix diameter constant at 50 mm and with different helix diameter at 40 mm, 50 mm, 60 mm, and keeping the helix pitch constant at 20 mm. The lateral load applied at

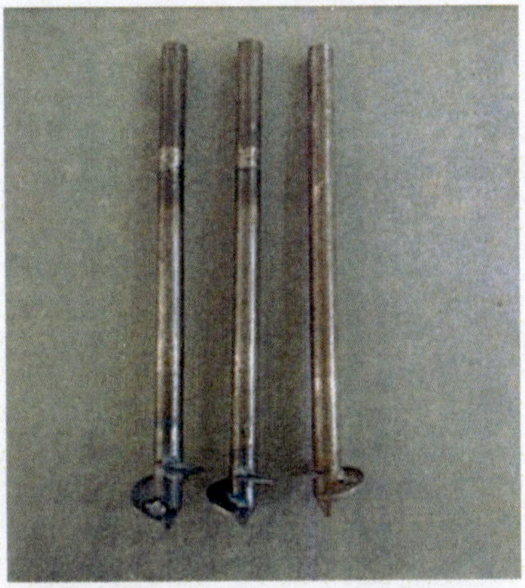

**Fig. 4** Helical pile (Dia-50 mm) Pitch—10 mm, 20 mm, 30 mm

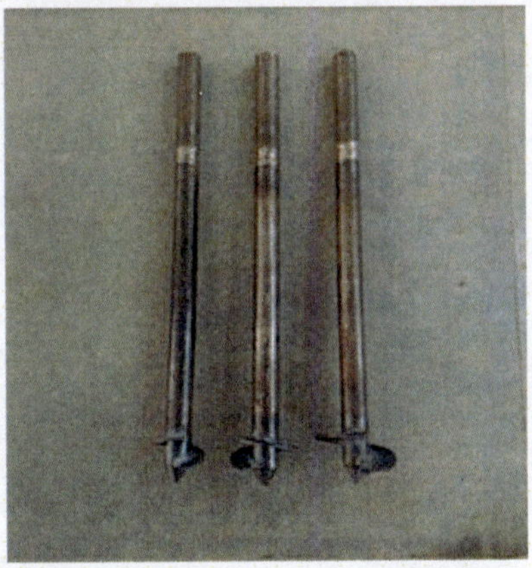

**Fig. 5** Helical pile (Pitch-20 mm) Diameter—40 mm, 50 mm, 60 mm

an eccentricity of 100 mm from the soil surface. All the model piles were having conical end of an apex of 60°. Model pile having various helix pitch and various helix diameter as shown in Figs. 4 and 5.

## 2.4 Pile Installation

After the soil bed had been prepared, the pile was slowly inserted into the soil bed by applying torque until the required depth of embedment was not achieved. The pile head was free and kept 100 mm above the sand surface to make the provision for application of lateral load. Helix pile having length 400 mm was driven at a depth 300 mm from the sand surface corresponding to L/d ratio 15. Figure 6 shows the schematic diagram of model experiment.

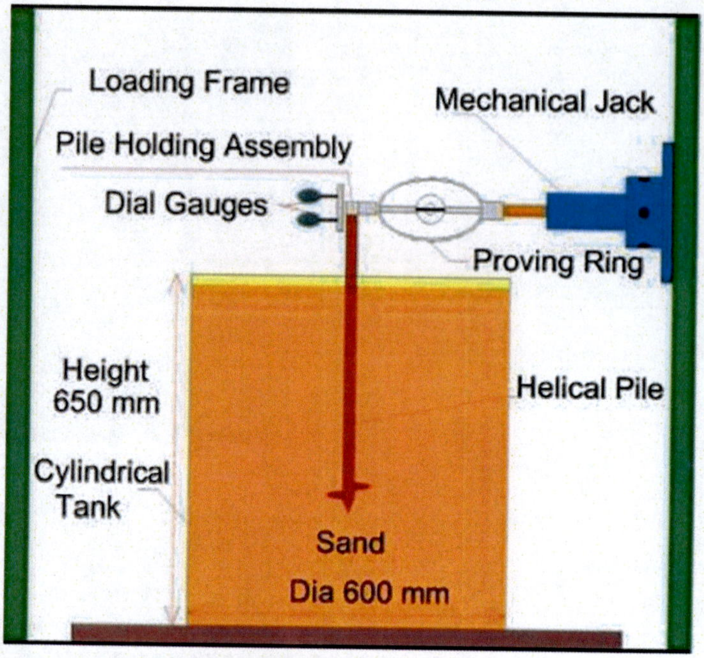


**Fig. 6** Schematic diagram of model experiment

### 2.5 *Test Procedure*

The lateral load tests were performed as per IS 2911 Part-4. Lateral load was applied at the top of the pile by rotating the mechanical jack. The observation of mechanical dial gauge records increase in lateral displacement corresponding to applied load. The ultimate lateral load was calculated at 12 mm displacement.

## 3 Results and Discussion

Series of model lateral load tests were carried out on piles having variation in helix pitch, keeping the helix diameter constant and variation in helix diameter, keeping the helix pitch constant. Load versus displacement curves were drawn for each test and comparison of both were shown in Figs. 7 and 8. These figures concluded that ultimate lateral load capacity increases with increase in helix diameter. In case of helical pile, helices are perpendicular to the pile shaft, so it resists the moment due to lateral load. Thus, increase in helix diameter increases the resisting moment. It is observed that the ultimate lateral load of helical pile decreases with the increase in helix pitch. The helix with minimal pitch behaves like base plate for the helical pile

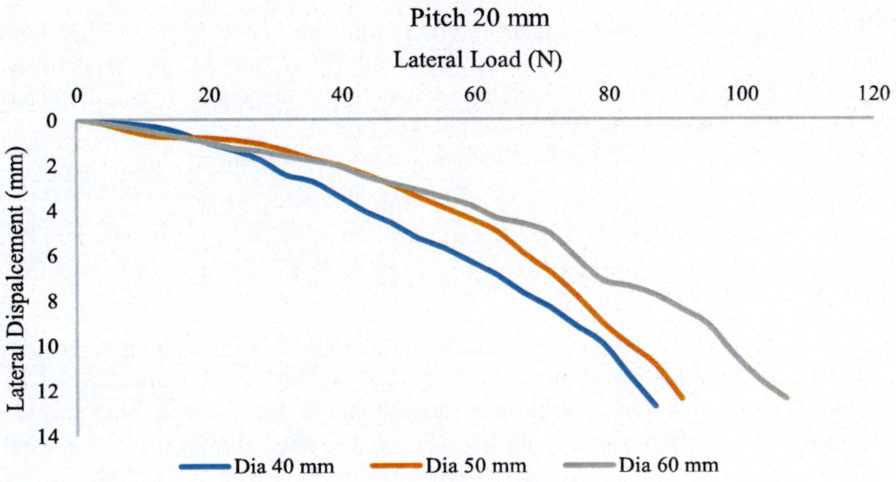


**Fig. 7** Load displacement curves for different helix diameter

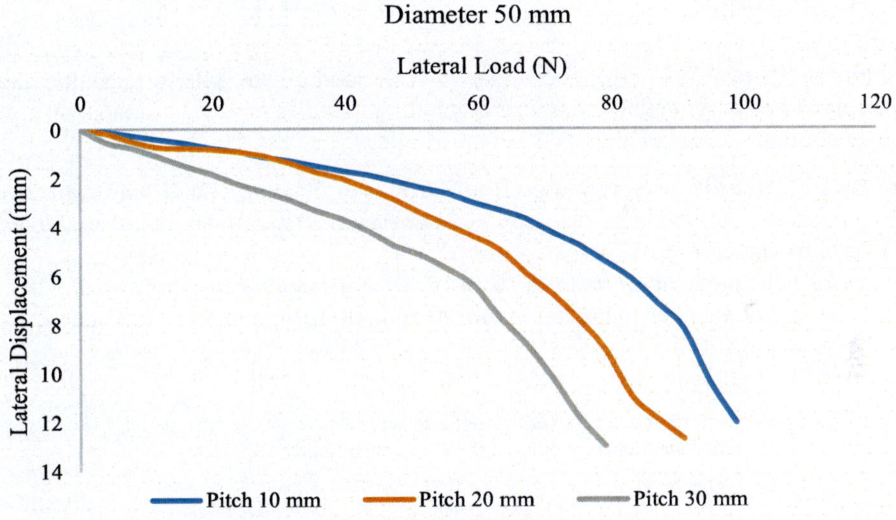


**Fig. 8** Load displacement curves for different helix diameter

and when this pitch increases, it loses its behavior of base plate. Thus, increase in pitch decreases the ultimate lateral capacity of helical pile.

Mittal et al. [5] proposed an empirical formula for predicting the ultimate lateral load of helical pile in cohesionless soil.

$$Q_u = \gamma.d^3.A.\left(\frac{H}{d}\right)^{2.431}.\left(\frac{h}{D}\right)^{-0.6284}$$

According to above empirical formula, the ultimate lateral loads of piles with diameter 40 mm, 50 mm, and 60 mm are 257 N, 291 N, and 327 N, respectively. The results by this formula are much higher than that in this research, and the D has significant effect on the ultimate lateral loads of pile.

Perko [6] developed an empirical formula for calculating ultimate lateral capacity, considering the influence of p.

$$Q_u = \frac{\gamma.d.h^3.K_p}{2(H+h)} \cdot \left(1 - \frac{p}{9d}\right)$$

According to above empirical formula, the ultimate lateral loads of piles with pitch 10 mm, 20 mm, and 30 mm are 132 N, 124 N, and 117 N, respectively. The results by this formula are quite close to the test data in this research. This formula could be useful to determine the ultimate lateral capacity of helical pile for wind farms.

## 4 Conclusion

In this study, test was performed on small-scale model helix pile to find ultimate lateral load by varying helix diameter and helix pitch distance. The findings of this study are summarized below:

- As the helix diameter increases from 40 mm to 50 mm, lateral load capacity increases to 9% and helix diameter increases from 50 mm to 60 mm, lateral load capacity increases to 17%.
- As the helix pitch increases from 10 mm to 20 mm, lateral load capacity decreases 10% and helix pitch increases from 20 mm to 30 mm, lateral load capacity decreases to 15%.

**Acknowledgements** The authors are very thankful to Prof (Dr.) R K Gajjar, Principal, LD College of Engineering, Ahmedabad for providing all the necessary research facilities.

## References

1. Abdrabbo FM, El Wakil AZ (2016) "Laterally loaded helical piles in sand" ELSEVIER—Alexandria Eng J 55:3239–3245. https://doi.org/10.1016/j.aej.2016.08.020
2. Sakr M (2009) "Performance of helical piles in oil sand" Canadian Geotech J 46:1046–1061. https://doi.org/10.1139/t09-044
3. Sakr M (2018) "Performance of laterally loaded helical piles in clayey soils established from field experience". DFI Journal—J Deep Found Inst, ISSN: 1937-5247. https://doi.org/10.1080/19375247.2018.1430481

4. Narasimha S, Prasad Y (1993) "Estimation of uplift capacity of helical pile anchors in clay". J Geotech Eng ASCE 119(2):352–357. https://doi.org/10.1061/(asce)0733-9410(1993)119:2(352)
5. Mittal S, Ganjoo B, Shekhar S (2010) Static equilibrium of screw anchor pile under lateral load in sands. Geotech Geol Eng 28(5):717–725. https://doi.org/10.1007/s10706-010-9342-4
6. Perko HA (2009) Helical piles: a practical guide to design and installation. Wiley, New York